Bettina Menne ▍ Kristie L. E

Climate Change and Adaptation
for Human Health

BETTINA MENNE   KRISTIE L. EBI (Eds.)

Climate Change and Adaptation Strategies for Human Health

Published on behalf of the World Health Organization
Regional Office for Europe by Steinkopff Verlag, Darmstadt

WITH 57 FIGURES AND 62 TABLES

Dr. Bettina Menne
Global Change and Health

Dr. Kristie L. Ebi*
Global Change and Health
WHO Regional Office for Europe
*now Exponent, United States

ISBN 3-7985-1591-3 Steinkopff Verlag Darmstadt

Bibliographic information published by Die Deutsche Bibliothek
Die Deutsche Bibliothek lists this publication in the Deutsche Nationalbibliografie;
detailed bibliographic data is available in the Internet at <http://dnb.ddb.de>.

Steinkopff Verlag Darmstadt
a member of Springer Science+Business Media

www.steinkopff.springer.de

© World Health Organization 2006
Printed in Germany

Medical Editor: Sabine Ibkendanz   Production: Klemens Schwind
Cover Design: Erich Kirchner, Heidelberg
Typesetter: K+V Fotosatz GmbH, Beerfelden

SPIN 11577829   85/7231-5 4 3 2 1 0 – Printed on acid-free paper

This book is dedicated to ELIS, LUCA,
WILLOW, KATHY and JULIE
and all our children

Foreword

It is now five years since the IPCC concluded in 2001, on the basis of new and strong evidence, that not only was most of the global warming observed over the last 50 years attributable to human activities, but also that climate change could affect human health. The effects can be direct such as through increased heat stress, and loss of life in floods and storms, or indirect through changes in the ranges of disease vectors such as mosquitoes, water-borne pathogens and water and air quality, as well as food availability and quality.

Health authorities had already expressed their concern about climate change and its impact on human health. Three years earlier, at the World Health Assembly in 1998, they had recognized that climate change could be a potential threat to human health. In 1999, at the Third Ministerial Conference on Environment and Health, ministers of health and environment from the WHO European Region had acknowledged that "human-induced changes in the global climate system and in stratospheric ozone pose a range of severe health risks and potentially threaten economic development and social and political stability". They also called for national action by all countries to reduce and prevent as far as possible these environmental changes and to limit the exposure of human populations in Europe to climate change and increased ultraviolet irradiation, thus addressing the likely health risks over the coming decades.

These statements posed a great challenge to WHO and to Member States. Scientific evidence was showing very clearly that climate change was already starting to occur, and, even in the best scenario, the human population was going to face direct and indirect health effects over the coming decades. Adaptation strategies were needed, based on thinking about the types of risks that European populations might face. With this in mind and in order to fill this knowledge gap, we developed the network and the content of the cCASHh project. We aimed to describe with facts and figures the early observed effects of climate change on health and to identify the available public health measures to cope with the additional risks. We also aimed to assess the benefits from acting now, compared to acting late and to develop the necessary policies to support decision-makers in addressing these issues.

During the project (2001–2004), the WHO European Region was hit by a major flood in 2002 and by a severe heat-wave in 2003. Experience seemed to confirm what models had predicted. Although one record heat-wave and flood does not prove that Europe is getting hotter or the weather more extreme, the impacts made by these events highlighted shortcomings in the existing public health preparation and responses, particularly the lack of knowledge on effective preventive measures and the few mechanisms in place to predict or prevent the health effects, or even to detect them rapidly. I believe the cCASHh project has produced very important results, both in the content and in the methods used. It shows that the concurrent work of different disciplines in addressing public health issues can produce innovative and useful results, providing an approach that can be followed on other public health issues. The project has shown that information on potential threats and impacts can be developed and can be extremely useful in preparing the public for adverse events as well as facilitating the response when the events occur. This is a new dimension for public health which reverses the traditional thinking: from describing what has already occurred and identifying and reducing specific risk factors, to taking action on the basis of prediction and early warning to prevent health consequences in large populations. We hope this experience and approach will be further developed and tested, particularly where emerging environmental risks are concerned.

We would like to use this opportunity to express our gratitude to the many scientists and stakeholders, including policy-makers at different levels, who contributed to the development of this project. Without this constructive and extremely collaborative critical mass, the research results would have been less timely and perhaps less relevant. We would also like to thank the DGResearch for its generous contribution and the attention and support with which it has followed the implementation of the project.

This research has generated a number of conclusions and recommendations for action by Member States and the international community. The challenge is now to translate these actions into policy and to monitor their effectiveness and impact. With this in mind, we worked with Member States to include in the final Declaration of the Fourth Ministerial Conference on Environment and Health held in Budapest in June 2004 recommendations on the public health response to extreme weather events and a renewed commitment to address in a proactive and anticipatory approach the consequences for health of climate change.

We hope that the implementation of effective adaptation policies together with that of effective mitigation actions will limit the impact of climate change and protect the health of present and future generations. We believe this project has made a significant contribution to this vital endeavour.

ROBERTO BERTOLLINI

Director, Special Programme on Health and Environment,
WHO Regional Office for Europe

Foreword

It is an honour for me to write a few introductory lines for the book *"Climate Change and Adaptation Strategies for Human Health"*, which represents the results of the study "Climate Change and Adaptation Strategies for Human Health in Europe" (cCASHh) (May 2001–July 2004), co-ordinated by WHO and supported by the "Energy, Environment and Sustainable Development Programme" in the frame of the Fifth EU Framework Programme for Research and Development.

cCASHh aimed to
- identify the vulnerability to adverse impacts of climate change of human health
- review current measures, technologies, policies and barriers to improving the adaptive capacity of populations to climate change
- identify for European populations the most appropriate measures, technologies and policies to successfully adapt to climate change and
- provide estimates of the health benefits of specific strategies or combinations of strategies for adaptation under different climate and socioeconomic scenarios.

The flood events in 2002 and the heat-wave of August 2003 in Europe had given evidence that no one is on the safe side when it comes to the impacts of climate change. Though some may argue whether these extreme weather events are linked to global change or not, these events revealed in a rather drastic way our vulnerability and our non-preparedness. The prevention and preparedness to extreme weather events need cooperation at all levels and throughout disciplines. The cCASHh project was able to contribute timely results on both occasions. I would like to take the opportunity to congratulate the consortium for this successful endeavour.

This type of research activities needs an interdisciplinary approach. The cCASHh project was a good example for this approach. Projects supported during the Sixth Framework Programme and hopefully also during the future Seventh Framework Programme continue to further develop this important work.

KARIN ZAUNBERGER
Project Officer
European Commission
DG RTD

Acknowledgements

This book is the summary of the final results of the project "Climate change and adaptation strategies for human health" (cCASHh-EVK2-2000-00070), funded by the European Commission.

We would like to warmly thank Anna Alberini, Aline Chiabai, Milan Daniel, Vlasta Danielova, Hans-Martin Füssel, Gerd Jendritzky, Christina Koppe, Tom Kosatsky, Sari Kovats, Bohumir Kriz, Elisabet Lindgren and Michael van Lieshout. Without their continuous commitment and deep dedication to working in this project, this publication would not have been possible.

Many experts have directly contributed to this project by steering the work and writing background reports and articles, and have provided information and comments essential to the development of the project. We would like to thank very much for this: Ben Armstrong, Matteo Albrizio, Martha Anker, Annmaria Asp, Jürgen Baumüller, Cestmir Benes, Arieh Bitan, Ian Burton, Diarmid Campbell-Lendrum, Carlo Carraro, Dominique Charron, John Cowden, Clive Davies, Philippe Desjeux, Julio Díaz, Sally Edwards, Michael Ejov, Jan Erhart, Michele Faberi, Marzio Galeotti, Gretel Gambarelli, Andy Haines, Shakoor Hajat, George Havenith, Gloria Hernández, Jaroslava Holubova, Zdenek Hubalek, Charmaine Gauci, Peter Gerner-Smidt, Norman Gratz, Simon Hales, Daniela Janovska, Thomas G.T. Jaenson, Adam Jirsa, Ricardo Jorge, Richard J.T. Klein, Ivan Kott, Jan Kopecky, Zuzana Kristufkova, Katrin Kuhn, Kuulo Kutsar, Milan Labuda, Gudrun Laschewski, Alberto Longo, César López, Lena Malmström, Wiesław Magdzik, Andreas Matzarakis, Pim Martens, Pierre Marty, Jan Materna, Fergus Nicol, Glenn McGregor, Anthony McMichael, Paul McKeown, Rennie M. D'Souza, Kassiani Mellou, Torsten Naucke, Antonio Navarra, Fergus Nicol, Paulo Jorge Nogueira, Sarah O'Brien, Anna Paldy, Milan Pejcoch, Edmund Penning Rowsell, Hans Schmid, Scott Sheridan, Paul Sockett, Sue Tapsell, Hiroko Takasawa, Cristina Tirado, Jaroslav Valter, Antti Vaheri, Theresa Wilson, Linda Wirén, Wilfrid van Pelt, Teresa Wilson, Tanja Wolf and Kamil Zitek.

Many persons participated in the several workshops, reviewed documents and through their contributions allowed the project to grow. The coordinators would like to thank Lucien Abenhaim, Roger Aertgeerts, Bastien Affeltranger, Ingvar Andersson, Peter Baxter, Elena Borisova, Nick Brooks, Rui Calado, Sergio Castellari, Tanja Cegnar, Claude Chastel, Jean-Claude

Cohen, Susanna Conti, Carlos Corvalan, Thomas Downing, Peter Duchaj, Andrea Ellis, Pascal Empereur-Bissonnet, Agustín Estrada-Peña, Veronique Ezratty, Vytautas Gailius, Benedek Goncz, Duane Gubler, Paolo Guglielmetti, Debarati Guha-Sapir, Cağatay Güler, Katarina Halzlova, Juhani Hassi, Madeleen Helmer, Marika Hjertqvist, Sona Horvathova, Michael Hübel, Lyubomir Ivanov, Ilze Jansone, Geoff Jenkins, Anne-Marie Kaesbohrer, Wilhelm Kirch, Victor Kislitsin, Silvia Kostelna, Zbigniew W. Kundzewicz, Jan Kyncl, Marco Leonardi, Otto Malek, Alexander Malyavin, Merylyn McKenzie Hedger, Paola Michelozzi, Thierry Michelon, Matthias Niedrig, Buruhani Nyenzi, Mikko Paunio, Armin Petrascheck, Günter Pfaff, Florin Popovici, Paul Reiter, Stefania Salmaso, Darina Sedlakova, John Simpson, Jolanta Skrule, Alfred Spira, Jochen Süss, Richard Tol, Jaroslav Valter, Els Van Cleemput, Thomas Voigt, Jaroslav Volf, Gary Yohe and Rudolf Zajac.

We also would like to thank in particular Roberto Bertollini, Director of the Special Programme on Health and Environment, WHO Regional Office for Europe, and Karin Zaunberger, Project Officer of the European Commission for their trust in our work and the support provided.

Blessy Corda deserves particular thanks for taking care of the heavy burden of EC and WHO project administration and the high professionalism and daily smile with which this was carried out.

Nicoletta di Tanno took care of developing and updating the website http://www.eurowho.int/globalchange.

Table of Contents

Information about the Authors

ANNA ALBERINI is an associate professor at the University of Maryland, College Park, and a Programme Coordinator for Fondazione Eni Enrico Mattei, Venice, Italy. She is an economist and statistician specializing in environmental economics. Her research currently focuses on non-market valuation methods, with a special emphasis in the valuation of safety and life, the effects of environmental and safety regulations, and land use. She is also a member of the Science Advisory Board of the US Environmental Protection Agency, a member of the economists' network of the European Environmental Agency, and her work has appeared in the *Review of Economics and Statistics, Journal of Environmental Economics and Management, Journal of Urban Economics, Journal of Risk and Uncertainty, Risk Analysis, Land Economics,* and other peer-reviewed journals. Anna Alberini coordinated the economic work package of cCASHh.

ČESTMÍR BENEŠ is an epidemiologist specialized in epidemiological data analyses and is head of the National Reference Centre for Analysis of Epidemiological Data, National Institute of Public Health, Prague. He has participated in several epidemiological projects and studies.

IAN BURTON is Emeritus Professor at the University of Toronto and Scientist Emeritus with the Meteorological Service of Canada. He works internationally as an independent scholar and consultant and has recently served as an advisor/consultant to the World Bank, UNDP, the European Commission, the United States Agency for International Development, the Secretariat of the United Nations Framework Convention on Climate Change, several Canadian government departments and municipalities as well as other organizations including the World Health Organization. He plays an active role in the interaction between science and policy especially in the fields of environment and risk assessment. He specializes in the management and mitigation of natural hazards and disasters, and adaptation to climate change and variability. He is a Fellow of the Royal Society of Canada and the World Academy of Arts and Sciences, and recently completed a three year term as President of the International Society of Biometeorology.

CLIVE DAVIES is a medical entomologist and epidemiologist with a particular interest in the control of vector-borne and zoonotic diseases. After arriving at the London School of Hygiene and Tropical Medicine in 1987, he focused his research activities on the ecology, epidemiology and control of leishmaniasis. This involved a number of large scale field projects in Peru, Brazil, Colombia, Bolivia and Iran.

MARKÉTA BRAUN KOHLOVÁ is junior researcher at Charles University in Prague. She has a Masters in sociology. During recent years she received several scholarships, including one from the Robert Bosch Stiftung für Nachwuchsführungskräfte aus der Länder MOE. She was involved in the economic valuation assessment in the Czech Republic.

ALINE CHIABAI is a researcher at the Fondazione Eni Enrico Mattei and graduated with a degree in environmental economics. She has worked on "Air Pollution and Respiratory Diseases: A Contingent Valuation Study" as well as "Estimating the Cost of Air Pollution from Road Transport in Italy". Within cCASHh she carried out together with Anna Alberini, the conjoint choice questionnaire and the contingent valuation study for Italy.

MILAN DANIEL, PhD, DSc is a parasitologist specialized in medical entomology and zoology, ecology of natural foci of vector-borne diseases, landscape epidemiology using GIS and RS methodologies; he is the senior scientific worker at the Centre of Epidemiology and Microbiology, National Institute of Public Health, Prague, and senior lecturer at the Institute of Postgraduate Medical Education, School of Public Health, Prague, having been engaged in the research of natural foci diseases in Europe, Asia, Africa and Cuba. He participated in the EUCALB project and has published more than 220 peer-reviewed articles and book chapters.

VLASTA DANIELOVÁ, PhD, DSc is a virologist specialized in arbovirology and in the ecology of arboviruses in laboratory and field experiments. She is a senior scientific worker at the Centre of Epidemiology and Microbiology, and head of the Department of Tissue Cultures, National Institute of Public Health, Prague. She studied arboviruses in the Czech Republic and abroad and recovered new mosquito- and tick-borne viruses. She has published more than 150 peer-reviewed articles.

PHILIPPE DESJEUX currently works with the Institute for OneWorld Health. He will join the team that will implement a paromomycin pre-launch programme in Bihar, India. Prior to joining OneWorld Health, he worked for 17 years as the Leishmaniasis Research Coordinator for the Special Programme for Research and Training in Tropical Diseases of the World Health Organization (WHO/TDR). He received a medical doctorate from the University of Paris, and has certificates in special studies in gastroenterology and parasitology from the Odeon Faculty of Medicine and a certificate in systematic microbiology from the Pasteur Institute, both in Paris.

SALLY EDWARDS worked as a research assistant to Sari Kovats in the assessment of food-borne diseases and climate change. Sally is now working at the WHO Headquarters in Geneva.

Kristie L. Ebi, PhD, MPH is an epidemiologist who has conducted research on the potential health impacts of climate variability and change, including impacts associated with extreme events, thermal stress, and food- and vector-borne diseases, and research on the design of adaptation response options to reduce negative impacts. She is a senior managing scientist at Exponent and has worked on climate change issues at the WHO Regional Office for Europe and the Electric Power Research Institute. She is a lead author in the Fourth Assessment Report of the Intergovernmental Panel on Climate Change. She was a lead author for three chapters/reports for the Millennium Ecosystem Assessment, the Health Sector Analysis Team for the U.S. National Assessment of the Potential Consequences of Climate Variability and Change, and other assessment activities. She had a leading role in the implementation of the cCASHh project.

Hans-Martin Füssel is a research scientist at the Potsdam Institute for Climate Impact Research (PIK) in Potsdam, Germany. From June 2004 until June 2006, he was on secondment to the Center for Environmental Science and Policy at Stanford University. He holds master's degrees in computer science (Johann Wolfgang Goethe-University, Frankfurt am Main, 1992) and in applied systems science (University of Osnabrück, 1995). He received a PhD in physics for his thesis "Impacts Analysis for Inverse Integrated Assessments of Climate Change" (University of Potsdam, 2003). He worked on the ICLIPS and Eva project, and is co-founder of the ToPIK 2004 and project PREVENT, which aims to provide science-based support for interpreting the "ultimate objective" of the United Nations Framework Convention on Climate Change (UNFCCC). In cCASHh he assessed existing frameworks of impact, vulnerability and adaptation assessment.

Glenn McGregor is a reader in synoptic climatology. His research interests are climate and health, climatic variability and change. He is involved in a number of projects on climate and health and has the responsibility of developing heat stress watch warning systems for five European cities in the context of the EU PHEWE project and is PI for the NERC COAPEC project Climate Information for the Health Sector. He is also an advisor to the EuroHEAT project.

Shakoor Hajat is a lecturer in epidemiology and statistics at the London School of Hygiene and Tropical Medicine, Public and Environmental Health Research Unit (PEHRU). After studying at UCL and the University of Leicester, he began work in the Department of Primary Care and Population Sciences at the Royal Free and University College Medical School. There he received a PhD after investigating the short-term effects of air pollution on health in London. Since 2001, he has continued his research interests in environmental epidemiology at LSHTM, first in the Epidemiology Unit in EPH, and now in PEHRU. In cCASHh he reviewed the impacts of floods on human health.

GEORGE HAVENITH has been working for 22 years in the research areas of human thermal physiology, environmental ergonomics and clothing science. He has worked in research laboratories in the Netherlands, the United States, and since 1998 in the United Kingdom at Loughborough. There he is leading the Environmental Ergonomics Research Group, which deals with interactions of the human body with its environment (climate, noise, lighting etc.). Many of his research projects and publications involve partners from abroad. Besides his involvement in European Union projects, he has an active collaboration with Osaka Int. Women's University and Osaka Shin-Ai College in Japan and is also involved in a prestigious project on ageing with partners in the United States of America (NIH funding). He has published widely (>250 publications and reports) and research methods that he developed have been adopted by other laboratories and his models in International Standards (ISO, CEN). He advises a number of renowned international companies on thermal physiology and clothing science.

ZDENĚK HUBÁLEK, PhD, DSc is a microbiologist specialized in the ecology of arthropod-borne viruses and bacteria. He is also the principal research worker (in the Medical Zoology Laboratory) and deputy director at the Institute of Vertebrate Biology, Academy of Sciences of the Czech Republic, Brno. He has published more than 230 scientific papers. In cCASHh he reviewed the impacts of climate change on West Nile fever.

THOMAS G.T. JAENSON, PhD, DTMH, FRES is head of the Medical Entomology Unit, Evolutionary Biology Centre, Uppsala University, Sweden. He teaches medical entomology and arthropod ecology at Uppsala University, has written a textbook in medical entomology and has carried out extensive research on the ecology of malaria mosquitoes, tsetse flies, Chagas disease and Lyme disease vectors, and other vectors of human and animal diseases in Africa, Latin America and Sweden. He has directed several research projects, including two large-scale projects in Guinea-Bissau, West Africa, aiming at controlling human African sleeping sickness and human malaria. In the latter project, anti-mosquito bed nets drastically reduced malarial disease among a human population of about 50 000 people. He is presently directing research in Sweden, Guinea-Bissau and Guatemala evaluating the potential of plant-derived products for control of ticks, mosquitoes and other blood-feeding arthropods. In cCASHh he reviewed the literature of Lyme disease.

RICHARD J.T. KLEIN is a senior researcher at the Potsdam Institute for Climate Impact Research (PIK) in Germany, an associate fellow of the Flood Hazard Research Centre of Middlesex University and a fellow of the Oxford office of the Stockholm Environment Institute. At PIK he is joint head of the interdisciplinary research group Environmental Vulnerability Assessment and leader of the research platform Development and the Management Transition. He is a lead scientist in a number of international research projects on societal vulnerability and adaptation to climate variability and change.

TOM KOSATSKY, MD, MPH is a Montreal (Canada)-based community medicine specialist. He maintains a clinical practice in occupational and environmental medicine, conducts research in environmental health for the Montreal Public Health Department and is Associate Professor of Epidemiology and Occupational Health at McGill University. During 2004, he was engaged as an epidemiologist at the WHO European Centre for Environment and Health, with responsibilities in the area of climate change and health. Much of his Montreal research is community-based: current projects include an assessment of health-protective measures taken by persons with chronic disease during hot or smoggy days, the construction of a geographic information system to represent the susceptibility of city dwellers to extreme weather events, and an evaluation of the effect of living downwind of petroleum refineries on the respiratory health of children resident nearby.

CHRISTINA KOPPE, PhD is currently working with the Deutscher Wetterdienst (German Weather Service) at the Human Biometeorology Department. In her PhD thesis she developed a method that allows a health relevant assessment of the thermal environment taking the short-term adaptation of the population to the local meteorological conditions of the past 4 weeks into account. From 1996 to 2001, she studied hydrology at the Albert-Ludwigs-University in Freiburg. In addition, she has training in economy. Her research interests include the impacts of heat-waves on human health in current and future climates and the possibilities to predict and prevent these impacts.

R. SARI KOVATS is a lecturer in Environmental Epidemiology, at the London School of Hygiene and Tropical Medicine, Public and Environmental Health Research Unit (PEHRU). Recent work has focused on the epidemiology of heat-waves for the EC-funded cCASHh and EuroHEAT projects on preventing the health impacts of weather extremes. She has been an expert advisor since 1996 on climate variability, climate change and health for WHO Geneva, and the WHO European Centre for Environment and Health (Rome). She is also a member of the WMO Commission on Climatology Expert Team 3.8 on Health-related Climate Indices and their Use in Early Warning Systems, and was a member of the Technical Working Group on Research Needs for the EC Environment and Health Strategy in 2003/4. She is currently a lead author in the Fourth Assessment Report of the Intergovernmental Panel on Climate Change, and has worked extensively on previous assessments for the IPCC.

KATHRIN KUHN carried out the literature review on malaria within this project. She previously worked in the LSHTM where she carried out research on malaria and climate change in Europe. Currently she works at the Danish Cancer Institute.

Bohumír Kříž, MD, PhD is an epidemiologist and microbiologist, head of the Centre of Epidemiology and Microbiology, National Institute of Public Health Prague, former head of Chair of Epidemiology, Charles University, 3rd Medical Faculty, Prague, head of the National Surveillance Laboratory for Diphtheria. As a specialist for tropical diseases, he has worked as a WHO and UNICEF staff member and expert in Asia, Africa and Indonesia. His particular dedication was important in the eradication of smallpox. He has participated in several epidemiological, microbiological and environmental studies and has coordinated the activities on vector-borne diseases of the National Institute of Public Health, Prague, Czech Republic.

Elisabet Lindgren is a medical doctor with a PhD in natural resources management, having specialized in global environmental changes and human health. After a clinical career she became linked to the Department of Epidemiology at the Karolinska Institute. Since 1994 she has been based at Stockholm University and has contributed to numerous transdisciplinary international research activities. She has been advisor to the WHO since 1998, and was contributing author to the Third Assessment Report of the IPCC and to the Millennium Ecosystem Assessment Report.

Irene Lorenzoni is currently based at the Centre for Environmental Risk (CER) where, as part of the Understanding Risk Programme funded by the Leverhulme Trust, her research focuses mainly on perceptions and institutional aspects of global climate change (see http://www.uea.ac.uk/env/cer/index.html). Before joining CER, she worked for CSERGE on the development of future scenarios to frame research on climate impacts in the United Kingdom and decision-making at national and regional policy levels; on energy and waste management; and on integrated coastal zone management in the East Anglian region.

Pim Martens is director of the International Centre for Integrative Studies (ICIS), Maastricht University. He holds the chair 'Sustainable Development' at the same university, and is project leader and principal investigator of several projects related to sustainable development, globalization, environmental change and society, funded by, among others, the Dutch National Research Programme, the United Nations Environment Programme and the European Community. Pim Martens is executive editor of the International Forum on Science and Technology for Sustainability, and co-editor-in-chief of the international journal *Ecohealth*. Finally, he is a Fulbright New Century Scholar within the programme 'Health in a Borderless World' and winner of the Friedrich Wilhelm Bessel-Forschungspreis.

Pierre Marty, MD, PhD works at the University of Nice-Sophia Antipolis, Faculty of Medicine and at the University Hospital of Nice, France. He is a specialist in parasitology, mycology, tropical medicine and travel medicine. With good experience in the field of tropical medicine particularly in Central Africa (Cameroon and Gabon), he is a French and an European expert on congenital toxoplasmosis. In 2003, he became director of ERLEISH

(Equipe de Recherche sur les Leishmanioses) at the Faculty of Medicine of Nice. Having worked on Leishmaniases for 25 years, he is a specialist in Mediterranean leishmaniases particularly focused on asymptomatic carriers of *Leishmania infantum* in the human population and in coinfection HIV-*Leishmania* (member of the WHO Network). He has developed many collaborations with manufacturers for in vitro diagnostics tests of parasitic diseases.

JAN MELICHAR, PhD graduated from the Faculty of National Economics, University of Economics Prague and at present, he continues his PhD study at the Department of Environmental Economics, University of Economics. In 2003, he became a member of the environmental economics unit at the Charles University Environment Center. He is mostly focused on the valuation of environmental goods and damages, assessment and methodology of externalities, experimental economics and environmental fiscal reform. He works on several research projects funded by European Commission and many nationally funded projects.

BETTINA MENNE has worked at the WHO Regional Office for Europe since 1997. She is a medical doctor and medical specialist in hygiene and public health. For the last five years she has coordinated the setting up of the WHO programme on global change and health (climate change, stratospheric ozone depletion, energy, extreme weather events and globalization) and coordinated several in- and outhouse research teams and policy-making processes. She has gained sound experience in assessing environmental health aspects, through her participation in Environmental Performance Reviews. In addition, she is the convening lead author of the health chapter in the 4th assessment report of the IPCC. Before joining WHO, she worked for UNICEF, UNOPS, the Italian Ministry of Foreign Affairs and the University of Perugia, mainly planning, implementing and evaluating projects on the prevention of communicable diseases in Mali, Albania, Nicaragua and Italy. She wrote and coordinated the cCASHh project.

FERGUS NICOL is best known for his work in the science of human thermal comfort, principally for the 'adaptive' approach to thermal comfort. He has run a number of major projects funded by United Kingdom and European funding agencies including the European Union project Smart Controls and Thermal Comfort (SCATS). He works at Oxford Brookes and London Metropolitan Universities. At both he was responsible for developing multidisciplinary Master of Science courses in energy efficient and sustainable buildings. He was recently awarded professorships by both universities and is deputy director of the Low Energy Architecture Research Unit (LEARN). Currently convenor of the Network for Comfort and Energy use in Buildings, he is organizing the conference of the same name due to take place in April 2006. He is the principal investigator in the project "Predicting the Effect of Occupant Behaviour on Thermal Comfort and Energy Use in Buildings" and is also working on "Adaptive Strategies for Climate Change in the Urban Environment" and the EU-funded project "Sustainable Architecture Applied to Replicable Public-Access Buildings (SARA)".

TORSTEN J. NAUCKE, PhD, DSc is a parasitologist and entomologist specialized in medical parasitology. Based since 1992 at the Institute for Medical Parasitology, University Bonn, Germany, he has managed main projects in Greece (1993–1997) and in Spain (2000–2001), while current field-projects are the discovery and distribution of diseases transmitting arthropods (such as *Phlebotomus* and *Dermacentor*). A recent topic is the diagnosis of 'new diseases' in people and animals, like Anaplasmosis, Babesiosis, Filariosis, Hepatozoonosis, Leishmaniasis, Mycoplasmosis imported into or present in Germany.

GIUSEPPE NOCELLA has a degree in agriculture, an MSc in Agricultural Economics (University of Aberdeen, United Kingdom) and is currently PhD candidate at the University of Newcastle upon Tyne (United Kingdom). Since 1995, he has held courses on statistics and has been responsible for IT at the University of Bologna's Faculty of Agriculture in Cesena. He has also taught consumer behaviour in the Master's programme in food quality and safety held at the University of Florence. His research interests are mainly based upon consumer demand, behaviour and trust, an area in which he has published several articles in international journals and on which he has lectured at conferences.

ANNA PALDY is a medical epidemiologist with 30 years experience in different fields of public health. She has been working at the National Institute of Public Health (in the legal predecessor of the institute) since 1981, from 1981–1985 in the Department of Toxicology, from 1985–1998 at the Department of Community Hygiene, in 1998 she was nominated to be the leader of Department of Biological Monitoring and the Group of Information and Statistics, in 2002 – the deputy director of the institute. She received a MSc in public health in 1996, from the University of Kuopio, Finland; she wrote her PhD in environmental epidemiology in 1990 with the title of the thesis being "The Effect of Pesticides on the Health of Rural Population". She has been involved in several multicentre as well Hungarian studies. Her research fields covered cytogenetics and environmental epidemiology focusing on the effect of pesticides, air and water pollutants. During last ten years she studied the health impact of environmental pollution on mortality and morbidity, among others the health impact of climate change.

EDMUND PENNING-ROWSELL, BSc, MA, PhD is professor of geography and Pro Vice-Chancellor for Research at Middlesex University and is also head of the Flood Hazard Research Centre. He is specialized in natural hazard assessment and policy, with special reference to water planning. He has published several books and many papers on his research, and acted as a consultant to numerous national and international environmental agencies.

MILAN PEJČOCH, MSc is a microbiologist studying the natural foci of infections, foremost on the hantaviruses. As an expert at the National Institute of Public Health, Prague and head of the Department of Vector-borne Diseases at the Institute of Public Health in Brno, he has been working in the field of medical parasitology and bacteriology, nosocomial diseases, vector-borne infections and their ecology, as well as in the control of vectors of infections.

MILAN ŠCAZNY graduated from the Institute of Economic Studies at the Faculty of Social Sciences, Charles University in Prague; he is a PhD student in environmental economics at Charles University Prague, and the leader of the Environmental Economics Unit at the Environment Centre, Charles University Prague. He worked for the Czech Ministry of the Environment, Department of Environmental Economy from 1997–2000. His research focuses on the economic aspects of sustainable development, material flows analysis, environmental accounting, environmental tax/fiscal reform and evaluation of environmental externalities.

CRISTINA TIRADO is Food Safety Regional Advisor at the WHO Regional Office for Europe in Rome, responsible for the provision of support to 52 Member States on the development/strengthening of food safety strategies, policies, legislation, surveillance and monitoring programmes. She is a doctor in veterinarian medicine specialized in food science, food technology and food sanitation, with a MSc/PhD on Environmental Sciences and Aquaculture from Cornell University. Before arriving in Rome she was coordinating the WHO Surveillance Programme for Control of Food-borne Diseases in Europe at the FAO/WHO Collaborating Centre for Research and Training in Food Hygiene and Zoonoses of the Federal Institute for Risk Assessment in Berlin. Previously, she worked on food safety and on environmental health issues for several private, governmental and non-governmental agencies and international organizations such as the FAO.

TERESA WILSON, BSc, MSc is a research fellow at the Flood Hazard Research Centre, Middlesex University, United Kingdom. She has worked on a variety of quantitative and qualitative research projects in the field of flood-risk management. These include the evaluation of social vulnerability to flooding, and the risk to life posed by flooding in the United Kingdom. Other activities include the mapping of flood risk and flood impacts, including the analysis and mapping of census data and the calculation, mapping, and interpretation of flood-incurred economic damages.

TANJA WOLF, a medical geographer, started working for the WHO after receiving her degree. She is mainly responsible for carrying out literature reviews, on sustainable development, heat-waves and health, as well as mapping information on the achievements of the MDGs in the European Region. In addition, she is a contributing author to the 4th assessment of the IPCC.

1 Introduction

Bettina Menne, Kristie L. Ebi

Weather is an ancient human health exposure, says Hippocrates, in "On Airs, Waters and Places", circa 400 B.C. Although, there is still considerable uncertainty about the rate and extent of climate change that can be expected, it is now clear that these changes will be increasingly manifest through increases in extremes of temperature and precipitation, decreases in seasonal and perennial snow and ice extent, and sea level rise (Karl and Trenberth, 2003).

The project "Climate Change and Adaptation Strategies for Human Health" (cCASHh) was funded by the European Commission within its Fifth Framework Programme (FP5) under the thematic programme: energy, environment and sustainable development (EESD-1999) and the key action: global change, climate and biodiversity. The project started on 1 May 2001 and ended on 31 July 2004. The assessment was carried out in the WHO European Region.

The project started with the assumption that irrespective of actions that have been taken to reduce or halt climate change, human populations in Europe will be exposed to some degree of climate change over the coming decades. If the assumption is correct this requires anticipatory thinking on

- What are the climatic risks Europe might be facing?
- What can be learned from observed health impacts and vulnerabilities?
- What strategies, policies, and measures are currently available to reduce impacts?
- What are the costs?
- What are the projected future health impacts?
- Which policy responses need to be strengthened or developed?

The public health and climate change communities share the goal of increasing the ability of countries, communities and individuals to effectively and efficiently cope with the risks and changes that are likely to arise because of climate variability and change (Yohe and Ebi, 2005). Many lessons have been learned in public health in Europe over the last century. The present state of public health reflects the success or otherwise of the policies and measures designed to improve lifestyle, education, the environment, health care services and systems, and the economy. However, this does not mean that Europe is immune to climate-related impacts. An example is the extended impacts of the 2002 floods and the 2003 heat-wave.

Although one record heat-wave and flood do not prove that Europe is becoming hotter or the weather more extreme, the impacts of these events highlight shortcomings in existing public health responses, such as the lack of mechanisms to predict the events, to detect rapidly the health effects, and the lack of knowledge on effective prevention measures.

Health outcomes do not depend solely on the actual exposure to weather or climate-related hazards, including the character, magnitude and rate of climate variation. Outcomes also depend on the natural or social systems on which health depends, the characteristics of the population, and the adaptation (prevention) measures and actions in place to reduce the burden of a specific adverse health outcome (the adaptation baseline).

The objectives of the cCASHh project were:

▌ to identify population vulnerability to the adverse health impacts of climate change

▌ to review current measures, technologies, policies and barriers to improving adaptive capacity

▌ to identify for European populations the most appropriate measures, technologies and policies to successfully adapt to climate change and

▌ to provide estimates of the health benefits of specific strategies or combinations of strategies for adaptation under different climate and socioeconomic scenarios.

The project focused on four areas of concern to public health: thermal stress, extreme weather events, vector- and rodent-borne diseases, and food- and waterborne diseases. Against expectations, little empirical evidence was available in Europe on observed health effects associated with climate and weather, including extreme events. Thus, during the first years of the project, empirical studies were undertaken in order to reach a suitable level of understanding of the observed health impacts of climate variability and change.

Assessment Methods

Extended literature assessments were carried out for all activities. As shown in Figure 1, epidemiological studies, policy reviews, and surveys were carried out for all health outcomes. In addition, informal information was obtained from experts during workshops. Contingent valuation and integrated assessment modelling were conducted for selected impacts. A detailed description of the methods is available from the published peer-reviewed papers (see Annex 1).

Fuessel and Klein (2004) reviewed guidelines on climate change vulnerability and adaptation assessment (see Chapter 3, Fuessel et al., in this book). Questions like what to adapt to, who to adapt, where to adapt, and how good the adaptation is were taken into consideration and are discussed in Chapters 3 to 7.

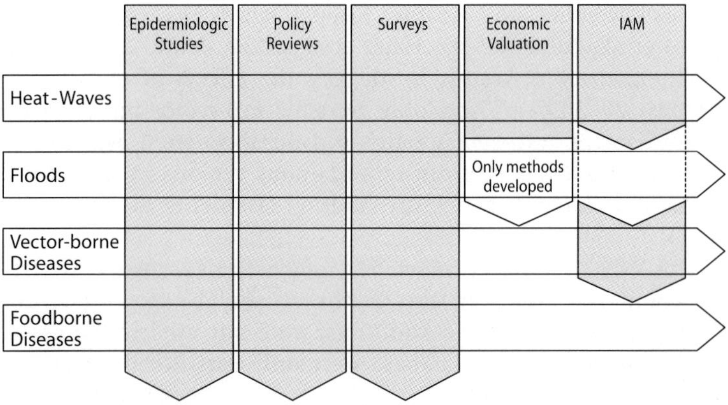

	Epidermiologic Studies	Policy Reviews	Surveys	Economic Valuation	IAM
Heat-Waves					
Floods				Only methods developed	
Vector-borne Diseases					
Foodborne Diseases					

Fig. 1. Methods used in the cCASHh study

The project faced several problems, including
- limited information on the current distribution and burden of climate sensitive diseases and associated population vulnerability;
- availability of data: the lack in many places of mortality and morbidity data over long enough time periods and at the sufficient spatial scale restricted the assessment of observed climate change-related impacts to few health outcomes;
- limited information on individual, social and environmental factors that modify health outcomes over time;
- lack of information on specific climate change related adaptation measures; and
- limited information on strategies, policies and measures to cope with the health effects of climate variability and change.

Quantitative and qualitative methods were used to obtain this information.

A severe heat-wave and flood occurred during the project that resulted in the loss of many lives. The project collected additional information on the impacts of these events as well as on available prevention measures. The adaptive capacity and country surveys were carried out after these events.

Epidemiological Studies

The main goals of the epidemiological studies were to assess the effects of weather and climate on health outcomes, and to better understand the current distribution and burden of climate sensitive diseases. Studies were conducted on (a) heat-related mortality, (b) temperature-related morbidity (hospital admissions), (c) foodborne disease (reported cases of salmonellosis and campylobacteriosis), and (d) vector-borne diseases. The ecological studies used time series methods to quantify the short-term associations between weather and health (Koppe, Jendritzky et al., 2004; Kovats, Edwards et al., 2004; Kovats, Hajat et al., 2004; Paldy, 2005).

The analysis of trends to identify the effects of observed warming over the past 20–30 years was only possible for vector-borne disease outcomes, where data of sufficient quality and duration were available: Seventy years of data were available from mountainous regions in the Czech Republic on altitudinal changes in tick distribution (Daniel et al., 2004; Danielova et al., 2004).

The WHO requested mortality information from the statistical offices of EU countries. However, the deaths attributable to floods and heat-waves are highly underreported and, thus, were not used in the assessment. Flood data from the EmDAT database were only used for illustrative and mapping purposes.

The WHO organized five expert and stakeholder workshops by thematic areas. The main scope of these expert and stakeholder consultations was to build consensus on the scientific content, on policy measures, and to identify knowledge gaps (Kirch and Menne, 2005; WHO, 2001; Hajat et al., 2003; Koppe, Jendritzky et al., 2004; WHO, 2004; WHO and EEA, 2004; WHO Regional Office for Europe, 2004).

Policy Review

In policy analysis a review on currently available measures was carried out, the effectiveness where possible reviewed and additional policies and measures where needed identified.

Qualitative approaches included:

- review of adaptation frameworks (Fuessel and Klein, 2004)
- review of existing policies and measures (Ebi in this book)
- several surveys were carried out that had the goal to
  - assess what is done by Ministries of Health to prevent the health consequences of extreme events (Kosatsky in this book)
  - assess the functioning and effectiveness of heat health warning systems. 44 meteorological services from 44 countries responded to the survey (Koppe, Jendritzky et al., 2004)
- discussions and interviews with scientists from vector-borne disease and foodborne disease surveillance networks.

The final choice of policies and measures to be adopted does not usually rest with the analyst (Scheraga et al., 2003). Rather it is the task of the analyst to clarify the choices and their implications or consequences. This information may then be used in the policy debate by the responsible authorities and by civil society, especially those likely to be affected by the policy choices.

Survey on Adaptive Capacity

Adaptive capacity is defined as the "potential, capability, or ability of a system to adapt to climate change stimuli or their effects of impacts" (IPCC, 2001), implying that at least in principle adaptation has the potential to reduce the damages of climate change, or to increase its benefits. The scope of the survey was to understand what were the most important determinants of adaptive capacity for the experts interviewed. Alberini and Chiabai (Chapter 8.2 in this book), developed a questionnaire (see Annex 4), that used the conjoint choice exercise as methodology (Louviere et al., 2000) and created an index of adaptive capacity by country.

Cost-benefit Analysis: Contingent Valuation Methodology

The original objective of the cCASHh study was to estimate the benefits of adaptation to the projected human health effects of climate change, and to discuss the challenges and difficulties associated with implementation. Alberini et al. (Chapters 9 and 10, Alberini et al. in this volume) conducted an economic valuation of the benefits of adaptation policies to mortality from heat events. It was necessary to place a monetary value on reductions in the risk of dying. To do so, economists estimate an individual' – and hence, society's – willingness to pay to reduce one's risk of dying. Willingness to pay is defined as the maximum amount that can be subtracted from an individual's income in the healthy state to keep his or her utility unchanged. One technique is Contingent Valuation, which is a survey-based method that relies on what individuals say they would do under specified circumstances. Two surveys were carried out in Italy and the Czech Republic.

Integrated Assessment

The integrated assessment work first reviewed existing scenarios, then developed causal pathways and models for thermal stress and some vector-borne diseases. To provide a rough indication of the spatial variation of health-related risks from climate change, climate-related European Union risk maps were developed. Chapter 11 (Martens et al.) describes the first step.

1.1   The Content of This Book

- GLENN MCGREGOR in Chapter 2 gives an overview of current and potential future climates in Europe.
- MARTIN FUESSEL et al. in Chapter 3 provide an overview of existing assessment frameworks on vulnerability, adaptation, risks, and impacts and reports on key steps necessary for adaptation assessment and planning.

The following four chapters and subchapters describe the observed associations between weather and climate with vector- and foodborne diseases, and describe the health effects of heat-waves and floods. The chapters also describe available prevention measures.

- Sari Kovats and Gerd Jendritzky in Chapter 4 assess the health impacts of heat-waves in Europe, including information on the impacts and responses to the 2003 heat-wave event.
- Kristie L. Ebi et al. in Chapter 5 describe the observed and projected impacts of floods in Europe. Adverse health outcomes can occur during or immediately after the event (such as injuries) or can occur later as a consequence of the event or its clean-up (such as mental health disorders).
- Chapter 6 is divided into six subchapters.
- Elisabeth Lindgren and Torsten Naucke (Chapter 6.1) describe the results of a literature review combined with input from experts on influences of climate on the epidemiology and ecology of Leishmaniasis.
- Elisabeth Lindgren and Thomas T. G. Jaenson (Chapter 6.2) describe the results of a literature review combined with input from experts on Lyme disease, including risk factors and the potential effects of climate change.
- Daniel et al. (Cchapter 6.3) describe the results of a literature review on tick-borne encephalitis, including an ecological assessment of ticks, disease risk factors, and the potential effects of climate change.
- Kuhn (Chapter 6.4) describes the current geographic distribution of the vector that can carry malaria, the possible effects of climate change on malaria in Europe, and future research needs.
- Zdenek Hubálek and Bohumír Kříž (Chapter 6.5) discuss the ecology, epidemiology, and prevention of West Nile virus in Europe. West Nile fever (WNF) is reemerging in Europe (since 1996) and has significant public health impact in affected areas.
- Milan Pejcoch and Bohumír Kříž (Chapter 6.6) assess the ecology, epidemiology and prevention of hantaviruses in Europe.
- Sari Kovats and Cristina Tirado in Chapter 7 review literature on the effects of ambient temperature on food-borne disease incidence, the methods used to evaluate the relationship between temperature and food-borne disease, and strategies for the surveillance and control of food- and water-borne disease in the context of climate change.

The next section considers current adaptation measures, such as preparedness for climate change, and an analysis of policies for and indicators of adaptive capacity.

- Ebi et al. in Chapter 8 describe policy processes and responses to climate change-related health risks at the level of communities, nations, and the European Union. Many measures are available to cope with current climate variability. However, additional adaptation measures will likely be needed in several settings to cope with climate change, while maintaining or improving current public health standards.

▌ TOM KOSATSKY (Chapter 8.1) assesses the results of a questionnaire sent to Ministries of Health in the WHO European Region to explore what is done to prevent health impacts associated with climate extremes. There is a wide variation in preparedness and response activities among the WHO European Region member states.

▌ ALBERINI and CHIABAI (Chapter 8.2) describe a conjoint choice exercise and a survey of public health officials and climate change expert to prioritize determinants of adaptive capacity for selected climate change-related health effects.

▌ The next two chapters focus on the economic valuation of temperature-related mortality and the benefits of adaptation policies. ALBERINI et al. in Chapters 9 and 10 describe the methods and report the results of two studies in the Czech Republic and Italy.

▌ PIM MARTENS et al. in Chapter 11 describe scenarios of population health in Europe based on the SRES scenarios.

In the conclusion (Chapter 12), we discuss key issues, such as the climatic risks that Europe might face, what can be learned from observed health impacts and vulnerabilities, adaptation options available to reduce projected health impacts, what are the costs, projected health impacts of climate change, measures that are likely to be needed to strengthen adaptive capacity, and research needs.

References

Daniel M, Danielova V et al (2004) An attempt to elucidate the increased incidence of tick-borne encephalitis and its spread to higher altitudes in the Czech Republic. Int J Med Microbiol 293(Suppl 37):55–62

Danielova V, Daniel M et al (2004) Prevalence of Borrelia burgdorferi sensu lato genospecies in host-seeking Ixodes ricinus ticks in selected South Bohemian locations (Czech Republic). Cent Eur J Public Health 12(3):151–156

Fuessel H, Klein R (2004) Conceptual Frameworks of Adaptation to Climate Change and their Applicability to Human Health. Potsdam Institute for Climate Impact Research (PIK), Potsdam

Hajat S, Ebi KL et al (2003) The human health consequences of flooding in Europe and the implications for public health: a review of the evidence. Applied Environmental Science and Public Health 1(1):13–21

IPCC (2001) Climate Change 2001. Impacts, Adaptations and Vulnerability. Contribution of Working Group II to the Third Assessment Report of the Intergovernmental Panel on Climate Change. Cambridge University Press, New York

Karl TR, Trenberth KE (2003) Modern global climate change. Science 302(5651): 1719–1723

Koppe C, Jendritzky G et al (2004) Heat-waves: impacts and responses. World Health Organization, Copenhagen

Koppe C, Jendritzky G et al (2004) Heat-waves: impacts and responses. Health and Global Environmental Change Series, No 2. W R O f Europe. World Health Organization, Copenhagen

Kovats RS, Edwards SJ et al (2004) The effect of temperature on food poisoning: a time-series analysis of salmonellosis in ten European countries. Epidemiol Infect 132(3):443–453

Kovats RS, Hajat S et al (2004) Contrasting patterns of mortality and hospital admissions during heat-waves in London, UK. Occup Environ Med 61(11):893–898

Louviere J, Hensher D et al (2000) Stated Choice Methods: Analysis and Applications. Cambridge University Press, Cambridge

Paldy A, Vamos A, Kovats RS, Hajat S (2005) The effect of temperature and heat-waves on daily mortality in Budapest. In: Kirch W, Menne B, Bertollini R (eds) Extreme weather events and public health responses. Springer, Berlin Heidelberg New York

Scheraga JS, Ebi KL et al (2003) From science to policy: developing responses to climate change. In: McMichael A, Campbell-Lendrum D, Corvalan C et al (eds) Climate change and health: risks and responses. WHO, Geneva, pp 237–266

Kirch W, Menne B, Bertollini R (ed) (2005) Extreme weather events and public health responses. Springer, Berlin Heidelberg New York, p 360

WHO (2001) Floods: Climate Change and Adaptation Strategies for Human Health. WHO, Copenhagen, EUR/01/503 6813

WHO (2004) Report of WHO meeting on Climate and Foodborne Disease. Rome, Italy, 27–28 February 2003. WHO, Rome (internal document)

WHO and EEA (2004) Report of the meeting "Extreme weather and climate events and public health responses" Bratislava, Slovakia, 9–10 February 2004. World Health Organization, Copenhagen

WHO Regional Office for Europe (2004) Declaration of the 4th Ministerial Conference on Environment and Health, Budapest, Hungary, 23–25 June 2004 (EUR/04/5046267/6):11

WHO (2004) Climate change and vector borne diseases. Internal WHO Report

Yohe G, Ebi K (2005) Approaching adaptation: parallels and contrasts between the climate and health communities. A Public Health Perspective on Adaptation to Climate Change. In: Ebi K, Burton I (eds)

2 Climatic Variability and Change across Europe

Glenn R. McGregor

2.1 Introduction

From a societal perspective, climate may be viewed a resource, a determinant or a hazard. Therefore, climatic variability and change are likely to have impacts on the resource base, alter the conditions under which a range of human activities can take place and present challenges to the safety of people and property. Understanding the nature and causes of climatic variability and change is, therefore, an important pre-requisite for developing effective adaptation strategies to cope with the vagaries of climate. The purpose of this chapter, therefore, is to give an overview of the nature of European climates in terms of their mean and variability characteristics, explain some of the mechanisms that contribute to climatic variability across Europe and assess how global warming might affect various aspects of Europe's climate.

2.2 The Climate System of Europe

The European climate system is part of the wider global climate system composed of the five interlinked components of the atmosphere, hydrosphere (oceans and land water bodies), the cyrosphere (snow and ice), the biosphere (flora and fauna) and the land surface (Fig. 1). These five components interact to produce the climate of a location while the changing relationship between them determines the nature of a location's climatic variability and climate change. The atmosphere is a fluid and, therefore, has a circulation, which manifests itself as the system of global and regional wind patterns. It also acts as an absorbent of heat emitted from the earth's surface in the form of terrestrial radiation, and a conduit for solar radiation as it makes its way from the sun to the earth's surface. Within the atmosphere a range of physical processes occur as well as chemical reactions that govern the levels of important gases such as methane and ozone. In the case of Europe, the main facets of the atmospheric circulation are the North Atlantic Oscillation (NAO), the upper level westerlies or Rossby waves, the Polar Front, the storm tracks and the Siberian high pres-

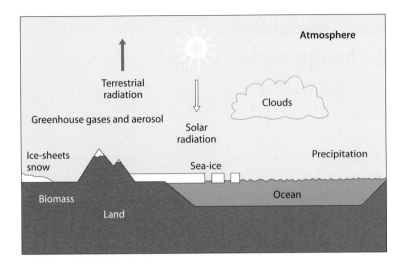

Fig. 1. The climate system

sure system. The NAO exerts an important influence on the inter-annual variability of climate and the trajectory of storm tracks, especially in winter, the upper westerlies "steer" weather systems across Europe, the Polar front represents the boundary between cold Arctic and warm moist tropical air masses and is where many weather systems that affect Europe develop, while the intensity and extent of the Siberian High, which is essentially a winter-time feature, is important for advecting very cold continental air masses over Europe from the east.

The oceans that surround Europe and their interaction with the atmosphere exert an important influence on climate. This is because there is continuous exchange of heat, momentum and water between the ocean and the atmosphere such that flows of air that cross the Atlantic Ocean or Mediterranean Sea often take on the characteristics of the underlying sea surface. The climate of locations immediately down-wind of such large water bodies are, therefore, generally warm and moist with a damped seasonal variation of temperature. Like the atmosphere, the North Atlantic ocean has a circulation; the main features of which are the large clockwise circulation of water referred to as the North Atlantic gyre. On the western side of the Atlantic basin, the northward flowing branch of the gyre is the Gulf Stream. This transports large amounts of heat to high latitudes in the Atlantic Basin, thus, keeping the climates of downwind regions, such as Scotland and Norway, much warmer than they would be otherwise. Transfer of heat from this warm current into the atmosphere also renders the atmosphere unstable, thus, leading to the occasional development of volatile storm systems and widespread rainfall producing weather systems that cross the Atlantic and influence Europe. The northern limit of the Gulf Stream coincides approximately with the region of formation of North

Table 1. General characteristics of main air mass types affecting Europe (modified from Robinson and Henderson-Sellers, 1999)

Air mass	Source region	Physical Properties at source
Polar maritime	Oceans north of approximately 50 degrees	Cool, moderately damp and unstable
Polar continental	Continental regions near or within Arctic circle such as Siberia and the position of the Siberian High	Cold, dry and very stable
Arctic maritime	Ocean regions within the Arctic circle	Cold, dry and stable
Tropical maritime	Subtropical North Atlantic in region of Azores high pressure system	Warm and moist. Unstable (stable) on western (eastern) side of North Atlantic basin
Tropical continental	Low latitude deserts over North Africa such as the Sahara	Hot, very dry and unstable

Atlantic Deep Water (NADW). This is because the northward flowing waters of the Gulf Stream cool and begin to sink to great depths within the ocean. This water then begins to flow south towards the equator and represents the beginning of the global thermohaline circulation, which is often referred to as the meridional overturning circulation (MOC). The position of the formation of NADW and the intensity of the MOC exert a fundamental control on Europe's climate at a variety of timescales. The counterpart of the Gulf Stream is the relatively cold southward flowing Canary Current on the eastern side of the Atlantic Basin. This transports cool water southwards off the coasts of Portugal and Northwest Africa creating stable atmospheric conditions that suppress rainfall formation resulting in dry climates in these subtropical coastal and maritime regions. Because of its high heat capacity, the ocean also acts as a heat sink to delay climate change.

The European land surface has an important influence on the flow of air over it, the absorption of solar energy, and the hydrological cycle. Mountains and high elevation areas such as the European Alps act to force the ascent of air, thus, creating unstable air masses or steer winds around their flanks. The nature of the land cover, whether it be urban, industrial, agricultural or natural, influences the absorption of heat at the earth's surface and the exchange of moisture with the overlying atmosphere. Seasonal snow cover also has similar effects on heat and moisture exchanges. The main elements of the cryosphere in the wider Atlantic-European sector are the Greenland ice sheet and Arctic sea ice. The former represents a high elevation surface that exerts an influence both on the atmospheric circulation and the absorption of energy in the high latitudes as well as a large source of fresh water for the North Atlantic and thus ocean salinity levels and the nature of the global thermohaline circulation. Sea ice cover is important as it decouples the ocean and atmosphere and halts the exchange of energy between these. Sea ice growth withdraws freshwater from the ocean, which leads to increases in ocean salinity. The terrestrial and marine biosphere, through sequestering and producing carbon in its gaseous or solid form, play a major role in the carbon cycle and, hence, the atmospheric concentration of carbon dioxide. In fact the summer greening of the European forests have a marked drawdown effect on global carbon dioxide levels as large amounts are consumed in the process of photosynthesis.

2.3   Climate Characteristics of Europe

The general controls on climate at a specific location in Europe are latitude, distance from the North Atlantic Ocean and the Mediterranean Sea (continentality), elevation and aspect. Furthermore a location's climate will be very much determined by its air mass climatology (Fig. 2) such that a high frequency of occurrence of polar continental air masses will result in a climate typified by a very cold winter and a warm to hot summer. In contrast locations that experience a high frequency of maritime tropical air masses will possess mild wet winters and moderately warm and humid summers.

A general impression of the climate characteristics of Europe can be gained by considering the climatological distribution of temperature and precipitation as portrayed in Figs. 3 and 4 for the winter (December, January and February) and summer months (June, July, August). In winter the distribution of isotherms, or lines of equal temperature, take on a more northwest to southeast orientation compared to summer especially in the case of Western Europe (Fig. 3a and 3b). This is a result of the dominating effect of the northward flow of warm water in the Gulf Stream of the North Atlantic that reaches about 70°N between Norway and Iceland and brings warm conditions to western Europe. The very low winter temperatures of eastern Europe are associated with little absorption of incoming solar radiation due to the high albedo or reflectivity of the snow covered land surface as well as the dominance of the Siberian high pressure system which maintains clear skies for many days and therefore allows large amounts of heat to escape to space. On occasions an increase in the intensity of the Siberian High results in the outflow of very cold dry continental air from Siberia all the way across Europe reaching the United Kingdom on occasions. In summer, the influence of the Gulf Stream is less and the temperature pattern across Europe is more a consequence of the astronomical position of the sun and the heating of the land surface so that the orientation of the isotherms is almost reversed. Comparing summer with winter also reveals that the seasonal temperature range increases from the west to east across

Fig. 2. The main air masses affecting Europe

Europe as the continentality or the distance from the Atlantic Ocean and Mediterranean Sea increases.

The regional variation in precipitation reflects the varying interaction between westerly flows of moist air off the Atlantic and the main mountain ranges of Europe. By forcing the flow of moist air up and over these physical barriers, precipitation formation is enhanced and, thus, some of the highest precipitation areas are associated with Europe's mountain regions such as the Highlands of Scotland, the west coast of Norway, the Central Massif of France, the French and Spanish Pyrennes, the Sierra Nevadas of Spain and the Alps, Caucus and Ural mountains of central and eastern Europe. During winter, much of the precipitation over these high elevation regions falls as snow, which in many cases may remain on the ground well into the early summer months. During summer, these same mountain ranges may initiate the development of thunderstorms with attendant in-

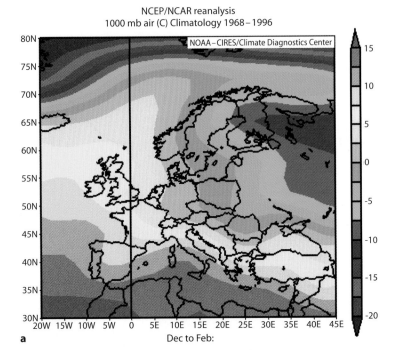

a   Dec to Feb:

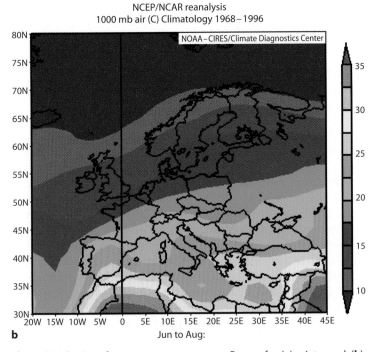

b   Jun to Aug:

Fig. 3. Distribution of mean temperature across Europe for (**a**) winter and (**b**) summer. Source: www.cdc.noaa.gov

tense falls of rain. While all of Europe may experience precipitation in excess of 1–2 mm/day during the winter months, there is a noticeable band of near zero rainfall in summer stretching across southern Europe from the Iberian peninsula in the west to southern Turkey in the east (Fig. 4a and 4b).

If the seasonal characteristics of temperature and precipitation across Europe are considered jointly, a number of major climate regions may be identified as described by the Koeppen climate classification (Fig. 5). This shows that southern Europe is dominated by the classic Mediterranean climates in which the summer is hot and dry with the average temperature of the warmest month being greater than 22 °C and the wettest month of winter possessing at least three times as much rainfall as in the driest month (less than 30 mm) of summer. Western Europe in general is dominated by a cool summer with the average temperature of the warmest month less than 22 °C and at least 30 mm of precipitation in the driest month. This matches the classic temperate humid climates found elsewhere in the world in mostly mid-latitude locations where the winters are wet and mild and the summers are drier than the winter and never overly warm. The continental climates of eastern Europe fall within the cold category of the Koeppen climate classification system. Such climates are typified by at least 30 mm of rain in the driest month with average temperature in the warmest month less than 22 °C while in the coldest month the average temperature is less than –3 °C. Northern Europe also falls within the cold climate category, as for eastern Europe, but differs from the latter in that the summer is short and cool with less than four months with an average temperature greater than 10 °C. Other less geographically important climates of Europe include the hot and cold dry steppe climates of the Iberian Peninsula and south eastern Europe respectively, the isolated high mountain climates of mainland Europe and Scandinavia and the warm wet climates of the northern coastal regions of the Ioannian and Black Seas (Fig. 5).

2.4 Climatic Variability

Although climate maps such as those presented in the previous section may create the impression that European climate varies little over time, this is not the situation in reality. This is because of the continuously changing relationship between the various components of the European climate system. Consequently a distinct characteristic of Europe's climate is its variability. Climate variability refers to variations in the mean state and other statistics (such as standard deviations, the occurrence of extremes etc.) of the climate on all temporal and spatial scales beyond that of individual weather events (e.g. inter-annual through decadal to millennial). Variability may be due to natural internal processes within the climate system (inter-

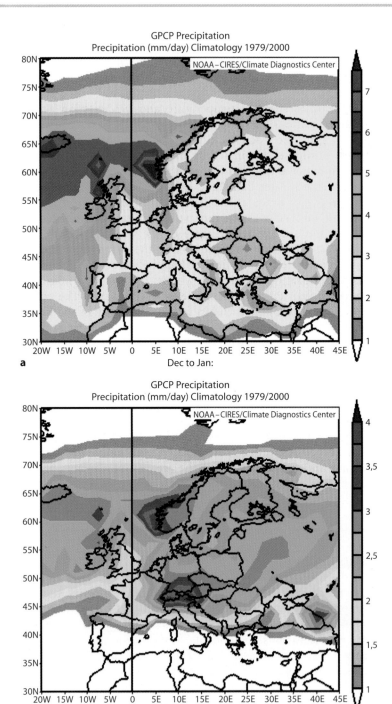

Fig. 4. Distribution of mean precipitation across Europe for (**a**) winter and (**b**) summer. Source: www.cdc.noaa.gov

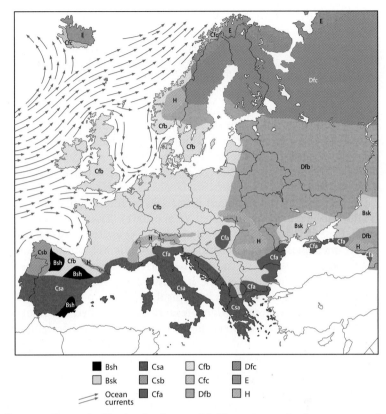

Fig. 5. The Koeppen climate classification for Europe. *Bsh* Warm steppe climate, mean annual temperature >18 °C; *Bsk* Cold steppe climate, mean annual temperature <18 °C; *Csa* Coldest month has temperature <18 °C but above −3 °C. Hot and dry summer period with mean precipitation of driest summer month less than one third of the wettest winter month. Driest month receives less than 30 mm. Mean temperature of the warmest month >22 °C; *Csb* Coldest month has temperature <18 °C but above −3 °C. Warm and dry summer period. Mean precipitation of driest summer month less than one third of the wettest winter month. Driest month receives less than 30 mm. Mean temperature of the warmest month <22 °C but at least four months with a mean temperature >10 °C; *Cfa* Coldest month has temperature <18 °C but above −3 °C. Moist with precipitation well distributed throughout the year and therefore no dry season. Mean temperature of the warmest month is >22 °C. *Cfb* Coldest month has temperature <18 °C but above −3 °C. Warm summer but moist with precipitation well distributed throughout the year and therefore no dry season. Mean temperature of the warmest month <22 °C but at least four months with a mean temperature >10 °C; *Cfc* Coldest month has temperature <18 °C but above −3 °C. Cool summer but moist with precipitation well distributed throughout the year and therefore no dry season. Mean temperature of the warmest month is <22 °C, but with one to three months with temperature >10 °C; *Dfb* Warmest month is >10 °C. Coldest month is <−3 °C. Moist with precipitation well distributed throughout the year. Warm summer with mean temperature of warmest month >22 °C but at least four months with a mean temperature >10 °C; *Dfc* Warmest month is >10 °C. Coldest month is <−3 °C. Moist with precipitation well distributed throughout the year. Cool summer, mean temperature of the warmest month is <22 °C, but with one to three months with temperature >10 °C

nal variability), or to variations in natural or anthropogenic external forcing.

Two notable variations in European climate over the last 1000 years are the Medieval Warm Period (MWP) and the Little Ice Age (LIA). The MWP occurred between the 11th and 14th when temperatures were about 0.2–0.3 °C warmer than the 15th to 19th centuries. However temperatures never achieved the levels they have in the 20th century. The LIA began in the 13th or 14th centuries and culminated somewhere between the mid 16th and 19th centuries. During this period temperatures across Europe were up to −1 °C cooler than present. Although these periods are of academic interest, variations at the inter-annual timescale bear greater implications for contemporary society.

Perhaps one of the strongest influences on inter-annual climatic variability in Europe is the NAO (Mitchell et al., 2001). The NAO can be simply described as an oscillation in atmospheric pressure between sub-polar and sub-tropical latitudes over the North Atlantic. This oscillation reaches its maximum expression in the Northern Hemisphere winter when an area of low pressure, referred to as the Icelandic Low, occurs in the vicinity of Iceland. At the same time over sub-tropical latitudes, in the region of the Azores, a semi-permanent area of high pressure occurs called the Azores High. The difference in pressure between the Azores High and the Icelandic Low creates a pressure gradient that determines the strength of the westerly winds over Europe such that a coincident strengthening or weakening of pressure in these two centres of action results in contrasting winter climates for western Europe. When atmospheric pressure in the Azores High region is anomalously high (low) and pressure in the Icelandic low is anomalously low (high), the NAO is said to be in a strong positive (negative) phase (Fig. 6). The cli-

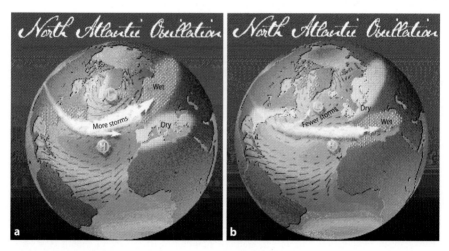

Fig. 6. Positive (**a**) and negative (**b**) phases of the North Atlantic Oscillation. Source: www.ldeo.columbia.edu/NAO/

matic consequence of a positive NAO phase is an intense winter atmospheric circulation over Western Europe with resultant mild and wet conditions; in the same phase, southern Europe and the Mediterranean experience warm and dry conditions. In contrast, a negative NAO phase, the result of the reversal of the "normal" pattern of atmospheric pressure, leads to a weakened westerly flow, the predominance of winds from a northerly direction and consequently cold dry conditions over Western Europe. Southern Europe and the Mediterranean, on the other hand, experience mild wet conditions. Unlike other low frequency variations in atmospheric circulation, such as ENSO and the Quasi Biennial Oscillation, which are important components of the climate system in subtropical to tropical regions, the NAO has no dominant periodicity although there is a hint of a weak periodicity at the 2–3 year timescale (Stephenson et al., 2000).

Over the last century, the NAO has demonstrated both persistence and variability (Fig. 7). For example the periods 1903–1914, 1920–1937 and 1973–95 were dominated by a NAO positive phase, which contributed to higher than normal temperatures during these periods. In contrast 1950–1960 was characterized by a negative phase when winter temperatures were frequently lower than normal. The recent positive phase of 1973–1995 also coincides with the period for which European surface air temperatures have shown rapid increases. Interestingly there is some evidence that the NAO may be returning to a period of more "normal" activity with no one particular phase dominating for a prolonged period.

The importance of the NAO for controlling winter climate at the interannual level is highlighted by the fact that variations in the NAO can explain 30–40% of the variation in temperature across Western Europe as well as have marked impacts on rainfall distribution (Hurrell, 1995; Marshall, 2001). Consequently, the social and economic consequences can be considerable. For example the persistent NAO positive phase in the last 20 years of the 20th century led to some of the lowest snow depths recorded in the Alps bringing hardship to those industries dependent on winter

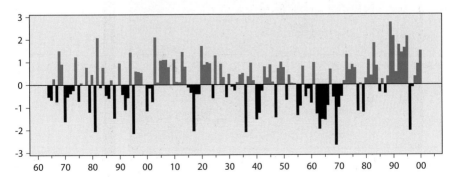

Fig. 7. Variation in North Atlantic Oscillation index 1865–2002. Source: www.cgd.ucar.edu/cas/jhurrel

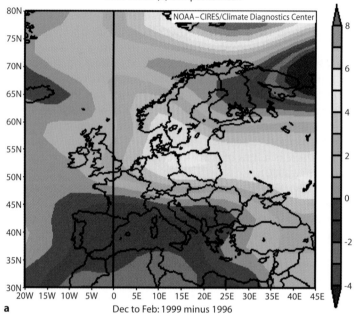

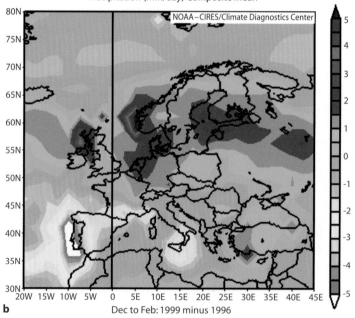

Fig. 8. Contrast in temperature anomalies across Europe for a strong negative (**a**) and positive (**b**) phase of the NAO. Source: www.cdc.noaa.gov

snowfall (Beniston, 1997). During the same period, severe drought prevailed across the Iberian Peninsula (Trigo et al., 2004) while Norway experienced a run of anomalously wet years and an advance of glaciers along its west coast. One notable anomaly during the predominantly positive phase of the NAO over the last 20 years was 1996 when the NAO became strongly negative; the outcome of which was below normal temperatures across most of western and northern Europe (Fig. 8). In contrast the Mediterranean region experienced anomalously warm conditions (the classic NAO negative phase pattern). In Norway, there were concerns about hydroelectric production because of below normal precipitation amounts, while in England there was unusually high winter mortality because of the anomalously low temperatures (McGregor, 2005).

In addition to influencing the general nature of winter conditions over Europe, the NAO plays an important role in determining the position of the winter storm tracks across the North Atlantic-European sector. When the NAO is in a positive phase, more North Atlantic storms traverse western Europe compared to the situation of a negative NAO phase, when the average storm track is displaced to the south so that more storms track through the Mediterranean Basin (Fig. 6). Furthermore, the trajectory taken by a particular storm can exert a strong influence on the geographical distribution of loss of life and damage to property, livelihoods and infrastructure. For example, cyclones Anatol (2–4 December 1999) and Janette (27–28 October 2002) brought damaging winds to northern England, Denmark, southern Sweden and northern Germany and Poland while cyclones Lothar (26 December 1999) and Martin (27–28 December 1999) caused extensive damage across France, southern Germany, Switzerland, northern Italy and parts of Austria. Because the mean trajectory of storms and coastal wave climates and sea levels are sensitive to storm related winds, the NAO also exerts an influence on the variability of sea level along the coast of north western Europe on a range of timescales (Yan et al., 2004).

Another important atmospheric centre of action for climatic variability across Europe is the Siberian High (SH). This is a semi-permanent area of high pressure, usually centred over northern Mongolia that is dominant in the winter. Associated with it are the coldest and densest air masses in the Northern Hemisphere. The SH forms mainly in October and can persist until the end of April. Although quasi-stationary, the SH can spread across Eastern Europe, occasionally extending into Western Europe. On such occasions outbreaks of very cold air push across Europe bringing falling temperatures, heavy snowfall down wind of major water bodies and a number of associated social and economic impacts (Walsh et al., 2001). Climatological analyses of the SH show that not only does it display marked inter-annual variability but also an unprecedented decline in its intensity over the last two to three decades, which may be related to increases in surface air temperature in the source region of the SH (Panagiotopoulos et al., 2005), a trend that might be expected with increasing concentrations of greenhouse gases (Gillette, 2003).

The atmosphere is not alone in influencing climatic variability over Europe as the ocean and atmosphere interact with each other mainly through variations in sea surface temperature (SST) exerting a control on the thermodynamics and dynamics of the overlying atmosphere. These interactions may occur over the short to long term such that changes in the intensity of the MOC and its surface component the Gulf Stream exert a fundamental control on the climate of northwest Europe. A weakening (strengthening) of the MOC brings about widespread cooling (warming) across western Europe. Paleo-climatic studies have also shown that changes in the nature of the MOC can occur within a decade resulting in rapid climate change in the Atlantic-European sector (Lockwood, 2001; Marshall, 2001; Seager et al., 2002; Taylor, 1966). In relation to the inter-annual timescale, recent investigations (Junge and Stephenson, 2003; Rodwell and Folland, 2002, 2003) have shown that a tripole anomaly pattern in North Atlantic SST has an influence on winter and spring seasonal variability in temperature and precipitation across Europe because this ocean climate pattern exerts an influence on the NAO. A positive tripole pattern, which is associated with a positive NAO index, comprises cooler than normal SST in the subtropical North Atlantic and to the south of Greenland, and anomalously warm seas off the east coast of North America. Variations in the tripole pattern, through its affect on rainfall patterns, have also been linked to flow variability in some of Europe's major rivers (Rimbu et al., 2002). North Atlantic variability in combination with tropical Pacific ENSO SST variability may also have an impact on seasonal climate variability over the North Atlantic/ European region (Sutton and Hodson, 2003). Evidence is also emerging for a link between extreme climate events such as the anomalously hot 2003 summer and ocean and atmosphere interactions in the tropical Atlantic related to wetter than average conditions in both the Caribbean and Sahel regions (Cassou et al., 2005).

Despite the theoretical importance for climate of the land surface (Dirmeyer, 2003; Pitman, 2003), little research has been conducted on the relationship between European climatic variability and land surface characteristics. Because changes in land cover occur beyond the timescale of inter-annual variability, the climatic response to natural or human-related alterations in surface characteristics is likely to manifest itself at decadal timescales or more (Schneider et al., 2004). However, on an annual basis, alterations of land surface characteristics occur across large tracts of Europe and the Northern Hemisphere because of seasonal snow cover. As snow cover variations influence the thermal and stability characteristics of the overlying atmospheric circulation, year-to-year variations in late winter Eurasian snow cover influence summer temperatures over Europe such that unusually cool summers across Europe follow winters with extensive Eurasian snow cover and vice versa (Qian and Saunders, 2004).

2.5 Climate Trends

As well as inter-annual variability, a noteworthy feature of climate across Europe is the trend in a number of facets of climate (Heino et al., 1999; Klein-Tank et al., 2002; Frich et al., 2002). For example, increases in minimum or night time temperatures across Europe have been noted for most locations over the period 1946–2004, with geographical contrasts apparent between the winter and summer and comparatively greater increases for summer compared to winter (Fig. 9a and 9b). Moreover night-time increases are greater than that recorded for maximum or day-time temperatures (Klein-Tank et al., 2002). Trends in warm spell duration (number of consecutive days above a given temperature threshold) have also been observed but unlike the situation for night-time temperatures, increases in this climate index are greater for winter compared to summer (Fig. 10a and 10b). Commensurate with warming winters have been decreases in snow depth and snow cover and the number of frost days across Europe (Bamzai, 2004; Bednorz, 2002, 2004; Brown, 2000; Farlaz, 2004; Hantel et al., 2000; Laternser and Schneebeli, 2003).

A feature of the period 1946–2004 is symmetric warming (Klein-Tank and Können, 2003) meaning that there has been an approximately equal increase in the occurrence of both cold and hot extremes and, thus, no change in temperature variability. However within this period, two "asymmetric" sub-periods, namely 1946–1975 and 1976–1999 may be identified due to contrasting relationships between the mean and extremes. For 1946–1975 a period of slight cooling occurred across Europe with an associated decrease in the number of warm extremes. However the annual number of cold extremes did not increase implying a reduction in temperature variability. In contrast, pronounced warming and an increase in the annual number of warm extremes at a rate two times faster than the expected change in cold extremes, which implies increased variability, characterized the period 1976–1999 (Klein-Tank and Können, 2003). From a biometeorological point of view such changes have had an impact on levels of human thermal comfort and the length of the discomfort season (McGregor et al., 2002), the phenology of a range of plant and animal species (Ahas et al., 2002; van Veilt and Schwartz, 2002) and the possible re-emergence of some tick-borne diseases (Randolph, 2004).

Over the last 50 years, statistically significant increases in total rainfall have been detected across Europe for winter but not so for summer. This is especially true in the case of northern Europe. Trends in measures of precipitation extremes partly mirror the observed trends for precipitation totals but the spatial coherence of trends is low (Fig. 11). However, where changes in annual amounts are significant there is a disproportionate change in the contribution of very wet days to precipitation totals, indicating an increase in precipitation extremes (Klein-Tank and Können, 2003). As well as precipitation totals and extremes, precipitation intensity is of in-

TN90p: Days with TN > 90th perc. of daily min. temp. (warm nights)
Trends 1946–2004 DJF

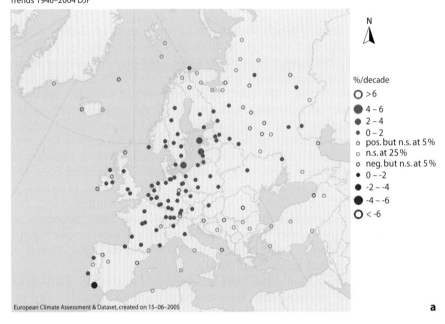

a

TN90p: Days with TN > 90th perc. of daily min. temp. (warm nights)
Trends 1946–2004 JJA

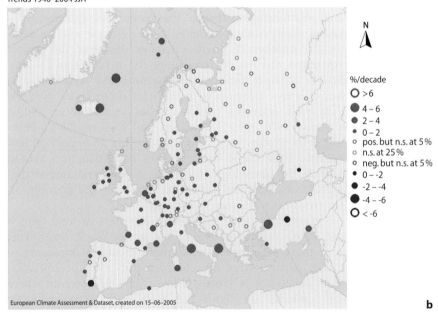

b

Fig. 9. Trend in minimum (night-time) temperature across Europe for (**a**) winter and (**b**) summer. Source: http://eca.knmi.nl

WSDI:Warm-spell duration index
Trends 1946-2004 DJF

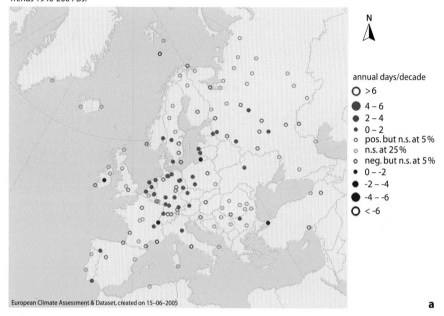

annual days/decade
- O >6
- ● 4 – 6
- ● 2 – 4
- ● 0 – 2
- o pos. but n.s. at 5%
- o n.s. at 25%
- o neg. but n.s. at 5%
- ● 0 – -2
- ● -2 – -4
- ● -4 – -6
- O < -6

European Climate Assessment & Dataset, created on 15–06–2005

a

WSDI:Warm-spell duration index
Trends 1946-2004 JJA

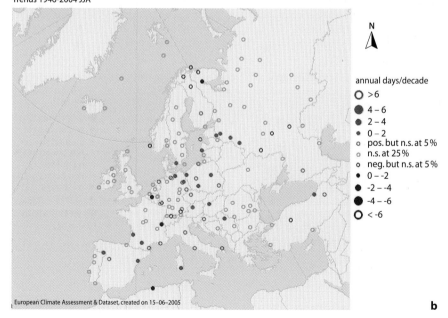

annual days/decade
- O >6
- ● 4 – 6
- ● 2 – 4
- ● 0 – 2
- o pos. but n.s. at 5%
- o n.s. at 25%
- o neg. but n.s. at 5%
- ● 0 – -2
- ● -2 – -4
- ● -4 – -6
- O < -6

European Climate Assessment & Dataset, created on 15–06–2005

b

Fig. 10. Trend in warm spell duration across Europe for (**a**) winter and (**b**) summer. Source: http://eca.knmi.nl

R95p: Days with RR > 95th perc. of daily an. (very wet days)
Trends 1946-2004 DJF

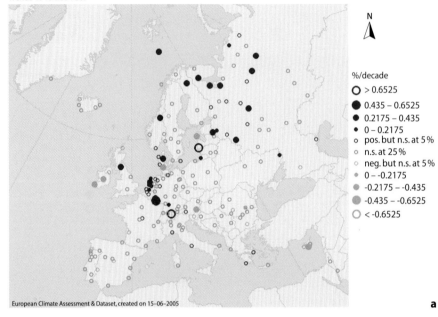

a

R95p: Days with RR > 95th perc. of daily an. (very wet days)
Trends 1946-2004 JJA

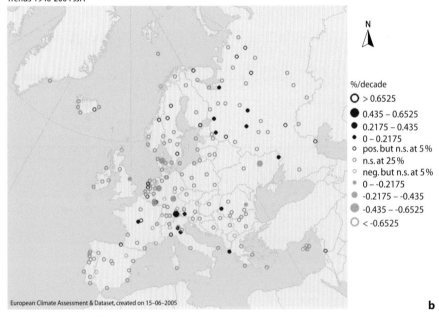

b

Fig. 11. Trend in 90th percentile of daily precipitation for (**a**) winter and (**b**) summer. Source: http://eca.knmi.nl

terest because potential increases in this precipitation characteristic have implications for flooding and soil erosion. For the United Kingdom, precipitation intensity increases have been noted but these are more so for the winter months (Osborn et al., 2000) as found for the European Alpine region (Frei and Schar, 2001). Elsewhere intensity increases have been associated with certain weather systems (Windman and Schar, 1998) and the changing relationships between wet day occurrence and wet day rainfall totals (Brunetti et al., 2000, 2001).

There have been few studies concerning the trends in humidity or atmospheric moisture over Europe. This is despite the fact that upward trends associated with increasing global mean temperatures are a distinct possibility (Trenberth, 1999). Like other facets of climate, atmospheric moisture over Europe displays marked variability from year to year (Phillips and McGregor, 2002). One measure of atmospheric moisture that has received some attention is precipitable water, the total depth of water produced if all atmospheric water was condensed and precipitated to the ground. As part of a study of precipitable water patterns across the northern hemisphere Ross and Elliot (2001) note that there are few detectable trends in this atmospheric moisture variable over Europe. For the atmosphere over southern Greece, Kassomenos and McGregor (2005) have observed moistening in maritime air masses and drying in their continental counterparts.

A limited number of studies have considered trends in storminess across Europe. Of the studies undertaken, most have focussed on Western Europe and have concluded that there are no discernable trends in storminess at this geographical scale. Rather inter-annual to decadal variability dominates and there is geographical variability in the temporal pattern of storminess (Bijl, 1999; Alexandersson et al., 2000; Pryor and Bathelmie, 2003; Schiesser et al., 1997; Stefanicki et al., 1998).

2.6   Climate Change

The Framework Convention on Climate Change (UNFCCC), in its Article 1, defines "climate change" as: "a change of climate which is attributed directly or indirectly to human activity that alters the composition of the global atmosphere and which is in addition to natural climate variability observed over comparable time periods". The UNFCCC, thus, makes a distinction between "climate change" attributable to human activities altering the atmospheric composition, and "climate variability" attributable to natural causes.

As there is mounting evidence that recent trends in European climate contain signatures of climate change (Klein Tank et al., 2005), much effort has been invested in undertaking climate change projections for Europe (Parry, 2000), which involves several steps. First, scenarios of energy production are used to construct global Greenhouse Gas Emission (GHGE) scenarios. These are then used as input into a global carbon cycle model that provides estimates of the sinks and sources of carbon. The balance between these provides an estimate of the increase in carbon dioxide concentrations in the atmosphere for a given GHGE scenario. Global climate models (GCM) are then run in order to establish how higher carbon dioxide concentrations may affect, for example, changes in temperature and precipitation. Estimated changes in climate variables form the input into climate change impact models, the results of which are used to assess the economic and societal consequences of a given change in climate.

Central to the development of climate change projections are GCMs. Although a variety of GCMs exist, associated with different climate modelling groups across the world, the general nature of them is the same. They are three-dimensional mathematical models that represent physical and dynamic processes that are responsible for climate. All models are first run for a control simulation that is representative of present-day climate or that of pre-industrial times. Next a perturbed run is made as the current levels of GHG bases are perturbed (increased) within the GCM to match some future level. Generally two types of perturbed climate model experiment can be run for estimating future climate namely equilibrium and transient-response experiments. In equilibrium experiments, the equilibrium response (new stable state) of the global climate following an instantaneous increase (e.g. doubling) of atmospheric CO_2 concentration is evaluated. Transient experiments are conducted with coupled atmosphere-ocean models (AOGCMs), which link detailed models of the ocean with those of the atmosphere. AOGCMs are able to simulate time lags between a given change in atmospheric composition and the response of climate. Most recent evaluations of climate impacts resulting from increases in GHG concentrations are based on scenarios formed from results of transient experiments as opposed to equilibrium experiments. Typically, the resolution of the output from GCMs is at a scale of 250×250 km. As this scale is often insufficient for making impact assessments, GCM model output is either downscaled, using statistical methods or dynamically by "nesting" a Regional Climate Model (RCM) within a GCM (Giorgi et al., 2004) so that higher resolution climate fields can be produced. In order to get an idea of the nature of the climate change resulting from increased GHG concentrations, the mean differences between the control run (current climate) and the perturbed run (future climate) are calculated. Generally comparisons are not made between current climate observations and climate fields for perturbed runs as GCM and RCM simulations of current climate may systematically overestimate or underestimate key climate characteristics (Moberg and Jones, 2004). Most climate change projections are made for fixed times (e.g. the 2020s, the 2050s, and the 2080s) or 30 year climatological periods (e.g. 2071–2100) in the future.

When analysing the differences between perturbed and control GCM and RCM runs, of interest is the nature of the climate change. Potentially the statistical characteristics of weather and climate events could change in a number of ways (McGregor et al., 2005). In the case of temperature, there simply could be a shift in the mean (location) towards higher (warmer) values. This would result in an increase in the number of extreme events at the hot end and a decrease at the cold end of the statistical distribution. Consequently, there would be not only more hot weather but also more record hot and less cold weather. As the probability of exceeding a fixed threshold may change non-linearly with shifts in the mean, a small change in the location could result in a large relative change in the probability of extremes. In addition to a change in location, there may be a change in the dispersion (scale) of the distribution. This would produce changes in the occurrence of extreme events at either end of the distribution and potentially have a greater effect on the frequency of extreme events than a simple change in location. In addition to changes in the location and scale, changes in the shape (skewness) of a distribution can be envisaged such that the distribution, of temperature for example, becomes skewed to the right. This would result in much more hot and record hot weather and fewer cold events. Many other types of change and combinations of changes in the distributions of climate variables are of course possible and will result in different climate outcomes (Ferro et al., 2005). For example, a pure shift in location is unlikely to occur in distributions of non-negative valued variables such as precipitation, for which location and scale often change simultaneously. This can alter disproportionately the occurrence of various aspects of precipitation extremes such as seasonal totals or daily intensities (Easterling, 2000).

A number of recent climate change experiments for Europe have used GCMs and RCMs 'forced' with the IPCC A2 and B2 GHG emission scenarios. While the B2 scenario represents a future typified by low emissions the A2 scenario assumes a high level of emissions throughout the 21st century, resulting from low priorities concerning greenhouse-gas abatement strategies and high population growth in the developing world (Naki'cenovi'c et al., 2000). Under this scenario, atmospheric CO_2 levels will reach about 800 ppmv by 2100 (three times their pre-industrial values). Projections based on this scenario, therefore, provide an estimate of the upper bound of climate futures discussed by the IPCC (Beniston, 2004).

Climate projections for the period 2071–2100 under A2 and B2 scenarios reveal considerable warming in all seasons ranging from 1–5.5 °C with temperatures generally 1–2 °C lower in the case of low emissions (Beniston et al., 2005; Ferro et al., 2005; Giorgi et al., 2004; Jones et al., 2001; Jones et al., 2004; Kjellstrom, 2004; Raisanen, 2004). Maximum warming occurs over northern and eastern Europe in winter. In summer, this is found over western and southern Europe. Changes in summer temperature extremes and periods of hot weather also occur under the A2 scenario. This can be seen in one projection of the impact of climate change on extreme events over Europe using the HIRHAM4 RCM developed at the Danish Meteorological Institute, which is one of the RCMs being used to investigate climatic change over Europe as part of the European Union project PRUDENCE

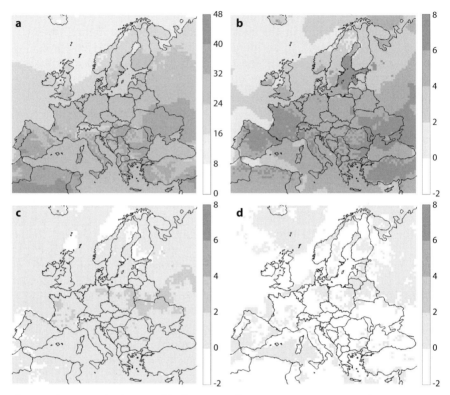

Fig. 12. Change in distribution of highest 10% of maximum temperatures under an A2 emissions scenario. (Source: McGregor et al., 2005)

(McGregor et al., 2005). Fig. 12 shows an analysis of the projected changes (A2 scenario minus control) in the 90th percentile for JJA maximum temperatures. The current situation is represented by the control run (Fig. 12a). This shows higher temperatures for the 90% quantile over regions possessing a Mediterranean climate and also central southeastern Europe where continental climates predominate (Fig. 5). The projected change in 90th percentile (Fig. 12b) is greatest in two clear regions, over southern France and central and northern regions of the Iberian Peninsula where the greatest temperature anomalies occurred during the August 2003 European heat-wave and regions bordering the Black Sea. Almost everywhere the change in the 90th percentile is greater than the change in median (Fig. 12c). Except in regions that experience an increase in skewness (not shown), the change in 90th percentile is mostly attributable to a change in both location and scale not just location alone indicating that a change in the variance, as well as the mean, will make a sizeable impact on future daily extreme temperatures. One other way in which changes in temperature extremes can be considered is by considering the change in

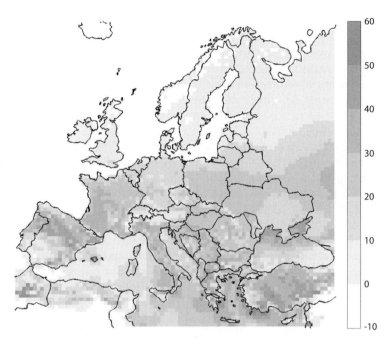

Fig. 13. Change in number of days above 30 °C across Europe under an A2 emissions scenario (Source: McGregor et al., 2005)

the number of days above 30 °C as this provides a crude approximation of the way in which heat-wave incidence and duration might change across Europe. According to the HIRHMAM RCM control simulation, the majority of Western Europe currently experiences about 5 to 10 days per summer with maximum temperatures in excess of 30 °C (McGregor et al., 2005). However, for the period 2071–2100 the situation could be quite different with the model simulation predicting increases of up to 60 days per summer in Mediterranean countries (Fig. 13). By 2100, countries such as France may experience temperatures above 30 °C as often as Spain and Sicily currently experience such events. Consequently, this climate model predicts increases in heat-wave frequency and duration across most of Europe, along with prolonged dry periods and increased probability of summer drought. Using a simple definition of a heat-wave, based on three successive days above 30 °C, a three- and ten-fold increase in the duration and frequency of heat-waves might be expected for many places across Europe by the end of the current century (Beniston et al., 2005).

In addition to alterations of the thermal climate, climate change is likely to bring about changes in precipitation amount and distribution. Climate models show the precipitation response to be far more variable than temperature. Apart from the south, winter precipitation increases are likely across most of Europe as a result of increases in atmospheric moisture content and storm activity (McCabe et al., 2001). In summer, Europe wide precipitation decreases have been predicted because of increased blocking activity over the northeastern parts of the North Atlantic. Summer storms off the Atlantic are, therefore, steered northwards with the result that in the far north summer wetness may increase. Although climate models indicate changes in winter and summer precipitation patterns and amounts, the situation is less clear for the transitional seasons of spring and autumn (Giorgi et al., 2004).

A consequence of the combined effects of increasing temperatures and changing precipitation patterns is an increased risk of drought across Europe in a warmer world (Jones et al., 2001). Predictions from three RCMs forced with the A2 emissions scenario show earlier onset and greater duration of drought. Using the annual maximum length of dry spell as an index of drought, Beniston et al. (2005) show that drought duration over the southwestern regions of the Iberian Peninsula is likely to increase by about 20–30 days compared to present. Other regions demonstrating marked changes in drought climatology include parts of central and western France as well as southern Italy and Greece. As well as changes in frequency and severity of meteorological drought, alterations to the distribution of hydrological drought are also likely with global warming (Strzepek and Yates, 1997; Graham et al., 2005).

Although summer precipitation amounts are likely to be substantially reduced over major parts of central and southern Europe, the opposite may hold true for trends in heavy precipitation amounts (Beniston et al., 2005; Giorgi et al., 2004; Semmler and Jacob, 2004). This is because climate change projections point to heavier/more intensive summertime precipitation events over some parts of Europe such as central and eastern Europe despite overall decreases in summer precipitation amounts. For Scandinavia and northeastern Europe increases in heavy precipitation return periods accompany like trends in precipitation amounts. Over large parts of southern Europe, where summer drying is likely, decreases in the return periods are expected. For winter, climate models indicate a clear geographical division between northern and southern Europe for changes in heavy precipitation with increases in return period in the north and the opposite in the south. This matches the projected changes in winter precipitation totals (Jones et al., 2001; Raisanen et al., 2004). Ferro et al. (2005) have also shown that climate change-related alterations in winter heavy precipitation characteristics over northern Europe, the Alps and around the Adriatic, as described by the 90% quantile, are mainly due to changes in the scale or dispersion of the statistical distribution.

Changes in summer precipitation intensity and return periods holds clear implications for flash flooding in summer (Christensen and Christensen, 2003) such that intensive precipitation events that led to recent flooding in the Elbe, Donau, Moldau and the Rhone are likely to become more frequent. As for summer, winter precipitation increases will have impacts on flood occurrence especially as increased precipitation will bring soil moisture capacities to, or close to, their maximum. Wet antecedent conditions are likely to increase the probability of wintertime flooding.

Future changes in storminess will have potential impacts on the occurrence of wind storms across Europe and storm surges in low lying coastal regions. Such changes are related to alterations in the distribution of atmospheric pressure and associated weather systems over the wider North Atlantic-European sector. Leckebusch and Ulbrich (2004) demonstrate changes in cyclone track density and extreme cyclone systems under an A2 scenario climate simulation using the HadCM3 GCM. The A2 scenario shows a tendency towards deepening cyclones over Spain, France, United Kingdom and Germany with associated increases in extreme wind events. Furthermore, an RCM (HadRM3H) simulation demonstrates an increase of the 95th percentile of the daily maximum wind speed over extended parts of Western Europe. Similarly Zweirs and Kharin (1998) have found evidence for a 3–25% increase in the 90th percentile for daily winter (DJF) wind speeds in a latitudinal band between 45 and 55 °N. These projections match those reported from the PRUDENCE project such that a 25% increase in the high percentiles of wind speed over western Europe, especially over the region between the Alps and the Baltic Sea, might be expected under an A2 climate change scenario (Beniston et al., 2005). Substantial impacts arising from changes in wind storm climatology may be expected as outlined by Dorland et al. (1999) for the Netherlands.

Storm surges represent a significant hazard for many low lying coastal regions of Europe (Trigo and Davies, 2002; Svennson and Jones, 2002). Future climate change-related rises in sea level would add to this risk. A combination of conditions conspires to create storm surges namely high water levels as a result of the tidal cycle and strong onshore winds (Holt, 1999). Simulations of climate change related storm surge indicate a greater frequency of surge events along the coasts of the Netherlands, Germany and Denmark (Beniston et al., 2005; Kass et al., 2001; Langenberg et al., 1999; Lowe et al., 2001; Lozano et al., 2004; Woth et al., 2004).

Compared to analyses of changes in temperature, precipitation and storms, the impacts of climate change on air quality has received little attention. Climate change could potentially impact local to regional air quality through shifting regional weather patterns and their associated statistics, increasing or decreasing anthropogenic emissions via changes in human behaviour and altering the levels of biogenic emissions as a result of higher temperatures and land cover change. However, establishing the scale (local, regional, global) and direction of change (improvement or deterioration) of air quality is a challenge. This is because future air quality, especially at the local to regional level, will be moderated by background levels of a range of pollutants at the global scale (Syri et al., 2002). Of the range of pollutants of concern for human health ozone has received most attention in the climate change literature (Derwent et al., 2001; Johnson et al., 2001; Jones and Davies, 2000; Langer et al., 2005). Assuming little change in background levels of ozone precursors, assessments indicate a strong increase in the mean daily maximum of ozone over southern and central Europe and a decrease in northern Europe in the summer. The trends in future summer air quality are a consequence of an increased frequency of stable anticyclonic conditions with little boundary layer ventilation and associated high temperature and cloud-free conditions.

2.7   Conclusions

The main controls on climate across Europe are the distribution of land and sea and the arrangement of major topographic barriers. Because of the relatively warm seas that border Europe, most of central and western Europe possesses a temperate climate, with mild winters and summers. Climate seasonality and therefore hot summers and cold winters, increases with distance from the Atlantic Ocean and Mediterranean Sea. A distinct characteristic of Europe's climate is its variability, which is controlled by the interplay between important elements comprising the atmospheric, ocean and land surface components of the European climate system. Chief among these is the North Atlantic Oscillation, which exerts a strong control on inter-annual climatic variability especially through altering winter temperature, precipitation and storm patterns.

Over the last 50 years, a number of clear trends are evident in European climate. These include an increase in minimum and maximum temperatures, changes in precipitation characteristics at a range of temporal and spatial scales and increases in the number of extreme events such as high temperatures, heavy precipitation and persistent dryness. These changes have occurred over the same period as the increase in global temperature. While there is still debate concerning attribution of these trends to human-related global warming, there is mounting evidence that Europe's climate may be responding to increases in the intensity of the global greenhouse effect.

Climate change holds a number of implications for Europe. In a warmer Europe, the current climate patterns will change with a possible expansion of Mediterranean and maritime temperate climates with a commensurate reduction in continental temperate climates. There is likely to be drying over southern Europe while northern Europe becomes milder and wetter. Not only will there be a change in mean climate conditions but there will be a shift in the climatology of extreme events with a reduction in return intervals for heat-waves, periods of prolonged dryness and drought, heavy precipitation, storminess and storm surges. While changes in mean climate may be within the coping range of society, future extreme events are likely to test coping strategies and require society to adapt to the changing European climate.

References

Ahas R, Aasa A, Menzel A, Fedotova G, Scheifinger H (2000) Changes in European spring phenology. International Journal of Climatology 22:1727–1738

Alexander L, Tett S, Jonsson T (2005). Recent observed changes in severe storms over the United Kingdom and Iceland. Geophysical Research Letters (in press)

Alexandersson A, Tuomenvirta H, Schmith T, Iden K (2000) Trends of storms in NW Europe derived from an updated pressure data set. Climate Research 14:71–73

Bamzai AS (2003) Relationship between snow cover variability and Arctic oscillation index on a hierarchy of time scales. International Journal of Climatology 23:131–142

Barring L, von Storch H (2004) Northern European storminess since about 1800. Geophysical Research Letters 31:L20202, doi 10.1029/2004GL020441

Bednorz E (2002) Snow cover in western Poland and macro-scale circulation conditions. International Journal of Climatology 22:533–541

Bednorz E (2004) Snow cover in eastern Europe in relation to temperature, precipitation and circulation. International Journal of Climatology 24:591–601

Beniston M (1997) Variations of snow depth and duration in the Swiss Alps over the last 50 years: links to large-scale forcings. Climatic Change 36:281–300

Beniston M (2004) The 2003 heat wave in Europe: A shape of things to come? An analysis based on Swiss climatological data and model simulations. Geophysical Research Letters 31:L02202, Doi:10.1029/2003gl018857

Beniston M et al. (2005) Future extreme events in European climate: An exploration of regional climate model projections. Climatic Change (submitted)

Bijl W, Flather R, de Ronde JG, Schmith T (1999) Changing storminess? An analysis of long-term sea level data sets. Climate Research 11:161–172

Black E, Blackburn M, Harrison G, Hoskins B, Methven J (2004) Factors contributing to the summer 2003 European heatwave. Weather 59:217–223

Bronstert A (2003) Floods and climate change: Interactions and impacts. Risk Analysis 23:545–557

Brown RD (2000) Northern hemisphere snow cover variability and change, 1915–97. Journal of Climate 13:2339–2355

Brunetti M, Buffoni L, Maugeri M, Nanni T (2000) Precipitation intensity trends in northern Italy. International Journal of Climatology 20:1017–1031

Brunetti M, Colacino M, Maugeri M, Nanni T (2001) Trends in the daily intensity of precipitation in Italy from 1951 to 1996. International Journal of Climatology 21:299–316

Cassou C, Terray L, Phillips AS (2005) Tropical Atlantic influence on European heat waves. Journal of Climate (in press)

Christensen JH, Christensen OB (2003) Severe summer-time flooding in Europe. Nature 421:805–806

Derwent RG, Collins WJ, Johnson CE, Stevenson DS (2001) Transient behaviour of tropospheric ozone precursors in a global 3-D Ctm and their indirect greenhouse effects. Climatic Change 49:463–487

Dirmeyer PA (2003) The role of the land surface background state in climate predictability. Journal of Hydrometeorology 4: 599–610

Dise M, Engel H (2001) Flood events in the Rhine basin: Genesis, influences and mitigation. Natural Hazards 23:271–290

Dorland C, Tol RSJ, Palutikof JP (1999) Vulnerability of the Netherlands and Northwest Europe to storm damage under climate change. Climatic Change 43:513–535

Easterling DR, Evans JL, Groisman PYa, Karl TR, Kunkel KE, Ambenje P (2000) Observed variability and trends in extreme climate events: A brief review. Bulletin of the American Meteorological Society 81:417–425

Estrela MJ, Peñarrocha D, Millán M (2000) Multi-annual drought episodes in the Mediterranean (Valencia region) from 1950–1996. A spatio-temporal analysis. International Journal of Climatology 20:1599–1618

Falarz M (2004) Variability and trends in the duration and depth of snow cover in Poland in the 20th century. International Journal of Climatology 24:1713–1727

Ferro CAT, Hannachi A, Stephenson DB (2005) Simple techniques for describing changes in probability distributions of weather and climate. Journal of climate (in press)

Fink AH, Brucher T, Kruger A, Leckebusch GC, Pinto JG, Ulbrich U (2004) The 2003 European summer heatwaves and drought – synoptic diagnosis and impacts. Weather 59:209–216

Frei C, Schär C (2001) Detection Probability of Trends in Rare Events: Theory and Application to Heavy Precipitation in the Alpine Region. Journal of Climate 14:1568–1584.

Frich P, Alexander LV, Della-Marta P, Gleason B, Haylock M, Klein-Tank AMG, Peterson T (2002) Observed coherent changes in climatic extremes during the second half of the twentieth century. Climate Research 19:193–212

Gillett NP, Zwiers FW, Weaver AJ, Stott PA (2003) Detection of human influence on sea-level pressure. Nature 422:292–294

Giorgi F, Bi XQ, Pal J (2004) Mean, interannual variability and trends in a regional climate change experiment over Europe. II: Climate change scenarios (2071-2100). Climate Dynamics 23:839–858

Graham et al (2005) Assessing climate change impacts on hydrology from an ensemble of regional climate models, model scales and linking methods – a case study on the Lule River Basin. Climatic Change (submitted)

Grazzini F, Ferranti L, Lalaurette F, Vitart F (2003) The exceptional warm anomalies of summer 2003. ECMWF Newsletter 90:2–8

Hantel M, Ehrendorfer M, Haslinger A (2000) Climate sensitivity of snow cover duration in Austria. International Journal of Climatology 20:615–640

Heino R, Brázdil R, Førland E, Tuomenvirta H, Alexandersson H, Beniston M, Pfister C, Rebetez M, Rosenhagen G, Rösner S, Wibig J (1999) Progress in the study of climatic extremes in Northern and Central Europe. Climatic Change 42:151–181

Hisdal H, Stahl K, Tallaksen LM, Demuth S (2001) Have streamflow droughts in Europe become more severe or frequent? International Journal of Climatology 21:317–333

Holt T (1999) A classification of ambient climatic conditions during extreme surge events off Western Europe International Journal of Climatology 19:725–744

Hurrell JW (1995) Decadal trends in the North Atlantic Oscillation: Regional temperatures and precipitation. Science 269:676–679

Johns TC, et al. (2003) Anthropogenic climate change for 1860 to 2100 simulated with the HadCM3 model under updated emission scenarios. Climate Dynamics 20:583–612

Johnson CE, Stevenson DS, Collins WJ, Derwent RG (2001) Role of climate feedback on methane and ozone studied with a coupled ocean-atmosphere-chemistry model. Geophysical Research Letters 28:1723–1726

Jones JM, Davies TD (2000) The influence of climate on air and precipitation chemistry over Europe and downscaling applications to future acidic deposition. Climate Research 14:7–24

Jones PD, Hulme M, Briffa KR, Jones CG, Mitchell JFB, Murphy JM (1996) Summer moisture availability over Europe in the Hadley Centre general circulation model based on the Palmer Drought Severity Index. International Journal of Climatology 16:155–172

Jones R, Murphy J, Hassell D, Taylor R (2001) Ensemble mean changes in a simulation of the European climate of 2071/2100 using the new Hadley Centre regional modelling system HadAM3H/HadRM3H. Hadley Centre Report 2001

Jones CG, Willen U, Ullerstig A, Hansson U (2004) The Rossby Centre Regional Atmospheric Climate Model part 1: Model climatology and performance for the present climate over Europe. Ambio 33:199–210

Junge MM, Stephenson DB (2003) Mediated and direct effects of the North Atlantic Ocean on winter temperatures in northwest Europe. International Journal of Climatology 23:245–261

Kassomenos P, McGregor GR (2005) The Inter-annual variability and trend of precipitable water over Southern Greece. Journal of Hydrometeorology (in press)

Kjellstrom E (2004) Recent and future signatures of climate change in Europe. Ambio 33:193–198

Kistler R, Kalnay E, Collins W, Saha S, White G, Woollen J, Chelliah M, Ebisuzaki W, Kanamitsu M, Kousky V, van den Dool H, Jenne R, Fiorino M (2001) The NCEP–NCAR 50 year reanalysis: monthly mean CD-ROM and documentation. Bulletin of the American Meteorological Society 82:247–-267

Klein-Tank AMG, plus 35 others (2002) Daily dataset of 20th-century surface air temperature and precipitation series for the European Climate Assessment. International Journal of Climatology 22:1441–1453

Klein-Tank AMG, Können GP (2003) Trends in Indices of Daily Temperature and Precipitation Extremes in Europe, 1946–99. Journal of Climate 16:3665–3680

Klein-Tank AMG, Können GP, Selten FM (2005) Signals of anthropogenic influence on European warming as seen in the trend patterns of daily temperature variance. International Journal of Climatology 25:1–16

Langenberg HA, Pfizenmayer A, von Storch H, Sundermann J (1999) Storm related sea level variations along the North Sea coast: natural variability and anthropogenic change. Continental Shelf Research 19:821–842

Langner J, Bergstrom R, Foltescu V (2005) Impact of climate change on surface ozone and deposition of sulphur and nitrogen in Europe. Atmospheric Environment 39:1129–1141

Laternser M, Schneebeli M (2003) Long-term snow climate trends of the Swiss Alps (1931–99). International Journal of Climatology 23:733–750

Leckebusch GC, Ulbrich U (2004) On the relationship between cyclones and extreme windstorm events over Europe under climate change. Global and Planetary Change 44:181–193

Linnerooth-Bayer J, Amendola A (2003) Introduction to special issue on flood risks in Europe. Risk Analysis 23:537–543

Lloyd-Hughes B, Saunders MA (2002) A drought climatology for Europe. International Journal of Climatology 22:1571–1592

Lockwood JG (2001) Abrupt and sudden climatic transitions and fluctuations: A review. International Journal of Climatology 21:1153–1179

Lozano I, Devoy RJN, May W, Andersen U (2004) Storminess and vulnerability along the Atlantic coastlines of Europe: analysis of storm records and of a greenhouse gases induced climate scenario. Marine Geology 210:205–225

McGregor GR, Walters S, Wordley J (1999) Investigating the relationship between daily hospital respiratory admissions and weather using winter air mass types. International Journal of Biometeorology 43:21–30

McGregor GR (1999) Winter ischaemic heart disease deaths in Birmingham, UK: A synoptic climatological analysis. Climate Research 13:17–31

McGregor GR, Markou MT, Bartzokas A, Katsoulis BD (2002) An evaluation of the nature and timing of summer human thermal discomfort in Athens, Greece. Climate Research: 20:83–94

McGregor GR (2004) Winter North Atlantic Oscillation, temperature and ischaemic heart disease mortality in three English counties. International Journal of Biometeorology 49:197–204

McGregor GR, Ferro CAT, Stephenson DB (2005) Projected changes in extreme weather and climate events in Europe. In: Kirch W, Menne B, Bertollini R. Extreme Weather and Climate Events: Public Health Responses. Springer, Heidelberg, pp 13–23

Marshall RT, Kushnir Y, Battisti D, Chang P, Czaja A, Dickson R, Hurrell J, McCartney M, Saravanan R, Visbeck M (2001) North Atlantic climate variability: Phenomena, impacts and mechanisms. International Journal of Climatology 21:1863–1898

Mitchell JK (2003) European river floods in a changing world. Risk Analysis 23:567–574

Moberg A, Jones PD (2004) Regional climate model simulations of daily maximum and minimum near-surface temperatures across Europe compared with observed station data 1961–1990. Climate Dynamics 23:695–715

Mudelsee M, Borngen M, Tetzlaff G, Grunewald U (2003) No upward trend in the occurrence of extreme floods in central Europe. Nature 425:166–169

Nakicenovic N, et al (2000) IPCC Special Report on Emission Scenarios. Cambridge Univ Press, Cambridge, UK and New York, USA, 599 pages

Osborn TJ, Hulme M, Jones PD, Basnett T (2000) Observed trends in the daily intensity of United Kingdom precipitation. International Journal of Climatology 20:347–364.

Panagiotopoulos F, Shahgedanova M, Hannachi A, Stephenson DB (2005) Observed trends and teleconnections of the Siberian High: a tecently declining center of action. Journal of Climate 18:1411–1422

Parry ML (ed) (2000) Assessment of the Potential Effects and Adaptations for Climate Change in Europe: The Europe ACACIA Project. Jackson Environment Institute, University of East Anglia, Norwich, UK, 320 pages

Phillips ID, McGregor GR (2001) Western European Water Vapor Flux–Southwest England Rainfall Associations. Journal of Hydrometeorology 2:505–524

Piccarreta M, Capolongo D, Boenzi F (2004) Trend analysis of precipitation and drought in Basilicata from 1923 to 2000 within a southern Italy context. International Journal of Climatology 24:907–922

Pirazzoli PA, Tomasin A (2003) Recent near-surface wind changes in the central Mediterranean and Adriatic areas. International Journal of Climatology 23:963–973

Pitman AJ (2003) The evolution of, and revolution in, land surface schemes designed for climate models. International Journal of Climatology 23:479–510

Plate EJ (2002) Flood risk and flood management. Journal of Hydrology 267:2–11

Prather M, Gauss M, Berntsen T, Isaksen I, Sundet J, Bey I, Brasseur G, Dentener F, Derwent R, Stevenson D, Grenfell L, Hauglustaine D, Horowitz L, Jacob D, Mickley L, Lawrence M, von Kuhlmann R, Muller JF, Pitari G, Rogers H, Johnson M, Pyle J, Law K, van Weele M, Wild O (2003) Fresh air in the 21st century? Geophysical Research Letters 30(2): art. no. 1100

Pryor SC, Barthelmie RJ (2003) Long-term trends in near-surface flow over the Baltic. International Journal of Climatology 23:271–289

Qian B, Saunders MA (2003) Summer U.K. Temperature and its links to preceding Eurasian snow cover, North Atlantic SSTs, and the NAO. Journal of Climate 16:4108–4120

Raisanen J, Hansson U, Doscher R, Graham LP, Jones C, Meier M, Samuelsson P, Willen U (2004) European climate in the later 21st century: regional simulations with two driving global models and two forcing scenarios. Climate Dynamics 22:13–31

Randolph SE (2004) Evidence that climate change has caused 'emergence' of tick-borne diseases in Europe? International Journal of Medical Microbiology 293 (Suppl 37):5–15

Rmbu N, Boronean C, Bu C, Dima M (2002) Decadal variability of the Danube river flow in the lower basin and its relation with the North Atlantic Oscillation International Journal of Climatology 22:1169–1179

Robinson PJ, Henderson-Sellers A (1999) Contemporary Climatology. Pearson Education, Harlow, 317 pages

Robson AJ, Jones TK, Reed DW, Bayliss AC (1998) A study of national trend and variation in UK floods. International Journal of Climatology 18:165–182

Rodwell MJ, Folland CK (2002) Atlantic air-sea interaction and seasonal predictability. Quarterly Journal of the Royal Meteorological Society 128:1413–1443

Rodwell MJ, Folland CK (2003) Atlantic air-sea interaction and model validation. Annals of Geophysics 46:47–56

Ross RJ, Elliott WP (2001) Radiosonde-based Northern Hemisphere tropospheric water vapour trends. Journal of Climate 14:1602–1612

Schiesser HH, Pfister C, Bader J (1997) Winter storms in Switzerland North of the Alps 1864/65–1993/94. Theoretical and Applied Climatology 58:1–19

Schneider N, Eugster W, Schichler B (2004) The impact of historical land-use changes on the near-surface atmospheric conditions on the Swiss Plateau. Earth Interactions 8:1–27

Seager DS, Battisti J, Yin J, Gordon N, Naik N, Clement AC, Cane MA (2002) Is the Gulf Stream responsible for Europe's mild winters? Quarterly Journal of the Royal Meteorological Society 128:2563–2586

Semmler T, Jacob D (2004) Modeling extreme precipitation events – a climate change simulation for Europe. Global and Planetary Change 44:119–127

Simmons AJ, Gibson JK (eds) (2000) The ECMWF project plan. ECMWF Project Report Series, vol 1. ECMWF, Reading, UK

Smits A, Klein-Tank AMG, Können GR (2005) Trends in storminess over the Netherlands. International Journal of Climatology (in press)

Stefanicki G, Talkner P, Weber RO (1998) Frequency changes of weather types in the Alpine region since 1945. Theoretical and Applied Climatology 60:47–61

Stephenson DB, Pavan V, Bojariu R (2000) Is the North Atlantic Oscillation a random walk? International Journal of Climatology 20:1–18

Stevenson DS, Johnson CE, Collins WG, Derwent RG, Edwards JM (2000) Future estimates of tropospheric ozone radiative forcing and methane turnover – the impact of climate change. Geophysical Research Letters 27:2073–2076

Stott PA, Stone DA, Allen MR (2004) Human contribution to the European heatwave of 2003. Nature 432:610–613

Sutton RT, Hodson DLR (2003) Influence of the ocean on North Atlantic climate variability 1871–1999. Journal of Climate 16:3296–3313

Sutton RT, Hodson DLR (2005) Atlantic ocean forcing of North American and European summer climate. Science 309:115–118

Svensson C, Jones DA (2002) Dependence between extreme sea surge, river flow and precipitation in eastern. Britain International Journal of Climatology 22:1149–1168

Sweeney J (2000) A three-century storm climatology for Dublin 1715–2000. Irish Geography 33:1–14

Syri S, Karvosenoja N, Lehtila A, Laurila T, Lindfors V, Tuovinen JP (2002) Modeling the impacts of the Finnish Climate Strategy on air pollution. Atmospheric Environment 36:3059–3069

Taylor AH (1996) North-South shifts of the Gulf Stream: Ocean-atmosphere interactions in the North Atlantic. International Journal of Climatology 16:559–584

Trenberth KE (1999) Conceptual framework for changes of extremes of the hydrological cycle with climate change. Climatic Change 42:327–339

Trigo IF, Davies TD (2002) Meteorological conditions associated with sea surges in Venice: a 40 year climatology. International Journal of Climatology 22:787–803

Trigo IF, Davies TD, Bigg GR (2000) Decline in Mediterranean rainfall caused by weakening of Mediterranean cyclones. Geophysical Research Letters 27:2913–2916

Trigo RM, Pozo-Vázquez D, Osborn TJ, Castro-Díez Y, Gámiz-Fortis S, Esteban-Parra MJ (2004) North Atlantic oscillation influence on precipitation, river flow and water resources in the Iberian Peninsula. International Journal of Climatology 24:925–944

van Vliet AJH, Schwartz MD (2002) Phenology and climate: the timing of life cycle events as indicators of climatic variability and change. International Journal of Climatology 22:1713–1714

Walsh JE, Phillips AS, Portis DH, Chapman DL (2001) Extreme Cold Outbreaks in the United States and Europe, 1948–99. Journal of Climate 14:2642–2658

Widmann M, Schär C (1997) A principal component and long-term trend analysis of daily precipitation in Switzerland. International Journal of Climatology 17:1333–1356

Woth K, Weisse R, von Storch H (2005) Dynamical modelling of North Sea storm surge climate: An ensemble study of storm surge extremes expected under possible future conditions as projected by four different Regional Climate Models. International Journal of Climatology (in press)

Yan Z, Bate S, Chandler RE, Isham V, Wheater H (2002) An analysis of daily maximum wind speed in northwestern Europe using generalized linear models. Journal of Climate 15:2073–2088

Yan Z, Tsimplis MN, Woolf D (2004) Analysis of the relationship between the North Atlantic oscillation and sea-level changes in northwest Europe. International Journal of Climatology 24:743–758

Zwiers FW, Karin VV (1998) Changes in the extremes of the climate simulated by CCC GCM2 under CO_2 doubling. Journal of Climate 11:2200–2222

3 Adaptation Assessment for Public Health

HANS-MARTIN FÜSSEL, RICHARD J. T. KLEIN, KRISTIE L. EBI

The main objective of the EU-funded research project cCASHh (Climate Change Adaptation Strategies for Human Health) has been to conduct an assessment of possible ways of adapting of populations to climate change in Europe, based on projections of the potential impacts of climate change on heat stress, health effects of riverine floods, vector-borne diseases, and food- and waterborne diseases. This objective requires the integration of public health and climate change approaches and perspectives. For the integration to be successful, the climate change community must understand in which way "health is different" (i.e. what distinguishes human health from other climate-sensitive systems or sectors), and the public health community must understand in which way "climate change is different" (i.e. what distinguishes anthropogenic climate change from other risk factors to human population health).

The primary goal of this chapter is to facilitate the mutual understanding between experts from public health, on the one hand, and climate change and adaptation, on the other. Because the main audience of this book is stakeholders with a background in public health, we will present the main concepts of climate change adaptation science, introduce the relevant terminology developed by the climate change community, discuss the different goals and methods to climate impact and adaptation assessment, present the key questions for adaptation to climate variability and change, and suggest key steps for adaptation planning to reduce climate-related health impacts.

3.1 Climate Change and Public Health

3.1.1 Introduction

For the public health community, climate has always been an important consideration. Every epidemiologist knows that climatic factors are important determinants of human health and well-being. Ambient temperatures outside the comfort range that a population is acclimatized to are associated with thermal stress; weather-related disasters, such as floods and storm surges, cause significant loss of life; and many infectious diseases are limited to certain climatic zones. Two broad categories of climate-related impacts are generally distinguished: *Direct* impacts, such as the effects of thermal stress and extreme weather events, are directly caused by the exposure of humans to hazardous meteorological conditions. *Indirect* impacts, such as vector-, rodent-, water-, and foodborne diseases, and aeroallergens, involve a mediation of climatic factors through a climate-sensitive environmental system.

It is now well established that humankind is affecting climate on a global scale. The rate of projected climate change will likely be outside the range experienced during at least the past 10000 years (Houghton et al., 2001). As described in more detail in Chapter 2 [this book], anthropogenic climate change will entail changes in *average climate* conditions (e.g. annual temperature and sea level), in *interannual climate variability* (e.g. NAO and drought spells), and in *extreme weather events* (e.g. wind storms and floods). These changes will not be evenly distributed around Europe. Thus, future weather conditions will be less predictable, although the trend will be for increasing average annual temperatures and more extreme weather events.

Given the close link between climatic factors and human health, it is obvious that anthropogenic climate change can affect human health and well-being, by altering the urgency of current health risks and by introducing new health risks into previously unaffected areas. Consequently, there is now considerable scientific and policy interest in the potential impacts of anthropogenic climate change on human health, and in effective response measures.

3.1.2 Human Adaptation to Climatic Conditions

Humans have always adapted to the climate of their homeland, including its natural variability. This process comprises physiological acclimatization, behavioural strategies (such as clothing, scheduling daily work, and seasonal migration), technical measures (such as building design and air-conditioning), and institutional mechanisms (such as establishing disaster preparedness schemes). The history of human adaptation to climatic factors comprises great successes as well as disastrous failures. On the one hand, humans have successfully managed to live in nearly every climatic zone of the Earth, from the Arctic ice to the hot deserts. On the other hand, the rise and fall of many great civilizations has been linked to regional climatic shifts, including the establishment of the first Chinese dynasty, the collapse of the ancient civilizations in Egypt, Indus, and the Mesopotamian, the discontinuity in ancient Greek civilization, and the decline of the Maya culture (e.g. Haug et al., 2003).

Societies can respond to the threat of climate change by two fundamental response mechanisms: *mitigation of climate change* and *adaptation to climate change.* In the terminology of the climate change community, mitigation refers to actions that limit the amount and rate of climate change (the "exposure") by constraining the emissions of greenhouse gases or enhancing their sinks. Adaptation, in contrast, refers to any actions that are undertaken to avoid, prepare for or respond to the detrimental impacts of observed or anticipated climate change. Mitigation and adaptation vary significantly in their scope, type of actions, characteristic spatiotemporal scales, and principal actors. Mitigation is the only strategy that can reduce impacts of climate change on all systems and on a global scale but it requires international cooperation and takes a long time to become fully effective because of the inherent inertia of the climate system. Adaptation can address climate-related risks in human-managed systems on a local or regional scale and on a shorter time scale but its scope is generally limited to specific systems and risk types.

Planned adaptation to the health impacts of climate change comprises a wide range of preventive public health measures. Eventually, behavioural changes, medical interventions, or the use of technologies will be required to reduce climate-related health effects. The public health sector and other relevant sectors may facilitate these actions by appropriate educational, institutional, legal and financial measures, and other policy changes. Because the measures considered in adapting to future climate change are, in general, not new, most of them also reduce the vulnerability to current climate variability. However, adaptation to climate change may require action by people who have not considered climate an important factor for their decisions in the past.

Adaptation also refers to the process by which adaptive measures are implemented: it can be immediate and intuitive (e.g. buying a fan to cope with the heat), but it can also involve a long process of information collection, planning, implementation and monitoring (e.g. setting up an early-warning system for heat stress). The terms "autonomous adaptation" and "planned adaptation" are generally used to distinguish between these two types of adaptations, even though the distinction is not always sharp. Planned adaptation may be a response either to perceived changes in climate and associated health risks ("reactive adaptation") or to anticipated risk changes in the future ("proactive" or "anticipatory adaptation"). In view of the recent literature on adaptation to climate change, to which this book is a contribution, the term "(planned) adaptation" will be used here to describe any public health intervention aimed at avoiding, preparing for, or responding to the potential impacts of climate change on population health, as well as the process by which these interventions are implemented. This is consistent with recent work by McMichael and Kovats (2000) and Kovats et al. (2003).

3.1.3 A Conceptual Framework for Climate Impact and Adaptation Assessment

Two institutions (in a broad sense) are of great importance for framing international climate policy and science: The United Nations Framework Convention on Climate Change, agreed upon in 1992, provides the legal framework for international climate policy. The Intergovernmental Panel on Climate Change (IPCC), which was jointly established by the United Nations Environment Programme (UNEP) and the World Meteorological Organization (WMO) in 1989, unites thousands of scientists worldwide in an effort to provide impartial scientific advice to the international policy community. These two institutions have developed specific approaches to address and assess the climate change problem, and they have also shaped a common, internally consistent terminology (McCarthy et al., 2001; Glossary). However, some key terms are used differently by the climate change community than by other communities, such as those concerned with environmental health and risk management. The deviations are largely explainable by the fact that the focus of the UNFCCC is on the *incremental* impacts of anthropogenic climate change and that key terms were initially defined to facilitate assessments of *mitigation* rather than *adaptation*. Box 1 introduces key terms that will be used throughout this book. Where necessary, the pertinent definitions agreed upon by the IPCC were adapted so that they apply more specifically to health risks associated with anthropogenic climate change and natural climate variability.

▌ Box 1. Key concepts and terms in climate change assessments according to the IPCC (adapted from McCarthy et al., 2001).

▌ **Anthropogenic climate change:** A statistically significant variation in either the mean state of the climate or in its variability, persisting for an extended period (typically decades or longer) that is due to persistent anthropogenic changes in the composition of the atmosphere or in land use.

▌ **Mitigation (of climate change):** Anthropogenic interventions that constrain the amount and rate of global climate change by reducing the sources and enhancing the sinks of greenhouse gases.

▌ **Adaptation (to climate variability and change):** An adjustment in human systems in response to actual or expected climatic stimuli or their effects, which reduces the associated risks to population health through preventive measures.

▌ **Climate impacts:** Consequences of climate change for human health, i.e. the disease burden attributable to climate change. Estimations of future climate impacts are contingent on a variety of assumptions. The terms *potential* and *expected* impacts are often used to distinguish between assessments that neglect adaptation, and those that do consider it. (Anthropogenic climate change is inseparably overlaid with natural climate variability. Since the counterfactual distribution of exposure, i.e. the undisturbed climate, is only partly known and itself highly variable, the attribution of certain health effects to anthropogenic climate change is often difficult in practice).

▌ **Exposure:** The nature and degree to which a population, or an individual, is exposed to climatic stimuli. (The individual-level interpretation of exposure is only meaningful for *direct* impacts of climatic stimuli, such as thermal stress. For *indirect* impacts, such as food borne diseases, exposure to *climatic stimuli* can only be defined at the population level.)

▌ **Sensitivity:** The degree to which the health of a population, or an individual, is affected by climate-related stimuli. The effects may be direct or indirect. (The overall sensitivity of population health to climate change is determined by a complex causal web involving various climatic and non-climatic risk factors. Because anthropogenic climate change entails changes in multiple climate characteristics, the use of a one-dimensional exposure-response relationship is often insufficient for assessing potential health effects of future climate change.)

▌ **Adaptive capacity (or adaptability):** The ability of a population to successfully adapt to climate variability and change in the future.

▌ **Vulnerability (to climate change):** The expected level of adverse effects for a given level of global climate change. It is determined by the regional manifestation of global climate change, the sensitivity of the system to these changes, and its adaptive capacity. (The IPCC definition of vulnerability combines the biophysical and social dimensions of vulnerability. Since the use of this term varies so much between, and sometimes even within, the climate change and health communities, we will not use it here, except in circumstances where there is no risk of misunderstanding.)

▌ **Climate impact assessment:** The practice of identifying and evaluating the detrimental and beneficial consequences of climate change on human health.

▌ **Adaptation (policy) assessment:** The practice of identifying options to adapt to climate change and evaluating them in terms of criteria such as availability, benefits, costs, effectiveness, efficiency, and feasibility.

Figure 1 embeds the main concepts defined in Box 1, and their analytical relationships, in a broader framework that includes non-climatic risk factors and their drivers. Figure 2 then applies this framework to the four categories of diseases investigated in cCASHh. Although we do not want to discuss this diagram in detail here, a few things shall be noted. The depicted diagram combines features of a system-dynamics diagram with those of an influence diagram. Bold arrows denote causal relationships as-

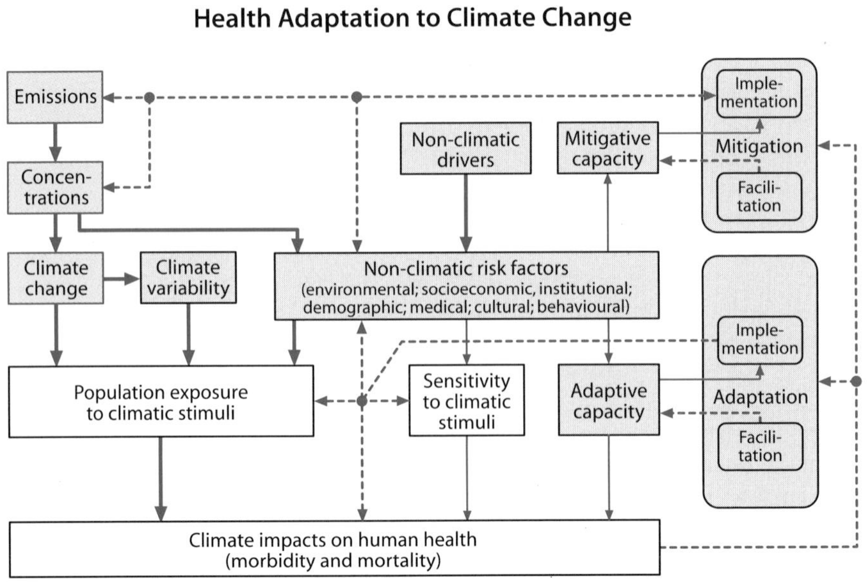

Health Adaptation to Climate Change

Fig. 1. Conceptual framework for adaptation to the health effects of climate change. Bold arrows mean "A causes (changes in) B", thin arrows stand for "A influences B", and dashed arrows denote purposeful human responses to reduce adverse health effects. Blue-shaded, non-shaded, and grey-shaded boxes denote concepts that are analysed at the global level, at the regional level, and at different levels, respectively. Source: adapted from Füssel and Klein (2002)

sociated with physical flows, which can be largely investigated with formal mathematical models. Thin arrows denote general functional relationships between certain elements of the framework, which are not always amenable to quantitative analysis. Finally, the dashed arrows depict a feedback loop where experienced or anticipated climate impacts trigger societal response strategies (i.e. mitigation and adaptation).

The diagram shows that the impacts of climate change on human population health are a function of the exposure of a population to changing climate conditions and of its sensitivity to these changes, which depends on environmental, social, economic, institutional, technical, demographic, and behavioural factors. The exposure to climatic and non-climatic factors as well as the sensitivity to them changes over time under the influence of various driving forces. The ability of a population to reduce its sensitivity to experienced or anticipated climate change is described as its "adaptive capacity" in the climate change adaptation literature. Even though details of this concept are still debated, there is general agreement that the availability of financial resources and their distribution across the population, the stock of human and social capital, the level of infrastructure, the structure and efficacy of critical institutions, and the responsiveness and capability of the governance and decision-making process are important deter-

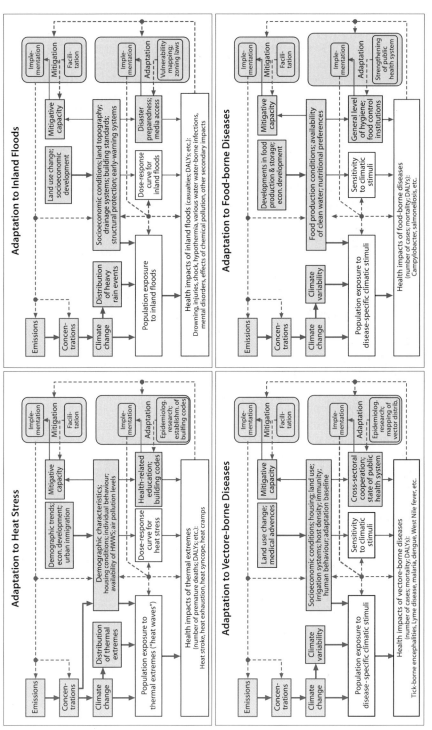

Fig. 2. Application of the conceptual framework from Fig. 1 to four different diseases

minants of the ability of a social system to adapt to climatic changes (Smit and Pilifosova, 2001; Yohe and Tol, 2002).

Adaptation can reduce climate impacts in different ways. It may involve the reduction of population exposure to climatic stimuli (e.g. through adapted urban planning), the reduction of population sensitivity (e.g. through vaccination campaigns), the modification of non-climatic risk factors (e.g. through environmental control of disease vectors), and the direct reduction of the disease impact (e.g. through improved medical planning that improves the early treatment of cases).

Figure 1 shows the intricate relationship between adaptive capacity and adaptation. On the one hand, the adaptive capacity of a community determines the range of feasible adaptation options; on the other hand, increasing adaptive capacity is the goal of certain types of adaptation options. The terms "Implementation" and "Facilitation" are used to distinguish adaptations that directly reduce the vulnerability of a community to climatic stressors from those that do so indirectly, by increasing adaptive capacity, respectively.

3.1.4 Assessing the Health Impacts of Climate Change

In the past fifteen years, a range of studies have been conducted on the relationship between climate, weather, and health, either using mathematical models or taking empirical observations (McMichael et al., 1996, 2001). The purpose of these studies has been to produce information that helps to understand how population health is potentially affected by a change in climatic conditions. However, climate impact and adaptation assessments for human health face a variety of challenges. The cause-and-effect chain from climate change to changing disease patterns is extremely complex and includes many non-climatic 'moderating' factors, such as environmental and social conditions, access to health care, demographics and behaviour. Hence, future health impacts are not only determined by the climatic development, but also by concurrent changes in these non-climatic factors (see Fig. 3). Epidemiologists are empiricists by training, who have traditionally focussed on comparatively straightforward relationships between individual risk factors and health outcomes. Hence, they are not used to deal with highly uncertain future scenarios of dynamical changes in multiple interacting risk factors.

The complex causal structure, and the uncertainties associated with future projections of relevant climatic and non-climatic factors, poses severe constraints on the accuracy with which future health impacts can be assessed. The scantiness of reliable epidemiological data on the current relationship between climatic factors and health outcomes, in particular in developing countries, further hampers the assessment of future health risks. For some indirect health effects, such as certain vector-borne diseases, there may not even be certainty about the direction of risk changes at the

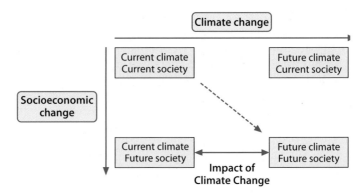

Fig. 3. Impacts of climate change in a dynamical systems perspective

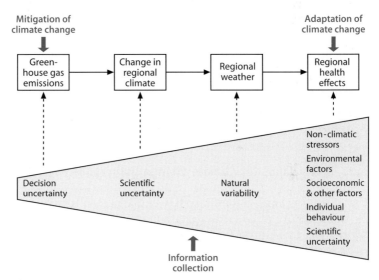

Fig. 4. Cascade of uncertainties in the assessment of climate impacts on human health. Top boxes: simplified causal chain; bottom box: sources of uncertainty; blue arrows: fundamental response options. Source: Füssel (2004)

local scale. Figure 4 shows the *cascade of uncertainties* along the causal chain that underlies the climate-health relationship, and lists its main causes. Prognostic uncertainty increases from greenhouse gas emissions to global climate change, to its regional manifestation, and further to the effects on human health.

In public health, anthropogenic climate change is often regarded as an environmental health topic. There are obvious arguments in favour of this classification, because climate can be viewed as part of a wider set of environmental conditions. However, there are also important differences between climate change and other environmental health topics. Most importantly, climate change is not hazardous *per se*. Hence, the classical toxicological model applied in environmental epidemiology, where a defined exposure to a specific agent causes an adverse health effect to identifiable exposed populations, is generally not adequate for assessing indirect or multi-causal impacts of climate change on human health. These health issues rather call for the application of ecological and systems-based approaches (Patz and Balbus, 1996; Bernard and Ebi, 2001).

3.2   Adaptation Assessment: Why and How?

Adaptation policy assessment for climate change and human health, as done in cCASHh, combines ideas from a variety of approaches, most importantly climate impact and adaptation assessment, risk management, health impact assessment, comparative risk assessment, and effectiveness of intervention studies. We will focus here on the contribution of climate impact and adaptation assessment for the development of policies to reduce the level of climate-related diseases now and in the future because these approaches are expected to be the least familiar to the target audience of this book.

3.2.1 Policy Goals for Climate Change Vulnerability Assessment

The climate change community has embarked on a major effort to assess the vulnerability of countries, communities, economic sectors and ecological systems to anthropogenic climate change (McCarthy et al., 2001). The assessment of vulnerability can serve two principal purposes. The *academic purpose* is to produce information that helps to understand how a system (i.e., country, community, sector, ecosystem etc.) is potentially affected by and responds to a change in climatic conditions. The *policy purpose* is then to evaluate and interpret this information and present it to decision-makers in a way that improves their decision-making.

The results of vulnerability assessments are applied in three related but distinct decision contexts:

▌ *Specification of long-term targets for the mitigation of global climate change*
   Projections of potential or expected impacts of climate change for different climate scenarios are combined with (subjective) assessments of their acceptability.
▌ *Identification of particularly vulnerable communities to prioritize international aid*
   Assessments of the exposure of a region to climate change, its sensitivity, and its adaptive capacity are combined to describe the overall vul-

nerability of that region to global climate change. Priority for international aid is then given to "particularly vulnerable" countries or regions.

▌ *Recommendation of adaptation measures for specific regions and sectors* Current climate-related risks and anticipated risk changes in the future are evaluated to identify the need for additional policies and measures. Available adaptation options are assessed according to their expected benefits, costs (in a broad sense), and feasibility. Selected options are recommended for implementation.

Different approaches and methods are applied in climate change assessments, depending on the specific goal of the assessment. Assessments for the first, second, and third policy goal are denoted as *climate impact assessment*, *climate vulnerability assessment* and *adaptation policy assessment*, respectively. The first two assessment types aim at determining the expected impacts of climate change on human health. Their main difference is the greater consideration of socioeconomic factors and feasible adaptations in climate vulnerability assessments. The third assessment type, in contrast, aims at recommending specific adaptation policies that reduce the vulnerability to all climatic stressors, regardless of their attribution to anthropogenic causes or natural variability. This purpose requires close interaction with relevant stakeholders, consideration of wider policy goals and a focus on policies that are robust enough given the unavoidable uncertainties in projections of future climate change and its impacts. For a more detailed review of the different approaches to climate change assessment and their evolution, see Füssel and Klein (2002).

3.2.2 Rationales for Adaptation Assessment

Well until the second half of the 1990s, climate change assessment was aimed primarily at understanding system behaviour under different scenarios of climate change, in order to project potential impacts of climate change. Many of these climate impact assessments have been criticized because they either ignored adaptation, or they assumed unrealistic levels of adaptation (Smithers and Smit, 1997). The metaphorical names "dumb farmer" and "clairvoyant farmer" have been used to characterize such extreme assumptions about adaptive behaviour. The former does not change their behaviour at all under changing climate conditions, whereas the latter has perfect foresight of future climate conditions and faces no restrictions in implementing adaptation measures. Obviously, actual behaviour will fall somewhere between these two extremes, depending on the detection and recognition of climatic changes, the existence of potential response measures, and their feasibility in terms of economic and other resources. Recognizing the limited consideration of adaptation in early climate impact assessments, scientists and policy-makers alike have increasingly called for a more thorough assessment of adaptation.

The rationale for adaptation assessment is twofold. First, it contributes to more realistic climate impact and vulnerability assessments by considering what adaptations are likely in a specific context (points 1 and 2 above). Extending the list of metaphorical names, these assessments might employ an "average farmer" who adjusts their practice in reaction to persistent climate change only. Second, adaptation policy assessment directly contributes to policymaking by assessing what adaptations are recommended under given social and climatic conditions, and when (point 3 above). This purpose is about turning an "average farmer" into a "smart farmer", who uses available information about future climate conditions to adjust to them proactively. This book is concerned with the latter purpose of adaptation assessment.

The two purposes of adaptation assessment have important consequences for the assessment practice, in particular for the appropriate spatiotemporal scales, the handling of uncertainty, and the consideration of non-climatic factors. Climate impact and vulnerability assessments are often conducted at a coarse spatial resolution and for the long time horizon that is characteristic for the specification of targets for mitigation policy. In addition, most impact assessments focus on the expected level of climate impacts rather than on the distribution of uncertainty around that figure. Adaptation policy assessments, in contrast, need to be relevant at the generally finer spatial and shorter temporal scales of operation of adaptation measures. Furthermore, uncertainty lies at the very heart of any policy-focussed adaptation assessment because the degree of confidence in future risk changes determines the specificity with which adaptation strategies can be designed.

3.2.3 Climate Change Assessments and Comparative Risk Assessment

Several authors have suggested to link climate impact and adaptation assessment with comparative risk assessment (CRA; e.g. Kay et al., 2000; Campbell-Lendrum et al., 2001). These approaches have many things in common, but there are also important differences. The main difference of climate *impact* assessments, compared to CRA, is that they are generally concerned with the cumulative impacts of *future* exposure scenarios. In addition, climate-related health impacts generally have a complex *multifactorial* causal structure that makes attribution to single factors difficult. The methodological differences between *adaptation* assessments and CRA are more fundamental. Most importantly, adaptation does *not* define a counterfactual distribution of population exposure to climate change. In addition, many adaptation measures have beneficial health effects *independent* of anthropogenic climate change.

Figure 5 links the different assessment approaches by extending the concepts "attributable" and "avoidable" burden of disease. These terms are used in CRA to characterize fractions of the current disease burden that

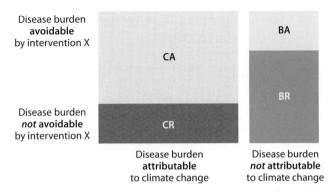

Fig. 5. Attributable and avoidable burden of disease in the context of adaptation to climate change. *CA*: climate-attributable burden, *a*voidable fraction; *CR*: climate-attributable burden, residual fraction; *BA*: baseline burden, *a*voidable fraction; *BR*: baseline burden, residual fraction

are attributable to a specific risk factor and that could be avoided by a counterfactual distribution of exposure to this factor, respectively (Murray and Lopez, 1999). The blue bar (BA+BR) states the burden of a specific disease for some *future* time period, assuming a constant baseline climate and no interventions. The grey bar (CA+CR) represents the *additional* future burden of disease for a specific climate change scenario. This figure would be termed "disease burden attributable to climate change" in CRA and "potential impacts of climate change" in climate impact assessments. It is also shown how a specific intervention avoids a certain fraction of the disease burden attributable and not attributable to climate change (CA and BA, respectively). Vulnerability assessments that account for feasible adaptations assess the "residual fraction of the attributable disease burden" (CR), which is called "expected impacts of climate change" in IPCC terminology. Adaptation policy assessments are typically interested in the health benefits of an intervention independent of their attribution, which corresponds to the "total avoidable disease burden" (CA+BA).

In this conceptualization, the total *attributable* disease burden and the total *avoidable* disease burden at a specific point in time are independent, and the latter may even exceed the former. It is important to note that reality is considerably more complex than the diagram shown in Figure 5, due to the dynamics of the system and the different levels of uncertainty associated with the different "boxes". In the case of climate change, the relative size of the attributable burden (the grey bar) typically increases over time, and the associated prognostic uncertainty is larger than for the baseline burden (the blue bar).

3.3   Designing Adaptation to the Health Impacts of Climate Change

Substantial guidance exists for assessing the expected impacts of climate change on population health and a range of other systems. In contrast, guidance for adaptation policy assessment is still under development. Therefore, cCASHh has produced its own analytical framework for adaptation assessment, thereby contributing to the methodological development process.

3.3.1 Main Approaches to Climate Impact and Adaptation Assessment

In 1996, climate scientists were able to pinpoint human activities as being at least partly responsible for the global warming that was observed during the twentieth century (Houghton et al., 1996). It has also become clear that even with the most drastic measures to curb emissions of greenhouse gases, climate will continue to change for many decades on account of past emissions. As confidence about anthropogenic climate change increased, and the long time scales involved in mitigating climate change were acknowledged, the adaptation dimension of vulnerability assessment became more important. In addition, there has been increasing recognition of the importance of current and future climate variability, including extreme weather events, for many climate-sensitive impact sectors.

Earlier approaches to climate impact and adaptation assessment, such as suggested in the IPCC Technical Guidelines (Carter et al., 1994) and the UNEP Handbook on Methods for Climate Change Impact and Adaptation Strategies (Feenstra et al., 1998) follow a top-down approach, in which the adaptation assessment commences only after the model-based impact assessment has been completed. The adaptation assessment typically focusses on technical options for reducing adverse impacts of climate change. These types of adaptation assessments shed light on the importance of adaptive behaviour in general, and on the large diversity of potential adaptation measures and stakeholders. However, they provided only limited guidance to stakeholders concerned about adaptation. The main reasons are the extensive reliance on model-based impact assessments, which are not equally available across all world regions and climate-sensitive sectors, the mismatch in spatial and temporal scales between climate impact projections and typical adaptation decisions, and the limited consideration of the implications of scientific uncertainties, existing climate-related risks and socioeconomic factors for adaptation planning.

Recognizing the limitations of top-down approaches to adaptation assessment in providing policy-relevant knowledge to adaptation decision-makers, bottom-up approaches have recently been developed. A common feature of these approaches is that they start from the current hazards to a particular system, and from the available experience in constraining the associated risks, rather than from model-based scenarios of future climate impacts. Thus, projections of potential future climate change and its impacts are now used along with information on today's climate variability to identify priorities for adapting to climate variability and change.

A major initiative under the guidance of the UNDP (United Nations Development Programme) and the GEF (Global Environment Facility) is currently developing an Adaptation Policy Framework that focusses on the specific needs and constraints for adaptation in developing countries (UNDP, 2003). It is by combining current and future climate-related risks that this approach brings the issue of long-term climate change also to decision-makers with a shorter time horizon.

Similar to the shift in purpose that can be observed for climate vulnerability assessments in other sectors, assessments of the expected health impacts of climate change are now being complemented with studies aimed at contributing to the decision-making of stakeholders. In particular, Kovats et al. (2003) recently developed "Methods of assessing human health vulnerability and public health adaptation to climate change".

No single methodological approach is appropriate to assess all types of climate-related health risks and to develop proposals on effective adaptation strategies. The relative importance of top-down and bottom-up methods in adaptation assessment depends on the purpose and scope of the study and the availability of relevant data and other resources. Factors in favour of bottom-up approaches are if current climate-related risks are unsatisfactorily controlled, the planning horizon is short, resources are limited, uncertainty about future risk changes is high and climatic stress factors are closely intertwined with non-climatic factors. We note that many of these factors are particularly important in developing countries. The cCASHh project combines quantitative model-based analyses with more qualitative assessment techniques, including expert elicitation.

3.3.2 Key Questions for Adaptation

The design of effective and efficient adaptation strategies, policies and measures requires consideration of the following questions (Smithers and Smit, 1997; Smit et al., 2000):

▌ *Adaptation to what?*
The starting point for planned adaptation to climate change is the recognition that the past may no longer be a good guide for the future. One goal of a vulnerability assessment is then to identify the types of climatic stresses to which a population will need to adapt. Precision in identifying the stressors will increase the specificity of the response strategy. If uncertainties are very large, then policy-makers may emphasize the implementation of adaptation measures that are robust across a range of possible future climate scenarios, such as additional research,

increased monitoring and surveillance, development of early warning systems, improved disaster response planning, and strengthening overall adaptive capacity.

▌ *Who or what adapts?*

Another goal of an adaptation assessment is to determine which actors should be involved in the development, implementation and evaluation of response options designed to address the health risks of concern. This goal is linked to the identification of those population subgroups that are particularly at risk for adverse health outcomes attributed to climate variability and change. The better the knowledge about the distribution of future changes in health risks (as well as current vulnerabilities to climate variability), the more specific action is possible.

▌ *How does adaptation occur?*

Adaptation is a process through which actors at individual, community, regional, national, and international scales implement strategies, policies, and measures to reduce current and future vulnerability. The scale of response depends on the health outcome of concern. For example, if the concern is increased morbidity and mortality in heat-waves, then individuals can buy fans and wear lighter clothing, communities can develop and implement early warning systems, regions and nations can modify urban planning regulations to decrease the urban heat island, and national and international organizations can coordinate and collaborate on gathering the necessary meteorological data. The question then is how to facilitate these processes in an effective and efficient manner.

▌ *How good is the adaptation?*

Evaluating the feasibility and effectiveness of adaptation options is of central importance in any adaptation policy assessment, whereby the criteria and methods depend on the context of a specific assessment. Most adaptation assessments for human health involve a comparison of the expected health benefit (in terms of lives saved, reduced disease incidence etc.) with the economic costs of a specific adaptation strategy, thus, its cost effectiveness. However, additional non-quantifiable criteria are often at least as important, such as the feasibility of a strategy in terms of required skills and infrastructure, its political and cultural acceptability, its compatibility with other policy goals and its robustness under different plausible climate scenarios.

3.3.3 Climate Change Adaptation Assessment as a Risk Management Strategy

Adaptation to climate variability and change is a strategy for managing climate-related risks. Hence, adaptation to climate change shares many features with other approaches to risk management. However, there are also some important differences. Most importantly, the long-term nature of climate change requires a dynamic framework for risk analysis. Such a framework needs to consider that climatic hazards as well as vulnerable populations are changing over time, and that response measures to reduce the vulnerability to the hazard will be implemented over time. In addition, the complexity of the exposure, the number and importance of non-climatic risk factors, and the uncertainty about key risk factors now and in the future are typically larger for climate-sensitive impacts than for other risk management issues.

The design of effective adaptation strategies, policies and measures in public health needs to be based on an understanding of current health risks caused by climatic and non-climatic factors as well as potential future risk changes. The assessment of future risks involves the combination of epidemiological knowledge about the relationship between health outcomes, climatic and non-climatic risk factors with future projections for these risk factors, including a description of key uncertainties. Potential response strategies are evaluated according to their effectiveness, efficiency, and feasibility. This integrative task requires the close collaboration across scientific disciplines, including epidemiologists, climatologists, economists, public health managers and other experts. Since most climate-related health risks are not completely new, cooperation between public health experts from different regions will facilitate knowledge transfer from areas that already have experience coping with a particular disease to areas that may become affected in the future.

The risk management process is typically divided into an initiation and scoping phase, risk assessment (which comprises risk analysis and risk evaluation), risk control, implementation of measures, and monitoring and evaluation (e.g. Canadian Standards Association, 1997). This basic structure is also reflected in a framework for adaptation decision-making in the face of climate change risk developed in the United Kingdom Climate Impacts Programme (UKCIP; Willows and Connell, 2003). One important feature of the UKCIP framework is the characterization of adaptation as a cyclical process instead of a one-shot approach, which is partly due to the dynamically changing hazard. It is also emphasized that the structure of the adaptation cycle needs to be followed in a flexible manner, as new knowledge and experience may require repeating earlier steps.

3.3.4 Key Steps for Adaptation Planning to Reduce Climate-Related Health Impacts

We present below the major steps of the adaptation cycle for the health effects of climate change. We highlight selected issues where adaptation to the health effects of climate change differs from more conventional topics for risk management. All steps should be performed in close collaboration with the stakeholders who will eventually implement the recommended adaptation actions.

▓ *Scoping the project*

The first step in the adaptation cycle is the determination of the scope of the assessment (such as the regional scope, the considered time period, the types and levels of actors addressed, and the resources available) and of the decision-making criteria. Benefit-related criteria for public health interventions may refer to the number of people who benefit, the number of lives saved, the reduction in disability-adjusted life years (DALYs), the reduction in risk for particularly vulnerable population groups, the compliance with established legal standards, and the equitable distribution of the benefits. Cost-related criteria typically include one-time costs for implementing a measure, running costs, and (unpaid) labour time. Because adaptation to climate change is often integrated into existing policies and programmes, additional decision criteria from this broader policy context may also be relevant, such as the expected contribution of a measure to sustainable economic development, to a strengthening of the rural regions, or to gender equality.

▓ *Risk screening*

This step identifies, based on expert judgement or based on questions established by policy-makers, the relevant climate-sensitive health impairments in the assessment region. The regional distribution of currently prevalent health impairments may provide important information about their dependence on climatic factors. However, it is also important to look at new health risks that could arise under future climatic conditions.

▓ *Examination of the adaptation baseline*

The adaptation baseline comprises all policies and measures that are already in place to reduce a specific health risk. In this step, these measures are determined and their efficacy under current climate conditions is assessed.

▓ *Analysis of future risk changes*

This step analyses potential future changes in climate-sensitive health risks on the spatiotemporal scale that is relevant for adaptation. The analysis is based on projections of relevant climatic as well as non-climatic factors, including likely impacts of climate change on other sectors. For example, a number of weather variables determine perceived temperature, including temperature, humidity, wind speed etc. In order to project possible changes in heat-related diseases and deaths, information about changes in these variables, in demographic characteristics, and in selected socioeconomic variables, would be needed for areas at future risk of heat-waves. An in-depth discussion of the available quantitative and qualitative assessment methods is provided in Kovats et al. (2003).

Available scenarios for future changes in relevant climatic and non-climatic risk factors are typically associated with substantial uncertainty. In some cases, such as heat-waves, future increases in risk levels are almost certain whereas in others, such as some vector-borne diseases with a complex etiology, there may even be uncertainty about the direction of local or regional risk changes. If climate change is expected to have only moderate effects on existing health risks, it is generally sufficient to maintain or strengthen current public health measures (the "adaptation baseline"). However, additional action is typically required in the case of newly emerging health risks or when existing health risks are not yet effectively controlled.

Evaluation of future risk changes

In this step, expected changes in risk levels are evaluated according to the criteria established in step 1. To this end, results from the previous step on the severity of the projected health impacts, their likelihood, and the associated uncertainties are combined with information about the relative importance of other health hazards. This step may also involve subjective evaluations of relevant stakeholders, such as the importance assigned to the reduction of risk levels in high-risk groups compared to the whole population, or a higher relevance assigned to specific diseases due to the social impacts associated with their prevalence.

Identification and evaluation of adaptation options

In this step, potential adaptation strategies are identified and evaluated according to their effectiveness, feasibility and acceptability for the respective population (cf. step 1). As discussed previously, there are a wide range of options to reduce climate-related health impacts, including technical, institutional, legal, behavioural and medical measures. Projections of future health risks, along with information on current adaptation measures, are the basis for generating possible adaptation strategies. It is expected that most of the health risks of tomorrow will be similar to the health risks of today. This makes the design of response strategies easier because interventions for a particular disease that are effective today are likely to be effective in the future if the disease moves to a new location. However, the transfer of response strategies from one location to another may be hindered by differences in social, economic, institutional, cultural, or environmental conditions.

Prioritization of adaptation options

In this step, effective and feasible adaptation measures are prioritized according to their urgency. The close collaboration between public health experts and policy-makers is needed here because this prioritization involves value judgements, such as about risk preferences, and consideration of the existing policy context. The timing of adaptation requires a careful balance of two types of risks. Postponing adaptation may be associated with the risk of considerable impacts that could have been avoided by earlier action. However, early action may involve the misallocation of resources due to limited knowledge of what to adapt to. The reliability of information about climate change and associated risks is expected to increase over time due to a larger body of observational data and better prognostic models.

The appropriate timing of adaptation depends on the severity and the degree of confidence in the projected risk changes as well as the lead-time required for and the costs of available response options. In general, early action is particularly important if the measure is already effective under current climate conditions, if severe impacts are possible (e.g. high mortality from heat-waves), if adaptation measures have a long lead time (e.g. changing infrastructure to reduce the extent of the urban heat island effect in a region), if decisions have long-term effects (e.g. building settlements in areas that are at risk of flooding), and to reverse trends that threaten future adaptive capacity. Delaying action can be a rational adaptation strategy if the risks are moderate and response measures can be introduced quickly when need is, or if the cost of adaptations are exceedingly high given the current level of uncertainty.

▌ *Decision about adaptation strategy*
▌ *Implementation of decision*
▌ *Monitoring and evaluation of effectiveness*

The last three steps of the adaptation cycle fall into the realm of policy-making and are discussed in more detail in Chapters 8 and 12 [this book]. However, it is important to emphasize that provisions for an ex-post evaluation of interventions must be made early in the planning phase. Currently, insufficient data on the cost and effectiveness of health interventions often limits the ability of decision-makers to make well-informed decisions.

3.4 Academic and Policy Relevance

The type and magnitude of health risks are always central factors for the design of intervention strategies. In the context of adaptation to climate change, three additional factors are particularly important for decision-makers. First, the *confidence* in future risk projections largely determines the specificity of planned adaptations. The higher the certainty about future changes in a health risk at the scale of adaptation measures, the more specific adaptation is possible. Second, the *familiarity* with the health risk largely determines the need for new and additional measures. If the health outcome of concern is already prevalent in the region, and if it is effectively controlled, then additional adaptation measures can be drawn from existing experience. In contrast, adaptation to new health risks typically require significant additional efforts, and the development of response measures is likely to be drawn from the experience of others. Prospects for adaptation are worst if health risks are concerned that have long been prevalent in a region but are not yet effectively controlled due to the lack of measures that are both effective and feasible. The most prominent example of a climate-sensitive disease is probably the high prevalence of malaria in many tropical and sub-tropical regions, particularly in Sub-Saharan Africa. Third, the *urgency* of response measures is an important consideration because adaptation to climate change and variability is concerned with a dynamically changing stressor. The major considerations for balancing the risks of acting early versus acting late have already been discussed above.

While planned adaptation has a large potential for constraining the health risks associated with global climate change, it is clearly no panacea. For some diseases, no effective response measures are known. Even if an effective response exists, the availability of financial and other resources and the cultural acceptability of required behavioural changes often pose insurmountable barriers to their implementation. Countries that already suffer from a considerable burden of climate-sensitive diseases today will generally be unable to successfully adapt to the increased risks associated with future climate change on their own. Hence, any comprehensive long-term strategy for minimizing the risks associated with global climate change requires the combination of planned adaptation to climate change and mitigation of climate change. Furthermore, international burden sharing is needed to distribute the costs of adaptation according to the differential vulnerability of countries to climate change and to their differential contribution to the problem.

References

Bernard SM, Ebi KL (2001) Comments on the process and product of the health impacts assessment component of the national assessment of the potential consequences of climate variability and change for the United States. Environmental Health Perspectives 109(Suppl 2):177–184

Campbell-Lendrum DH, Kovats RS, Edwards S, Wilkinson P, Menne B, Corvalan C, McMichael AJ (2001) Estimating the global burden of disease attributable to climate change. In Primera Feria del Agua de Centroamerica y El Caribe, Panama

Canadian Standards Association (1997) Risk management: Guideline for decision-makers. CAN/CSA-Q850-97, Canadian Standards Association, Etobicoke, Canada

Carter TR, Parry ML, Harasawa H, Nishioka S (1994) IPCC technical guidelines for assessing climate change impacts and adaptations. Part of the IPCC Special Report to the First Session of the Conference of the Parties to the UN Framework Convention on Climate Change, Department of Geography, University College London, UK

Feenstra JF, Burton I, Smith JB, Tol RSJ (1998) Handbook on Methods for Climate Change Impact Assessment and Adaptation Strategies. Version 2.0. United Nations Environment Programme, Nairobi

Füssel H-M (2003) Impacts of climate change on human health – opportunities and challenges for adaptation planning. EVA Working Paper No. 4, Potsdam Institute for Climate Impact Research, Potsdam, Germany

Füssel H-M, Klein RJT (2002) Vulnerability and adaptation assessments to climate change: An evolution of conceptual thinking. In UNDP Expert Group Meeting "Integrating Disaster Reduction and Adaptation to Climate Change", Havana, Cuba (http://www.onu.org.cu/havanarisk/EVENTOS/cchange3/evento.html, accessed September 29 2005)

Haug GH, Günther D, Peterson LC, Sigman DM, Hughen KA, Aeschlimann B (2003) Climate and the collapse of Maya civilization. Science 299:1731–1735

Houghton JT, Ding Y, Griggs DJ, Noguer M, van der Linden PJ, Xiaosu D (2001) Climate Change 2001: The Scientific Basis. Cambridge University Press, Cambridge

Kay D, Prüss A, Corvalan C (2000) Methodology for assessment of environmental burden of disease. Report on the ISEE session on environmental burden of disease, Buffo, 22 August 2000. WHO/SDE/WSH/00.7, World Health Organization, Geneva, Switzerland

Kovats S, Ebi KL, Menne B (2003) Methods of assessing human health vulnerability and public health adaptation to climate change. Number 1 in Health and Global Environmental Change Series. World Health Organization, Regional Office for Europe, Copenhagen, Denmark

McCarthy JJ, Canziani OF, Leary NA, Dokken DJ, White KS (2001) Climate Change 2001: Impacts, Adaptation and Vulnerability. Cambridge University Press, Cambridge

McMichael AJ, Githeko A (2001) Human health. In: McCartly, JJ Canziani OF, Leary NA, Dokken DJ, White KS (eds) Climate Change 2001: Impacts, Adaptation, and Vulnerability, chapter 9. Cambridge University Press, Cambridge, 451–485

McMichael AJ, Haines A, Slooff R, Kovats S (1996) Climate Change and Human Health. World Health Organization, Geneva

McMichael AJ, Kovats RS (2000) Climate change and climate variability: adaptations to reduce adverse health impacts. Environmental Monitoring and Assessment, 61:49–64

Murray CJL, Lopez AD (1999) On the comparable quantification of health risks: Lessons from the global burden of disease study. Epidemiology 10:594–605

Patz J, Balbus JM (1996) Methods for assessing public health vulnerability to global climate change. Climate Research 6:113–125

Smit B, Burton I, Klein RJT, Wandel J (2000) An anatomy of adaptation to climate change and variability. Climatic Change 45:223–251

Smit B, Pilifosova O (2001) Adaptation to climate change in the context of sustainable development and equity. In: McCarthy JJ, Canziani OF, Leary NA, Dokken DJ, White KS (eds) Climate Change 2001: Impacts, Adaptation and Vulnerability, chapter 18. Cambridge University Press, Cambridge, 877–912

Smithers J, Smit B (1997) Human adaptation to climatic variability and change. Global Environmental Change 7:129–146

UNDP (2003) User's Guidebook for the Adaptation Policy Framework. Final Draft. United Nations Environmental Programme, New York City, NY

Willows R, Connell R (2003) Climate adaptation: Risk, uncertainty and decision-making. UKCIP Technical Report, United Kingdom Climate Impacts Programme, Oxford, UK

Yohe G, Tol RSJ (2002) Indicators for social and economic coping capacity – moving toward a working definition of adaptive capacity. Global Environmental Change 12:25–40

Heat-waves and Human Health

R. Sari Kovats, Gerd Jendritzky

Contributing Authors: Shakoor Hajat, George Havenith, Christina Koppe, Fergus Nicol, Anna Páldy, Tanja Wolf

4.1 Introduction

Heat-waves are an emerging public health problem in Europe. Although major heat-wave events have occurred (Greece 1987, Chicago 1995), they appear to be quickly forgotten. This chapter aims to review the current literature about the burden of heat in Europe and review the adaptive strategies to limit further increases in heat-related mortality and morbidity due to climate change. A major heat-wave occurred in western Europe in August 2003. The chapter will include information on the impacts and responses to that event.

4.2 The Impact of Heat on Health

4.2.1 Thermophysiology

Heat production of the human being by activity and heat release to the environment must be balanced in order to keep body core temperature within healthy limits. The body temperature regulation centres normally perform this in a wide range of varying heat exchange conditions. The body can lose heat by evaporation of sweat, convection (warming of air or water around the body), respiration (air inhaled is usually cooler and dryer than exhaled air), long wave radiation, and by conduction (contact with solids, e.g. floor) (Table 1). When air temperature and/or air vapour pressure increase, the gradients between skin and environment required for these heat losses decrease and heat loss is reduced. When air temperature approaches skin temperature heat loss by convection approaches zero, and there may even be a heat gain when air temperature goes above skin temperature. The main (and sometimes only) avenue for heat loss left is by sweat production and evaporation. With increasing water vapour pressure, even this will be compromised. Direct and diffuse solar radiation and long wave radiation from the surroundings can also result in significant heat gain for the human body.

Table 1. Environmental factors that influence heat loss

Air temperature	– T(skin) > T(air) convective heat loss from the skin to the environment
	– T(skin) < T(air) convective heat gain
Radiant temperature	– Radiant heat exchange between skin and environment
	In the sun, radiant temperature can easily exceed air temperature which results in an radiant heat transfer from the environment to the skin.
Surface temperature	Conductive heat exchange (minor role), long-wave radiation
Air humidity	Evaporative heat loss/gain
	The amount of moisture (not relative humidity) in the air determines whether moisture (sweat) in vapour form flows from the skin to the environment or vice versa. Evaporation of sweat is the most important avenue for the body to dissipate its surplus heat under hot conditions
Wind speed	Convection and evaporation, i.e. turbulent, sensible and latent heat fluxes
	Heat exchange increases with increasing wind speed

Increased sweat production in heat can lead to two types of problems: dehydration and hyponatraemia (caused by drinking large quantities of water with low salt concentrations). A significant proportion of heat stroke cases are due to activity in hot weather, mainly due to outdoor work, sport or army training (Epstein et al., 2000). It is important to note that such cases of exertional heat stroke are not necessarily associated with extreme weather or heat-waves (Shanks and Papworth, 2001). For less fit subjects, heat illnesses can occur at low levels of activity, or even in the absence of exercise. Low fitness levels lead to a low cardiovascular reserve and, thus, to low heat tolerance (Havenith et al., 1995). Further evidence has emerged from studies in Saudi Arabia, after a severe heat-wave: it was shown that the production of heat stress proteins and immune system reactions have a high importance in causing deaths (Bouchama and Knochel, 2002).

Addressing the thermophysiology of the thermal environment requires the application of complete heat budget models of the human being that take all mechanisms of heat exchange into account. Important input variables include air temperature, water vapour pressure, wind velocity, mean radiant temperature including the short- and long-wave radiation fluxes of the atmosphere, in addition to metabolic rate and clothing insulation. Such models possess the essential attributes to be used operationally in most biometeorological applications, including epidemiology (Laschewski and Jendritzky, 2002) and thermal stress forecasts in all climates, regions, seasons, and scales.

Fanger's (1970) PMV (Predicted Mean Vote) equation can be considered among the advanced heat budget models if Gagge's et al. (1986) improvement in the description of latent heat fluxes by the introduction of PMV*

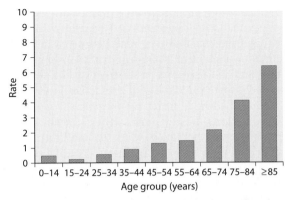

Fig. 1. Average annual rate (per 1 000 000 population) of heat-related deaths attributed to weather conditions (International Classification of Diseases, Ninth Revision (ICD-9), code E900.0) or exposure to excessive natural heat (ICD-10, code X30) – United States, 1979–2002 (after IoVecchio et al., 2005)

is applied (Fanger, 1970; Gagge et al., 1986). This approach is generally the basis for the widely used operational thermal assessment procedure "Klima-Michel-model" (Jendritzky et al., 1979; Jendritzky and Nuebler, 1981) of the German national weather service DWD (Deutscher Wetterdienst) with the output parameter "Perceived Temperature, PT" that considers a certain degree of adaptation by various clothing. The definition of the intensity of heat load or cold stress is based on PT thresholds. This procedure is run operationally, taking acclimatization into account (Koppe and Jendritzky, 2005). However, in most epidemiological studies on heat-related effects, air temperature has been used as the sole or main indicator of the thermal environment.

4.2.2 Heat-wave Events and Attributable Mortality

It is difficult to define the term "heat-wave" as it must meet both statistical (low frequency of occurrence) and social (an impact on the population) criteria. "Heat-waves" can be defined here as episodes with a sustained heat load that are known to effect human health. A further methodological difficulty is the lack of a standard definition of a "heat-wave". Some countries have local definitions, and it has proved difficult to identify a definition for international comparisons. The World Meteorological Organization does not have an operational definition, although a universal thermal climate index (UTCI) is in development (Jendritzky et al., 2000). Heat-waves can more readily be defined for the most extreme events (such as occurred in August 2003 in western Europe) than for less extreme events. In the scientific literature, the term is rather loosely applied. However, studies of heat-wave should consider both the intensity and duration of the event.

Heat-waves have a greater impact on mortality than the reported number of deaths or cases certified as due to classical heat illness. It is possible to quantify the excess if one estimates a baseline mortality, which is assumed would have occurred in the absence of the heat-wave. Attributable mortality is estimated by subtracting the "expected" mortality from the observed mortality during a pre-defined period. The expected mortality is calculated using a variety of measures, including moving averages, smoothing functions, and averages from similar time periods in previous years. Estimates can be sensitive to the method used to calculate the "expected" mortality (Whitman et al., 1997). Table 2 summarizes the impacts of heat-wave events on mortality that have been reported so far. The studies have used different methods and this makes comparison difficult. Where studies have looked at the mortality excess by cause of death, the impacts were greatest in mortality from respiratory and cardiovascular disease (Rooney et al., 1998; Huynen et al., 2001).

The heat-wave that occurred in August 2003 in Europe was unprecedented (Schar et al., 2004) and has caused the greatest impact on mortality ever reported in Europe. The heat-wave caused between 27 000 and 40 000 excess deaths in Europe, depending on the data sources, methodology, and the reference period. More than 14 800 excess deaths were observed in 13 major French cities from 4 to 15 August (Vandentorren et al. 2004). During the same period, excess mortality was also reported from Spain (Martinez-Navarro et al., 2004), Portugal (Falcao, Nogueira et al., 2003), Italy (Conti et al., 2005; Michelozzi et al., 2004), United Kingdom (Johnson et al., 2005), Germany (SBW, 2004), Switzerland (Grize et al., 2005), Belgium (Sartor, 2004), the Netherlands and the Czech Republic (Kysely, 2005). The Italian National Institute of Statistics reported a total of 19 200 excess deaths in comparison to the year 2002 in Italy over the period June-September (ISTAT, 2004).

Table 2 shows a range of estimates for mortality excess associated with individual heat-wave events that have been reported in the literature, including estimates published so far for the 2003 event. Comparison of these estimates should be made with caution as not only are different methods used to estimate the excess, but also the exposures are different. Even during August 2003, countries were affected by extreme temperatures of different magnitudes and duration. Similar temperatures can have different impacts depending on the duration of the event, or the acclimatization status of the population which is related to the time in the season when the heat-wave occurred as well as the long-term climate.

Heat-waves early in the summer (especially June) have been shown to be associated with greater impacts on mortality than heat-waves of comparable or hotter temperatures in the same population later in the summer (Hajat et al., 2002; Paldy et al., 2005). The impact of high temperatures later in the summer is sometimes diminished after an early heat-wave. In the summer of 2003 in Germany, several heat episodes occurred even before the August event and the impact of the August heat-wave was much high-

Table 2. Heat-wave events and attributed mortality in Europe

Heat-wave event	Attributable mortality (all cause)	Baseline measure	References
1976 – London, United Kingdom	9.7% increase England and Wales and 15.4% Greater London	31-day moving average of daily mortality in same year	McMichael and Kovats, 1998
1981– Portugal (month of July)	1906 excess deaths in Portugal, 406 in Lisbon	Predicted values	Garcia et al., 1981
1983 – Rome, Italy	35% increase in deaths in July 1983 in 65+ age group	Compared to deaths in same month in previous year	Todisco, 1987
1987 – Athens, Greece 21/7 to 31/7	estimated excess mortality >2000	Time trend regression adjusted	Katsouyanni et al., 1988
1991 – Portugal 12/7 to 21/7	997 excess deaths	Predicted values	Nogueira and Dias, 1999
1995 – London, United Kingdom 30/7 to 3/8	11.2% (768) in England and Wales, 23% (184) Greater London	31-day moving average of daily mortality in previous two years	Rooney et al., 1998
1994 – The Netherlands 19/7 to 31/7	24.4% increase, 1057 (95% CI 913, 1201)	31-day moving average of daily mortality in previous two years	Huynen et al., 2001
2003 – Italy 1/8 to 31/8	9704 (23.7%) throughout the country	Deaths in same period in 2002	Istituto Nazionale di Statistica (ISTAT) 2004
2003 – Italy 1/6 to 15/8	3134 (15%) in all Italian capitals	Deaths in same period in 2002	Conti et al., 2005
2003 – France 1/8 to 20/8	14802 (60%)	Average of deaths for same period in years 2000 to 2002	Hemon and Jougla, 2003
2003 – Portugal 1/8 to 31/8	1854 (40%)	Deaths in same period in 1997–2001	Botelho et al., 2005
2003 – Spain 1/8 to 31/8	3166 (8%)	Deaths in same period 1990–2002	Martinez-Navarro et al., 2004

Table 2 (continued)

Heat-wave event	Attributable mortality (all cause)	Baseline measure	References
2003 – Switzerland 1/6 to 31/8	975 deaths (6.9%)	Predicted values from Poisson regression model	Grize et al., 2005
2003 – The Netherlands 1/6–23/8	1400 deaths	Number of degrees above 22.3 °C multiplicated with the estimated number (25–35) of excess deaths per degree	CBS, 2003
2003 – Baden-Württemberg, Germany 1/8 – 24/8	1410 deaths	Calculations based on mortality of past five years	SBW, 2004
2003 – Belgium 15/5 to 15/9	1297 deaths for age group over 65	Average of deaths for same period in years 1985 to 2002	Sartor, 2004
2003 – England and Wales 4/8 to 13/8	2091 (17%). Mortality in London region: 616 deaths (42% excess)	Average of deaths for same period in years 1998 to 2002	Johnson et al., 2005

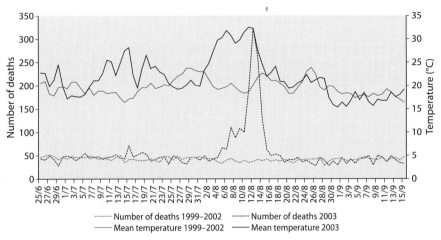

Fig. 2. Comparison of daily mortality and mean temperature in Paris for years 2003 and 1999–2002 (Empereur Bissonet, 2005)

er than expected. It has been suggested that this may be due to both short-term acclimatization and the loss of susceptibles from the population.

Lower than expected mortality can be seen following some of these events. It is likely that some of the mortality during the heat-wave is attributable to deaths brought forward by a matter of days or weeks (short-term mortality displacement). However, it can be seen that there is no apparent "dip" in reported mortality following the heat-wave in Paris in 2003 (Fig. 2). Undoubtedly some short-term mortality displacement did occur during this event, but it is also clear that many persons also died who would not have died in the following weeks. A study of a heat-wave in Belgium calculated that 15% of the excess mortality was due to deaths brought forward (Sartor et al., 1995), and other proportions have also been reported in the literature. However, robust methods have not yet been developed that quantify this effect, and the proportion of deaths due to short term mortality displacement during acute episodes remains uncertain.

4.2.3 Heat-waves and Morbidity

A few studies have investigated the impact of heat-waves on morbidity. Data on non-fatal outcomes are not routinely collected on a daily basis. Some studies have, therefore, used data on the use of health services, particularly hospital admissions, which are available as daily or weekly time series.

Heat-waves in the United States are associated with increases in emergency hospital admissions. The 1995 Chicago heat-wave was associated with an 11% increase in emergency hospital admissions and a 35% increase in the over 65 age group (Semenza et al., 1999). Of this excess, 59% were for heat-related illness (dehydration, heat exhaustion, and heat stroke) in persons with underlying chronic disease. A small increase was reported in hospitals in Birmingham during the 1976 heat-wave (Ellis et al., 1980), while no statistically significant excess was observed during the 1995 heat-wave in London (Kovats et al., 2004). In 2003, a 16% increase in admissions in those over 75 was detected in London (Johnson et al., 2005). An increase in admissions was reported in a hospital in Spain (Cajoto et al., 2005). Approximately 40% of the admissions during the heat-wave period were identified as heat related, but none were diagnosed as having heat stroke. In France, where the heat-wave was most intense, many hospitals were overwhelmed during the heat-wave, and a number of heat stroke cases were reported (Vanhems et al., 2003; Gremy et al., 2004; Lecomte and de Penanster, 2004).

Table 3. Thresholds and slopes for relationship between high temperature and emergency hospital admissions, by selected causes and age-groups, for Greater London, 1994–2000 (Kovats et al., 2004)

Cause of admission	Threshold (°C) (95% CI)	% increase per °C above threshold (95% CI)
All cause	12 (11)	−0.04 (−0.22, 0.13)
Diseases of circulatory system	24 (19)	1.71 (−2.70, 6.33)
Cerebrovascular disease	13 (1, 17)	−0.88 (−1.55, −0.21)
Diseases of respiratory system	23 (6)	5.44 (1.92, 9.09)
Diseases of the renal system	18 (16, 20)	1.30 (0.27, 2.35)
Renal failure, kidney stones	21 (5)	2.58 (−0.10, 5.32)
All other causes	6 (1, 9)	0.12 (−0.01, 0.24)
All causes, 0–4 years	12[a]	0.24 (0.02, 0.46)
All causes, 5–14 years	12[a]	0.20 (−0.21, 0.62)
All causes, 12–64 years	12[a]	−0.05 (−0.25, 0.15)
All causes, 65–74 years	12[a]	−0.14 (−0.46, 0.19)
All causes, 75+ years	12[a]	−0.22 (−0.49, 0.05)
Respiratory, 0–4 years	23[b]	3.91 (−3.33, 11.69)
Respiratory, 5–14 years	23[b]	5.20 (−3.04, 14.15)
Respiratory, 15–64 years	23[b]	3.34 (−1.49, 8.41)
Respiratory, 65–74 years	23[b]	7.71 (0.22, 15.76)
Respiratory, 75+ years	23[b]	10.86 (4.44, 17.67)

[a] Assumes same threshold as for all ages
[b] Assumes same threshold for respiratory disease across all ages

Time series studies of the effects of ambient temperature on hospital admissions across the whole temperature range have presented surprising results. A study in London for the cCASHh project found evidence for heat-related increases in emergency admissions for only a few specific outcomes: renal disease, and respiratory disease particularly in the 75+ age group (Kovats et al., 2004). Table 3 shows the effects of temperature for a range of cause and age-specific admissions. In Europe, higher temperatures do not appear to be associated with increases in admissions for cardiovascular disease (Kovats et al., 2004; Panagiotakos et al., 2004), although some effect is apparent in the US (Schwartz et al., 2004). Hospital admissions are not a perfect indicator of morbidity, as health system factors, such as admission thresholds, will vary between countries, and over time. However, the evidence so far indicates that increases in hospital admissions during heat-waves are not as dramatic as that seen in mortality. This suggests that the people who die during heat-waves are not reaching the attention of the medical services. This has important implications for prevention.

4.2.3 The Temperature-mortality Relationship

Seasonal patterns in mortality were described as soon as routine data on deaths became available (Farr, 1852). Since at least the 20th century, populations in temperate regions have deaths rates in winter higher than in the summer (Sakamoto-Momiyama, 1977; Eurowinter group, 1997). In particular, respiratory mortality is greatest in winter due to the seasonal occurrence of infections such as influenza. The degree of seasonality is not directly related to climate or latitude, however. The countries with the highest rates of "excess" winter mortality in Europe are Portugal and Spain (Healy, 2003). Excess winter mortality is lowest in the Scandinavian countries although their winters are much colder. The Scandinavians are well adapted to cold temperatures and it is only the mid-latitude countries that are badly effected by cold winters (Eurowinter group, 1997).

The shape of the "temperature"-mortality association can be described both qualitatively and quantitatively. The relationship is approximately U-shaped, with mortality increasing at both low and high temperatures (Kunst et al., 1993) (Fig. 3). The relationship between daily mortality and the thermal environment is much closer, and more short-lived in summer than in winter (Laschewski and Jendritzky, 2002; Pattenden et al., 2003). In order to quantify the effect of temperature on mortality, a linear relationship is assumed above the "threshold" or change point in the relationship. Table 4 lists the heat slopes for selected cause groups based on published studies in European cities and countries. These estimates are derived from models that take into account the seasonal and other long-term patterns in the outcome measure, in order to reveal any short-term effects of temperature. Many factors vary seasonally in addition to temperature, such as behaviour, diet and the occurrence of seasonal infections (Schwartz et al.,

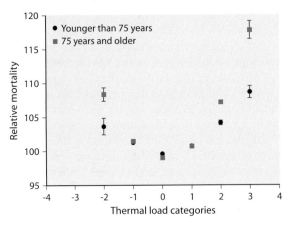

Fig. 3. Deaths relative to the expected mortality (in percent) for different thermal load categories in Baden-Württemberg (SW Germany) between 1968 and 2003 for two age groups. Thermal load categories: −4: extreme cold stress; −3: strong cold stress; −2: moderate cold stress; −1: slight cold stress; 0: thermal comfort; 1: slight heat load; 2: moderate heat load; 3: strong heat load; 4: extreme heat load

1996). The effect of temperature on mortality is greatest for respiratory diseases and cardiovascular diseases than for other causes of death.

Fig. 3 illustrates the impact of different thermal stress categories on relative mortality (defined as the number of deaths relative to the expected mortality) for two age groups (under 75 and over 75 years) in south-west Germany between 1968 and 2003. During days with thermal comfort or only a slight heat load or cold stress, there are only small deviations of mortality from the expected value. As the thermal stress (heat load and cold stress) becomes more extreme, mortality increases. Days with a strong or higher heat load can be classified as heat-wave days. Mortality during days with a strong heat load (category 3) increased by 18% in the older age group. Within the strong heat load category, the mortality increase is determined by the persistence of the situation. Relative mortality for both age groups increases with the number of days in a row with a strong heat load (Koppe, 2005).

The expected mortality was calculated on the basis of a Gaussian smoothing with a filter length of 365 days. The smoothed values are then corrected by applying a restoration factor in order to account for the course of the year (Laschewski and Jendritzky, 2002). Koppe (2005) estimated that in the age group over 75 years only about 20% of the increase in mortality after days with a strong heat load was attributable to short term mortality displacement (Koppe, 2005).

Table 4. The effect of temperature on daily mortality, for selected causes

Population	% change per increase 1 °C above cut point			Cut point (°C)	Lag (days)	Reference
	All cause	CVD	Resp			
Oslo, Norway	0.7 (−0.2, 1.5)	0.8 (−0.4, 2.0)	4.7 (2.2, 7.1)	10	7	Nafstad et al., 2001
London, UK	1.3 (1.0, 1.2)	–	–	18	0–1	Pattenden et al., 2003
The Netherlands	1.2	1.1	3.1	16.5	1–2	Kunst et al., 1993
Sofia, Bulgaria	2.2 (1.6, 2.9)	–	–	18	0–1	Pattenden et al., 2003
Madrid, Spain	1.0	–	–	20.3	1	Alberdi et al., 1998
Valencia, Spain	3.6 (1.2, 6.0)	2.3 (−1.5, 4.5)	5.7 (−2.9, 8.2)	24	1–2	Ballester et al., 1997
Barcelona, Spain	–	4.0 (IHD)	–	25		Saez et al., 2000

CVD cardiovascular disease, *Resp* respiratory disease, *IHD* ischaemic heart disease

Air Pollution

Stressful weather and high levels of air pollutants are assumed to have independent adverse effects on daily mortality. Few studies have simultaneously quantify weather-induced and pollution-induced effects on daily mortality. In studies of heat-related mortality, air pollutant exposure is a confounder because it is associated both with daily variations in temperature (exposure) and with daily variations in mortality (outcome). However, air pollutant exposure is also a probable interactor. There is some evidence of a physiological synergistic effect between temperature and pollutants (Shumway et al., 1988; Katsouyanni et al., 1993).

The same meteorological conditions that cause heat-waves also tend to produce high concentrations of air pollutants. For example, the concentrations of tropospheric ozone and particulates were very high in the South of England during the 2003 heat-wave (Stedman, 2004).

4.2.4 Determinants of Vulnerability

Physiological

The main predisposing factors for clinical heat illness include the following (Kilbourne, 1992): dehydration due to reduced food and liquid uptake, intestinal problems, use of diuretics or alcohol abuse; use of other drugs affecting the system: stimulants, β-blockers, anticholinergics, digitalis, barbiturates, especially combined with hypertension, age, fever, recent infections or skin burns; previous heat illness, low fitness, adiposity, fatigue, sleep deprivation, long-term high level exercise, and protective clothing. Age and illness are strong predictors in this sense as age highly correlates with increasing illness, disability, medication use, and reduced fitness.

Physical fitness tends to decrease with age due to a reduced physical activity level in the elderly. More strain is put on the cardiovascular system and less cardiovascular reserve is left, because any activity performed becomes more stressful. The cardiovascular reserve is especially relevant to the capacity for thermoregulation as it determines the capacity to move heat for dissipation from the body core to the skin by the skin blood flow (Havenith et al., 1995; Havenith, 1998). The fitness reduction can work like a vicious circle as the increased strain experienced with activity may in itself promote even further activity reduction which again may lead to a further reduction in fitness. In addition, heat and cold exposure is avoided which lead to a loss of heat and cold acclimatization. On a population level these and other changes lead to a reduced muscle strength, reduced work capacity, a reduced ability to transport heat from the body core to the skin, reduced hydration levels, reduced vascular-reactivity and a lower cardiovascular stability (blood pressure) in the elderly. These effects will put elderly people at a higher risk in extreme conditions, leading to an increase in morbidity and mortality (Havenith, 2001b).

Babies and infants are at higher risk of dehydration and therefore heat stress due to the relative higher volume of fluid in their bodies compared to an adult (King et al., 1981). There is anecdotal evidence of certified heat deaths (heat stroke) caused by children being left in cars on hot days, but as with exertional heat stroke, such deaths are not necessarily linked to extreme weather. There is currently little evidence for heat effects on mortality in young age groups (under 15 years) in European populations, as the baseline mortality in this age group is very low.

Socioeconomic Factors

Both individual- and population-level studies provide strong and consistent evidence that age is a risk factor for heat- and cold-related mortality. One of the reasons is that age is closely related to decreasing fitness. Studies, however, vary at the age at which the vulnerability is shown to increase. Most population-based time series studies show an effect in adult age groups (Sierra Pajares Ortiz et al., 1997) and the effect is larger in persons over 65 compared to other ages. As these studies use predetermined groups for the "elderly", there has not been a more detailed examination of the age at which vulnerability is increased between different populations.

The elderly in hospitals and institutions, such as residential care homes, may be vulnerable to heat-related illness and death (Faunt et al., 1995). An almost two-fold increase in mortality rate was reported in geriatric hospital inpatients (but not other inpatients) during the 1976 heat-wave in the United Kingdom (Lye and Kamal, 1977). In northern Europe, such institutions are unlikely to be air conditioned. In France, deaths rates in retirement homes doubled during the heat-wave in 2003 (Hemon and Jougla, 2003). A larger than expected excess was also reported in northern Italian nursing homes (Rozzini et al., 2004).

Impacts of heat-waves appear to be greater in urban than in rural populations. Such populations differ in their basic vulnerability to heat (underlying rates of cardiorespiratory disease) and in their exposures (different types of housing, and higher temperatures caused by the urban heat island effect (Clarke, 1972; Oke, 1997). Mapping of heat-wave deaths in St. Louis (1966) found the highest rates in inner city areas where population density was higher, open spaces were fewer, and where socioeconomic status was lower than in surrounding areas (Henschel et al., 1969; Schuman, 1972). The presence of an urban heat island will also tend to increase the heat exposure of urban populations (see below). Individuals are more vulnerable to heat stress if they are in poorly designed housing, with no access to air conditioning or cooler buildings. These risk factors are more likely to be present in urban than rural areas. During a heat-wave, mortality impacts are greater in cities than in the surrounding areas or in the country as a whole. This has been shown in the United Kingdom in 1995 (Rooney et al.,

1998), in Greece in 1987 (Katsouyanni et al., 1988) and in 1980 in Missouri (Jones et al., 1982). Rural populations in the Midwest in the United States are also affected by heat-waves (Sheridan, 2003), but there has been relatively little research on non-urban populations in Europe.

It has proved difficult to demonstrate clear socioeconomic gradients for heat-related mortality (Jones et al., 1982; Smoyer, 1998b; Hales et al., 2000; Gouveia et al., 2003). Such studies have used census-based, small area indicators of socioeconomic status. This method may be unable to detect real differences if the effects are small. In the United States, information on race and educational level is available on the death certificate. Time series analysis stratified by these indicators in seven United States cities, indicates that being black and having a high school education or less doubled the risk of heat-related mortality (O'Neill et al., 2003). Previous studies have shown that black people are more likely than whites to be living in impoverished neighbourhoods, even if they have similar incomes. These findings are consistent with the small area studies (Smoyer, 1998a), indicating that there is a socioeconomic gradient in the risk of heat-related death in the United States, where perhaps access to air conditioning is the most important protective factor. However, these methods cannot reveal whether the increased risk in groups with lower socioeconomic status is due to difference in housing, neighbourhood or the underlying prevalence of chronic disease. Case-control studies following the 1995 (Semenza et al., 1996) and 1999 (Naughton et al., 2002) heat-waves in Chicago both found that the strongest risk factor was living alone, and not leaving home daily.

4.3   Adaptation: Current Knowledge

The capacity of humans to adapt to varied climates and environments is considerable. Physiological, behavioural and architectural differences between cultures have developed over many millennia as a consequence of exposure to vastly different climatic regimes. Individuals undergo physiological acclimatization. At the population level, it can be seen that mortality responses are also adapted to the local climate. Populations in warmer climates show that thresholds for mortality responses to hot weather are higher than in populations with colder climates (Table 4).

Human beings undergo physiological adaptation to warmer climates. Many studies have illustrated this for people moving to hotter countries, and particularly for soldiers or athletes. Physiological adaptation can take place over a matter of days or weeks. The body responds by increasing the output of the sweat glands, and by improving cardiovascular stability upon heat exposure. Both responses lead to reduced thermal and cardiovascular strain.

Physiological adaptation is supported by a wide range of behavioural changes (clothing, drinking, food, siesta, activity). Housing design is also closely related to climate conditions (WHO, 1990; Roaf, 2005). This section reviews current knowledge about strategies that are currently used to reduce the impact of hot weather and heat-waves on human health and comfort.

Heat-wave emergencies have been reported predominantly in southern and south-eastern Europe (the Balkans) (CRED and OFDA, 2001). In Chicago in 1995, the water and power supplies failed in parts of the city, with serious implications for public health. The service responsible for dealing with deaths at home (often the fire service) may also become overwhelmed. Hospitals may become full, as well as the mortuaries, although the information for this is less clear. The heat-wave in July 1987 in Greece was associated with 2000 excess deaths in Athens (Katsouyanni et al., 1988), and an overwhelming of the hospitals. Non-health impacts, such as transport problems and forest fires, can also cause widespread disruption.

4.3.1 Health Education

Education to advise people of appropriate behaviour during hot weather is an essential component of heat-death prevention. Many governments in Europe have issued advice on how to avoid heat-related illness. However, there is evidence that perception of ambient temperature is poorer in the elderly (Collins et al., 1981). Therefore, an individual may not recognise that they need to change their behaviour at the first signs of heat stress. An understanding of human behaviour during heat events is needed before the most appropriate messages can be developed and targeted. There is a lack of qualitative research on behavioural responses to heat-waves and heat-wave warnings.

There have been several campaigns at getting people to be warmer in winter. Searches of the published literature found no studies of the effectiveness of heat- or cold-awareness campaigns.

4.3.2 Heat Health Warning Systems

Heat health warning systems (HHWS) are acute responses during heat-wave events (Koppe et al., 2003; Kovats and West, 2005). They should include the following meteorological and public health components:

- Sufficiently reliable heat-wave forecasts for the population of interest
- Robust understanding in the cause-effect relationships between thermal environment and health
- Effective response measures to implement within the window of lead time provided by the warning
- The community in question must be able to provide the needed infrastructure.

A range of methods are used in order to identify situations that adversely affect human health, which are described in detail elsewhere (Koppe et al., 2003).

Heat early warning systems when accompanied by specific health interventions are generally considered to be effective in reducing deaths during a heat-wave (Ebi et al., 2004). However, there have been no studies that evaluate the various community-based interventions associated with such systems. Heat-waves are infrequent events and vary in magnitude and in the time that they occur during the summer season. The criteria for the evaluation of the systems should include the number of deaths and hospital admissions prevented. There is very little evidence on the evaluation of individual systems. The effectiveness of intervention strategies should be investigated using formal epidemiological methods.

Public warnings should be aimed at the wider community in order to modify the behaviour of individuals and to raise awareness of the dangers that are connected with heat exposure in order to reduce heat-related impacts. Therefore, warnings need to be linked to specific advice on how people recognise the problem and what they should do to protect themselves and others.

Intervention plans should be best suited for local needs, through co-ordination between the local health agencies and meteorological officials. A comprehensive warning system should involve multiple agencies, such as city managers, public health and social services workers and emergency medical officers. Good co-ordination between the weather service, which issues the warning, and the local health agency is essential. In some systems, the Meteorological Services do not feel in control of the implementation of these warnings. Poor communication between the Meteorological Service and the Health Agency can prevent the implementation of an effective system.

The active care of susceptible individuals can only be achieved at the local level. It is therefore important that those who are susceptible can be easily identified if outreach services are to be attempted. One method may be through local community groups, or through active registration with a general practitioner. One system in the United States encourages relatives to register through a website. The outreach may be undertaken by professionals (health workers) or by volunteers, including friends or relatives.

Hospitals, primary care clinics and nursing homes should all be prepared for heat-waves. An emergency plan should be drafted and piloted. The plan should involve the education of doctors, nurses and other staff to identify heat problems and be familiar with the most appropriate treatments. A personnel plan could also be developed so that extra staff are in place if needed.

4.3.3 Heat in the Indoor Environment

Most homes have an indoor temperature between 17 °C and 31 °C. Humans cannot comfortably live in temperatures outside this range. The tolerance range of an individual is usually less than this and tends to become narrower with age or infirmity. The temperature of the outdoor air is a significant factor for human comfort, giving a good indication of the temperature occupants will find comfortable indoors (De Dear and Brager, 2001). There are three main factors that are associated with indoor heat exposures:

▌ *Thermal capacity of building:* a heavy building will warm up more slowly in a heat-wave, particularly if it is well insulated.

▌ *Position of apartment:* the upper floor of a building will generally be hotter than the other floors in summer and colder in the winter because the roof provides inadequate insulation and hot air may collect in upper storeys.

▌ *Behaviour and ventilation:* Occupants will respond as best they can to unusual weather conditions using the facilities provided by the building (windows, fans coolers etc.) but these can be less effective if used in a non-optimal manner. For instance, opening the windows to cool the building down at night can be effective if the building has a high thermal mass and can store the 'coolness' of the night air but this also requires that windows can be opened safely at night.

In developed countries, people spend the vast majority of their time indoors, at home and at work. The indoor environment has been investigated in relation to indices of thermal comfort. Perceptual scales have been developed to evaluate thermal comfort in an individual (e.g. ASHRAE scale). In temperate climates, the optimum indoor temperature for health is between 18 °C and 24 °C (WHO, 1987). In general, recommendations have focussed on maintaining a minimum indoor temperatures and the reduction in the impact of cold on health, with less emphasis on the potential impact of heat (WHO, 1990).

Because of heat gains from equipment and sun coming in through the windows, the mean temperature inside buildings is generally higher than that outdoors. In heat-wave conditions, a heavy-weight (hw) building will 'warm up' more slowly than a light-weight (lw) one (Fig. 4) and except in very long heat-waves, will not get quite so hot. In addition, the daily temperature swing is larger in lw building which heat up quicker in response to warming by the sun and occupant activities. Houses that have high thermal capacity are slow to warm up but once warm retain the heat. This was very true in Chicago where brown stone tenement blocks became like ovens. If the occupants are able to ventilate the building at night this helps keep daytime indoor temperatures down, especially in heavy-weight buildings. There has been relatively little health research done on the housing types in Europe that increase vulnerability to heat illness or heat-related mortality. Some work has been done in relation to exposures to cold (see

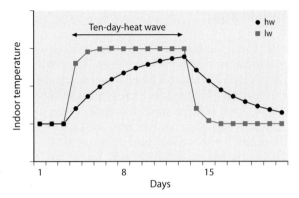

Fig. 4. The mean temperature in high-mass buildings (hw) rises (and drops) more slowly than in low-mass (lw) buildings with the onset of a heat-wav.

e.g. Rudge and Nicol, 2000). People are at higher risk of dying in a heat-wave if they live in certain types of apartment building (with high thermal load), and in a top floor apartment (Mirchandani et al., 1996; Semenza et al., 1996). The type of housing most vulnerable to heat could be dwellings with no cross ventilation, and no shading devices.

Buildings today are designed to have a long lifetime. Climate change (such as the increasing numbers of extreme hot days), changing lifestyles and new technologies all have implications for building design (Roaf, 2005). With respect to climate change, the design of comfortable, energy-efficient and safe buildings is a priority. In particular, the design should aim to limit both the frequency of occurrence of high temperature episodes inside the building and their intensity and duration. Traditional building designs have evolved in harmony with the environment and usually provide adequate protection against the heat. In recent decades, it seems that rapid urbanization has led to an increase in poor building design in relation to weather extremes in many cities. Thus, until populations in these dwellings develop methods to adapt, they will perhaps be more vulnerable to heat episodes (Oke, 1997). People living in informal dwellings in large cities may be very vulnerable to weather extremes as they are often constructed from materials which provide no thermal insulation. Further, these populations are already marginalized and suffer poor health (WRI, 2000).

There is very little information about how people behave in their houses, such as when they open windows or put on air conditioning in relation to outside temperatures. One strong trend over the past 20 years is a decrease in ventilation – people close their windows due to the fear of crime, outdoor noise, and air pollution, especially at night when ventilation is important. Indoor air pollution is related to outdoor air pollution. Indoor air may contain lower concentrations of outdoor pollutants (particulates, ozone and NO_x) but these may be supplemented by 'indoor pollutants' such as cooking by-products and chemical off-gases from furniture and building materials.

Air Conditioning

In developed countries, air conditioning or "comfort cooling" has a direct role in reducing temperature exposures in an individual when they are in the cooled environment. Air conditioning does not necessarily lower the humidity. There are three types of air conditioning in use in Europe:
▌ Central air conditioning
▌ Attached to hole in the wall
▌ Free standing unit.

Evidence from the United States indicates that air conditioning is an effective intervention to prevent heat stroke and heat-related illness (Marmor, 1975; Kilbourne et al., 1982; Kiernan, 1996; Semenza et al., 1996; O'Neill, 2003), both in dwellings and in nursing homes (Marmor, 1978). Half the cases of heat stroke admitted during the 1995 heat-wave in Madison, US, were attributed to indoor activity with no functioning air conditioning (Dixit et al., 1997). The benefits seem to be associated with inbuilt air conditioning, rather than free standing units (Rogot et al., 1992).

Although air conditioning reduces mortality during climatic extremes, it also takes away all the natural stimuli, which normally induce acclimatization to cold or heat and are beneficial in preventing climate-related illnesses (Havenith, 2001 a). In addition there are other illnesses (e.g. Sick buildings syndrome) which are exacerbated by air conditioning. Nor should we forget that the energy used by air conditioning is itself increasingly contributing to pollution, local warming by heat release and climate change. Continuous air conditioning is advised only in cases where ill health is present. For otherwise healthy people, it seems advisable to protect them from climatic extremes, but not to take away the normal stimuli in order not to compromise long-term protection.

4.3.4 Heat in the Outdoor Environment

Regional climate is modified at the local scale by human activities through changes in the land surface, such as building a city (Oke, 1997). Heat-waves present special problems in urban areas. Buildings retain the heat during the day and then re-radiate that energy during the night. In Chicago, a mean increase (urban-rural) of 2.8 °C was observed during the night in the summer which decreased to 1.8 °C during the daytime (Ackerman, 1985). In Athens, Greece, a mean maximum urban heat island of 4.6 °C in the summer (June–September) has been observed (Livada et al., 2002). These examples represent quite extreme values, and in most cities in Europe, heat islands are more apparent in winter than in summer.

Within a given city, the hottest areas are those with tallest buildings, with the greatest density of buildings, and without green spaces (Oke, 1997). This is often in the inner city, where population-density varies, depending on the level of out-migration to the suburbs.

Measures to reduce the urban heat island focus on:
- Increasing green spaces, planting trees in streets, greenery on walls and roofs (trees provide shade and can also improve air quality but reduce ventilation)
- Increasing ventilation and air flow between buildings (which also improves air quality)
- Increasing the number of courtyards and other open spaces
- Increasing the albedo of a city (e.g. painting roofs white)
- Decreasing anthropogenic heat production (e.g. natural space cooling, see above).

The urban heat island is an inevitable consequence of urban development. Appropriate and climate friendly urban planning, however, may help to reduce the magnitude of the urban heat island. For example, the benefits of tree planting projects include shading, cooling due to evapotranspiration, dust control, runoff control, consumption of carbon dioxide, water conservation. There are many competing priorities for urban planning. In practice, climate issues often have a low impact on urban design. Although urban planners are interested in climatic aspects, the use of climate information is unsystematic (Eliasson, 2000). Good building designs can provide effective measures to reduce heat stress of individuals living in cities.

4.4   Climate Change and Future Heat Stress

Global climate change is likely to be accompanied by an increase in the frequency and intensity of heat-waves, as well as by warmer summers and milder winters. The impact of extreme summer heat on human health may be exacerbated by increases in humidity (Gaffen and Ross, 1998).

Extreme weather events are, by definition, rare stochastic events. With climate change, even if the statistical distribution of such events remains the same, a shift in the mean will entail a non-linear response in the frequency of extreme events (Fig. 5). This was originally demonstrated using statistical theory by Katz and Brown (1992).

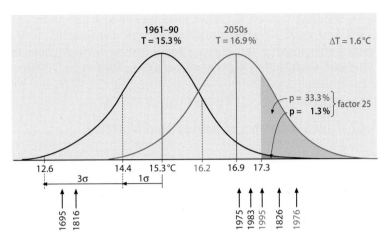

Fig. 5. Changes in temperature distributions with climate change. Increasing probabilities of extremes, example: summer temperatures in central England. Source: Climatic Research Unit, University of Norwich

4.4.1 Have Heat-waves Been Increasing in Europe?

Published papers on observed trends in daily, monthly or seasonal extremes describe an increase in trends over recent decades or longer (Meehl et al., 2000; Yan et al., 2002; Klein-Tank and Können, 2004; Luterbacher et al., 2004). The detection of changes in simple extremes based on climate statistics (e.g. hot days) requires long data series of sufficient quality (Easterling et al., 2000). In the United Kingdom, an increase in the frequency of very hot days in Central England has occurred since the 1960s, with a number of extreme summers in 1976, 1983, 1990 and 1995 (Hulme et al., 2002). Sustained hot days (heat-waves) have become more frequent, particularly in May and July (Hulme et al., 2002). The European Climate Assessment, 2003, confirms that Europe has experienced an unprecedented rate of warming in recent decades (Klein Tank et al., 2003). Studies of more-health relevant climate indices also describe observed increases in "heat stress days" in the United States (Gaffen and Ross, 1998, 1999).

The heat-wave that affected many parts of Europe during the course of summer 2003 produced record-breaking temperatures particularly during June and August (Beniston, 2004; Schar et al., 2004). Absolute maximum temperatures exceeded the record temperatures observed in the 1940s and early 1950s in many locations in France, Germany, Switzerland and the United Kingdom according to the information supplied by national weather agencies (WMO, 2003). Luterbacher et al. (2004) suggests that 2003 is likely to have been the warmest summer since 1500 based on documentary records. The 2003 heat-wave was associated with a very robust and persistent blocking high pressure system that some weather services suggested may be a manifestation of an exceptional northward extension of the Hadley Cell (Dalu et al., 2005). Climatologists now consider it "very likely" that human influence on the global climate has at least doubled the risk of a heat-wave such as that experienced in 2003 (Stott et al., 2004).

4.4.2 Future Estimates of Heat-related Mortality

The impacts of climate change on future heat-related mortality have been quantified for a range of populations. As the relationship between mortality and temperature has been quantified by several epidemiological studies, it is more readily incorporated into a risk assessment model than other health outcomes.

Scenarios are plausible and simplified descriptions of how the future climate may develop, based on a coherent and internally consistent set of assumptions about driving forces and key relationships (IPCC, 2001). Climate scenarios provide an important tool for estimating the potential impact of climate change on specific health outcomes in the future. Several studies have estimated future heat- or cold-related mortality using climate scenarios for specified times in the future (Table 5) using two main approaches:

▌ Increased frequency of air masses associated with mortality
▌ Extrapolation of a linear temperature-mortality relationship.

Predictions of temperature-related mortality in the United States have been derived from observed relationships with air masses or synoptic climatological classifications. The air mass/mortality relationship is evaluated using a regression model (Kalkstein and Smoyer, 1993). The Kalkstein and Greene study of 44 cities in the United States estimated that increases in heat-related deaths would be greater than decreases in cold-related deaths (by a factor of three) (Kalkstein and Greene, 1997). One of the main weaknesses of this study is the difficulty in using climate scenarios to estimate changes in the frequency of air masses. Furthermore, both heat and cold types are derived using a zero lag model, that is, delayed effects of the weather exposures are not taken into account. For cold, the delayed effects are more important and there this model underestimates the effect of low temperatures on mortality.

Extrapolation of a linear temperature mortality relationship has been undertaken by several studies (Bentham Langford, 1988) (Table 5). Several coefficients from studies of temperature and mortality can be combined using meta-analysis (Martens, 1998; McMichael et al., 2000). A final cause-specific 'V-shaped' dose-response relationship was applied to 20 cities in 17 countries. Additional deaths due to climate change were estimated for mortality due to cardiovascular and respiratory disease, and in the elderly.

There are several simplifications that need to be addressed in such modelling approaches:

▌ The application of a single temperature-mortality relationship to diverse populations, either from a single study or one derived from a meta-analysis

▌ Assumption that the temperature-mortality relationship does not change in the future

▌ Assumptions about changes in the "threshold" temperature due to changes in climate

▌ The "annualization" of a short-term (day-to-day) relationship to changes in annual (or decadal) means to generate an annual estimate for excess mortality

▌ Assumptions about climate changes, and changes in temperature at the local (population) level, including changes in the variability of daily temperatures as well as changes in the mean.

A central question in estimating future temperature-related mortality is acclimatization. Populations will acclimatize to warmer climates via a range of behavioural, physiological and technological adaptations. The initial physiological acclimatization to hot environments can occur over a few days, but complete acclimatization may take several years. Acclimatization will reduce the impacts of future increases in heat-related mortality. Estimates of rates of acclimatization over decades have been derived from observed short-term acclimatization during the summer (Kalkstein and Greene, 1997).

The World Health Organization comparative risk assessment methods were applied to global climate change in order to estimate the potential benefits of two stabilization scenarios on the global population (McMichael et al., 2004). Estimates were generated for future relative risks for the impact of climate change on cardiovascular and respiratory mortality via "direct temperature" effects. Populations were assumed to undergo physiological acclimatization which was reflected in a proportional shift in the threshold temperatures. No explicit adjustment was made for an effect of socioeconomic development on temperature-related mortality. However, the lower bound of the estimates assumed that 100% of the effect of climate change on heat-related mortality is avoided due to acclimatization.

Little work has been done to quantify the uncertainty around future projections. A Bayesian analysis of a study of heat-related deaths in Lisbon indicated that the greatest sources of uncertainty were climate projections (climate models), acclimatization assumption and then the temperature-mortality model methods (Dessai, 2003).

In addition to concern about the methodological approaches in risk assessment, there are many potential difficulties in attempting to project health outcomes in the future, particularly for the long time spans preferred in climate change impact assessments (i.e. to 2050 or 2100). Health outcomes result from the interaction of many factors such as the population's age structure and general health status, social and economic conditions, and environmental factors other than climate. Climate change is projected to progress slowly, giving populations time to adjust to the new conditions without major changes in health status. However, the potential for surprises and for increased climate variability exists.

4.5   What Strategies, Policies and Measures Are Needed?

Increases in the frequency and intensity of heat-waves is a virtually certain outcome of global climate change. The rate at which infrastructure changes will take place without specific advice is likely to be much slower, however. Neither the magnitude nor the time course of the various modifying factors for vulnerability can be predicted with any confidence. It is clear that preventive measures will be needed to counter the substantial initial adverse effects of heat, and long-term changes are required to housing and urban infrastructure.

The August 2003 heat-wave has revealed that capacity to dealing with extreme temperatures is low. France and other affected countries have developed national plans to avoid future health disasters (Bosch, 2004; Senat, 2004). The France plan, "Plan Canicule", has focussed on several key responses: a four level alert system to be triggered by adverse weather forecasts, a census of older people to create a register of those at risk, and the installation of air conditioning in the homes of elderly people. Response plans differ between the different European countries (see Chapter 8.1, Kosatsky). The risk of a similar event recurring within the next decade remains, and will increase as climate change accelerates.

Table 6 summarizes the policy measures in more detail in the context of an adaptation assessment that can be undertaken at the local, regional and national level (see Chapter 8, Ebi et al., on Policy Analysis). The table has been completed with reference to the United Kingdom and the answers will vary for other countries in Europe. Clearly, several options are available to policy-makers but a limiting factor seems to be evidence of their effectiveness in reducing heat-related mortality.

Educate, Warn and Protect

Clearly, a main strategy in reducing heat-related deaths is to educate people to be more aware of heat-related illness. It is especially important that persons responsible for the care of vulnerable people are educated, for example, people who work in residential care homes and other institutions. Education could be formally addressed within training and course curricula, and be part of the formal regulations of such institutions.

Education can be linked to warnings of imminent extreme weather. However, education is not sufficient on its own as many of the most vulnerable are unlikely to respond to health advice or change their behaviour in response to a warning. There is a lack of information on the most appropriate measures to educate people about heat avoidance.

Table 5. Predictive modelling – effects of climate change on deaths due to thermal extremes

Reference	Method	Inputs	Region	Key findings
Langford and Bentham, 1995	Regression model: Monthly mortality & temperature series [flu included]	Seasonal average temp from UK scenarios (Warrick and Barrow, 1991)	England and Wales	Winter deaths avoided: 2010: all cause 3301, Ihd 1308, Cvd 429 2030: all cause 6353, Ihd 2550, Cvd 836 2050: all cause 8922, Ihd 3631, Cvd 1187
Martens, 1998	MIASMA v1.0 – Empirical-statistical model, meta-analysis	ECHAM1-A, UKTR, GFDL89 [IPCC scenarios]	Cause + age specific Global – 20 cities	Changes in mortality rates for Cvd (<65), cvd (>65), respiratory, and total mortality Net reductions in mortality in all cities except respiratory mortality in a few cities
McMichael et al., 2000	MIASMA v.1.0 – same as Martens, 1998	HadCM2, ensemble mean +HadCM3	Global – 20 cities	Net reductions in mortality in all cities except Athens, due to decreases in winter mortality
Kalkstein and Smoyer, 1993	Model derived from observed relationship between synoptic air masses and mortality	GCM scenarios, no down-scaling	Global – 27 cities	Significant increases in heat-related mortality under various climate change scenarios, with or without acclimatization, projected for all cities
Kalkstein and Greene, 1997	Model derived from observed relationship between synoptic air masses and mortality	GCM scenarios, no down-scaling	44 US cities	Increases in heat-related mortality are much larger than decreases in cold-related mortality, overall, but cold relationships based on zero lags and therefore underestimated
Duncan et al., 1997	Model derived from observed relationship between synoptic air masses and mortality	climate scenarios for Canada	10 cities in Canada	240–1140 additional heat-related deaths by per year in Montreal by 2050, 230–1220 additional in Toronto, assuming no acclimatization No climate/mort relationship in some cities

Table 5 (continued)

Reference	Method	Inputs	Region	Key findings
Guest et al., 1999	Regression model: daily mortality and synoptic indices	CSIRO Mark 2 model, CO_2 doubling, high and low scenarios estimated	5 cities in Australia	Net decrease in TR mortality, particularly in 65+ age group, range 47–62 fewer deaths per year in all 5 cities. Significant increase in summer deaths in Sydney (76–239)
McMichael et al., 2003	Empirical-statistical model, derived from observed monthly mortality	High, medium and low emissions CSIROMk2, ECHAM4	6 cities in Australia (Adelaide, Brisbane, Hobart, Melbourne, Perth, Sydney)	Increases in heat-related mortality in over 65, increases large in temperature cities. Less reductions in cold-related mortality
Casimiro and Calheiros, 2002; Dessai, 2003	Empirical statistical model, observed-expected	2 Regional climate models – PROMES and HadRM2	Lisbon, Portugal	Heat-related death rates increase by 57 to 113% by 2020s, by 97–255% by 2050s, assuming no acclimatization. Acclimatization assumptions reduces estimates
McMichael et al., 2004	Empirical statistical model based on published temperature-mortality relationships	HadCM3–1S92a, and 2 stabilization scenarios	Global, 13 WHO regions	Relative risks. Decreased risk in Europe due to benefits of milder winters. Increased risk in some other regions
Department of Health, 2002	Empirical statistical model	UKCIP98 scenarios	England and Wales	Medium high scenario, by 2050, increase to 2793 heat-related deaths per year, and reduction of 60000 cold-related deaths per year

UK United Kingdom, *Cvd* cardiovascular disease, *Ihd* ischaemic heart disease, *TR* temperature related mortality, *US* United States

Table 6. Response options for the United Kingdom

Theoretical range of choice	Technically feasibility demonstrated?	Effective to address health outcome?	Economically feasible?	Socially and legally acceptable?	Closed/open (Practical range of choice)
Heat health warning systems with active care of vulnerable individuals	Yes (UK heat-wave plan launched 2004)	No direct evidence but good indirect evidence	Yes	Yes	Closed
Installation of limited air conditioning in care homes	Yes	No direct evidence, but good indirect evidence	Cost benefit analysis not done	Yes, but high energy use is bad for environment	Closed
Planning options to reduce heat islands	Yes	No direct evidence, but weak indirect evidence that planning is effective at reducing outdoor temperatures	Some alternatives shown to be cost-effective (Modelling studies)	Legally yes. Socially – usually very beneficial Co-benefits of planting trees	
Improved training and staffing levels for those responsible for care of elderly	Yes	No direct evidence, but good indirect evidence	Unknown	Yes	
Improved design of old and new buildings to avoid high indoor temperatures	No. Current emphasis on insulation and reducing cold, may in fact increase risk of heat.	No direct evidence, but good indirect evidence	Retro-fitting is very expensive	New housing guidance requires indoor temperatures to be below health risk – but not legally enforced	

UK United Kingdom, *DaH* Department of Health

Build and Design for Future Climate

In Europe, new buildings survive several decades. Therefore, the future climate should be taken into account by the construction of new buildings and planning of new parts of the city in order to provide as much thermal comfort to the citizens as possible even in the future. The emphasis should be on energy saving. Renewable energy [solar energy does not affect the urban heat island in the short term] should be increasingly used for heating and cooling purposes.

Changes in building design will help in new buildings. A large part of the building stock in European is old and designed for past climates. There is, therefore, a need to educate and train building professionals to improve climate friendly design.

4.6   Discussion and Conclusions

We must start from the fact that health risks by heat load are evident. The public, politicians and decision-makers are probably not completely aware about the environmental risks of hot weather. There is a high probability that thermal conditions like those in summer 2003 will occur more frequently and possibly with even higher intensity. Heat health warning systems in North America seem to indicate that lives can be saved by adequate warnings and interventions. Prevention measures are also necessary in urban planning and building design.

As with many environmental health problems, the technology to combat heat (air conditioning, natural ventilation, drinking fluids) is widely and readily available. However, many barriers remain to their effective use. The availability and distribution of resources may limit adaptive capacity in European populations. Social trends, such as increasing population of isolated old people, will increase the population at risk of heat-related mortality. The poor elderly have been identified as the population most vulnerable to heat-related mortality. As with all developed countries, the population is ageing, and unless inequities in the distribution of resources are reduced, then adaptive capacity will not be improved.

Adaptation strategies require further research to clarify those that are truly effective. The impact of heat-waves and climate change needs to be investigated further. In particular, the interactions between global and local environmental changes on health in cities. Measures and strategies for reducing the impact of heat stress in such populations need to be identified. Unfortunately, many modern houses are characterized by poor insulation and low thermal capacity that requires air conditioning. Air conditioning should not be necessary for climate-oriented European buildings, even when considering future climate conditions, when natural ventilation can be effective (Roaf, 2005). Clearly, we should be building cities that are more sustainable and energy efficient. An important component of this is to use optimum methods and materials for space cooling. The reliance on energy-intensive technology such as air conditioning is unsustainable and can be considered a maladaptation. The construction of poor insulated housing is also a form of maladaptation to the impacts of more extreme summer weather. Heat health warning systems are an important strategy to reduce heat-related deaths, providing they are accompanied by the active detection and care of vulnerable individuals.

References

Ackerman B (1985) The march of the Chicago urban heat island. Journal of Climate and Applied Meteorology 24:547–554

Alberdi JC, Diaz J, Montero Rubio JC, Miron Perez IJ (1998) Daily mortality in Madrid community, 1986–1992: relationship with meteorological variables. European Journal of Epidemiology 14:571–578

Ballester DF, Corella D, Perez-Hoyos S, Saez M, Hervas A (1997) Mortality as a function of temperature. A study in Valencia, Spain, 1991–1993. International Journal of Epidemiology 26:551–561

Bosch X (2004) France makes heat-wave plans to protect the elderly people. Lancet 363:1708

Botelho J, Caterino J, Calado R, Nogueira PJ, Paixao E, Falcao JM (2005) Onda de calor de Agosto de 2003. Os seus efeitos sobre a mortalidade da populacao portuguesa. Instituto Nacional de Saude Dr. Ricardo Jorge, Lisboa

Bouchama A, Knochel JP (2002) Heat stroke. N Engl J Med 346(25):1978–1988

Cajoto VI, Peromingo JA, Vicdeo GV, Leira JS, Frojan S (2005) Health impact of 2003 heat-wave at Hospital de Riveira (A Coruña) [Impacto de la ola de calor de 2003 en el Hospital de Riveira (A Coruña)]. Anales de Medicina Interna 22:15–20

Casimiro E, Calheiros JM (2002) Human health. In: Santos FD, Forbes K, Moita R (eds) Climate Change in Portugal: Scenarios, Impacts, and Adaptation Measures – SIAM project. Gradiva, Lisboa, pp 241–300

CBS (2003) Ruim duizend doden extra door warme zomer, Central Bureau of Statistics, The Netherlands, http://www.cbs.nl/nl/publicaties/artikelen/algemeen/webmagazine/artikelen/2003/1275k.htm, accessed 2.5.2005)

Clarke JF (1972) Some effects of the urban structure on heat mortality. Environmental Research 5:93–104

Collins KJ, Exton-Smith AN, Dore C (1981) Urban hypothermia: preferred temperature and thermal perception in old age. Br Med J (Clin Res Ed) 282:175–177

Conti S, Meli P, Minelli G, Solimini R, Toccaceli V, Vichi M, Beltrano C, Perini L (2005) Epidemiologic study of mortality during the Summer 2003 heat wave in Italy. Environmental Research 98:390–399

CRED/OFDA (2001) EM-DAT website (http://www.cred.be/emdat/intro.html, accessed 2 Jan 2004)

Dalu GA, Gaetani M, Pielke RA, Baldi M, Maracchi G (2005) Regional Variability of the ITCZ and of the Hadley Cell. Unpublished work. Colorado State University, Fort Collins

De Dear R, Brager GS (2001) The adaptive model of thermal comfort and energy conservation in the built environment. International Journal of Biometeorology 45:100–108

Department of Health (2002) Health Effects of Climate Change in the UK. Department of Health, London

Dessai SR (2003) Heat stress mortality in Lisbon. Part II: An assessment of the impacts of climate change. International Journal of Biometeorology 48:37–44

Dixit SN, Bushara KO, Brooks BR (1997) Epidemic heat stroke in a midwest community: risk factors, neurological complications and sequelae. Wis Med J 96:39–41

Duncan K. et al (1997) Health Sector, Canada Country Study: Impacts and Adaptation. Environment Canada, p 520–580

Easterling DR, Meehl GA, Parmesan C, Changnon SA, Karl TR, Mearns L (2000) Climate extremes: observations, modelling and impacts. Science 289:2068

Ebi KL, Teisberg TJ, Kalkstein LS, Robinson L, Weiher RF (2004) Heat Watch/Warning Systems save lives: estimated costs and benefits for Philadelphia 1995–1998. Bulletin of the American Meteorological Society 85:1067–1068

Eliasson I (2000) The use of climate knowledge in urban planning. Landscape and Urban Planning 48:31–44

Ellis FP, Prince HP, Lovatt G, Whittington RM (1980) Mortality and morbidity in Birmingham during the 1976 heatwave. Q J Med 49:1–8

Epstein Y, Shani Y, Moran DS, Shapiro Y (2000) Exertional heat stroke – the prevention of a medical emergency. J Basic Clin Physiol Pharmacol 11:395–401

Eurowinter group (1997) Cold exposure and winter mortality from Ischaemic heart disease, cerebrovascular disease, respiratory disease, and all causes in warm and cold regions of Europe. Lancet 349:1341–1345

Fanger PO (1970) Thermal Comfort. McGraw-Hill, New York Düsseldorf

Farr W (1852) Vital statistics: A memorial volume of selections from the reports and writings of William Farr, MD, DCL, CB, FRS

Faunt JD, Wilkinson TJ, Aplin P, Henschke P, Webb M, Penhall RK (1995) The effects in the heat: heat-related hospital presentations during a ten day heat wave. Aust NZ J Med 25:117–120

Gaffen DJ, Ross RJ (1998) Increased summertime heat stress in the US. Nature 396:529–530

Gaffen DJ, Ross RJ (1999) Climatology and trends of US surface humidity and temperature. J Climate 12:811–828

Gagge AP, Fobelets AP, Berglund PE (1986) A standard predictive index of human response to the thermal environment. ASHRAE Trans 92:709–731

Garcia AC, Nogueira PJ, Falcao JM (1981) Onda de calor de Junho de 1981 em Portugal: efeitos na mortalidade. Revista Portuguesa de Saude Publica 1:67–77

Gouveia N, Hajat S, Armstrong B (2003) Socio-economic differentials in the temperature-mortality relationship in Sao Paulo, Brazil. Int J Epidemiol 32:90–397

Gremy I, Lefranc A, Pepin P (2004) Consequences sanitaire de la canicule d'aout 2003 en Ile-de-France [Impact of the August 2003 heat wave: sanitary consequences in Ile-de-France]. Rev Epidemiol Sante Publique 52:93–98

Grize L, Hussa A, Thommena O, Schindlera C, Braun-Fahrlandera C (2005) Heat wave 2003 and mortality in Switzerland. Swiss Medicine Weekly 135:200–205

Guest C, Willson K, Woodward A, Hennessy K, Kalkstein LS, Skinner C, McMichael AJ (1999) Climate and mortality in Australia: retrospective study, 1970–1990, and predicted impacts in five major cities. Climate Research 13:1–15

Hajat S, Kovats RS, Atkinson RW, Haines A (2002) Impact of hot temperatures on death in London: a time series approach. Journal of Epidemiology and Community Health 56:367–372

Hales S, Kjellstrom T, Salmond C, Town GI, Woodward A (2000) Daily mortality in Christchurch New Zealand in relation to weather and air pollution. Aust NZ J Public Health 24:89–91

Havenith G (1998) Relevance of individual characteristics for human heat stress response is dependent on excercise intensity and climate type. European Journal of Applied Physiology 77:231–241

Havenith G (2001a) Individualised model of human thermoregulation for the simulation of heat stress response. Journal of Applied Physiology 90:1943–1954

Havenith G (2001b) Temperature regulation and technology. Gerontechnology 1:41–49

Havenith G, Inoue Y, Luttikholt V, Kenney WL (1995) Age predicts cardiovascular, but not thermoregulatory, responses to humid heat stress. European Journal of Applied Physiology 70:88–96

Healy JD (2003) Excess winter mortality in Europe: a cross country analysis identifying key risk factors. Journal of Epidemiology and Community Health 57:784–789

Hemon D, Jougla E (2003) Estimation de la surmotalite et principles caracteristics epidemiologues. Rapport d'etape 1/3. INSERM (Institut National de la santé et de la recherche medicale), Paris

Henschel A, Burton LL, Margolies L, Smith JE (1969) An analysis of the heat deaths in St Louis during July 1966. American Journal of Public Health 59:2232–2242

Hulme M et al (2002) Climate Change Scenarios for the United Kingdom. The UKCIP02 Scientific Report: Norwich, Tyndall Centre for Climate Change Research. School of Environmental Sciences, University of East Anglia

Huynen M, Martens P, Schram D, Weijenberg M, Kunst AE (2001) The impact of heat-waves and cold spells on mortality rates in the Dutch population. Environmental Health Perspectives 109:463–470

Insitut de Veille Sanitaire (2003) Impact sanitaire de la vague de chaleur en France survenue en août 2003. Département des maladies chroniques et traumatismes, Département santé environment, Paris

IPCC (2001) IPCC Climate Change 2001. Impacts, Adaptations and Vulnerability. Contribution of Working Group II to the Third Assessment Report of the Intergovernmental Panel on Climate Change. Cambridge University Press, New York

Istituto Nazionale distatistica (ISTAT) (2004) Bilancio demografico nazionale. Anno 2003. Comunicato Stampa Populazione, Roma

Jendritzky G, Maarouf AR, Fiala D, Staiger H (2000) An update on the development of a universal thermal climate index

Jendritzky G, Nuebler W (1981) A model analysing the urban thermal environment in physiologically significant terms. Arch Met Geoph Biokl B 29:313–326

Jendritzky G, Soenning W, Swantes HJ (1979) Ein objektives Bewertungsverfahren zur Beschreibung des thermischen Milieus in der Stadt- und Landschaftsplanung („Klima-Michel-Modell"). Beiträge d Akad f Raumforschung und Landesplanung 2:8

Johnson H, Kovats RS, McGregor GR, Stedman JR, Gibbs M, Walton H, Cook L, Black E (2005) The impact of the 2003 heatwave on mortality and hospital admissions in England. Health Statistics Quarterly 25:6–11

Jones TS et al (1982) Morbidity and mortality associated with the July 1980 heat wave in St Louis and Kansas City, Mo. JAMA 247:3327–3331

Kalkstein LS, Greene JS (1997) An evaluation of climate/mortality relationships in large US cities and the possible impacts of climate change. Environmental Health Perspectives 105:84–93

Kalkstein LS, Smoyer KE (1993) The impact of climate change on human health: some international implications. Experientia 49:969–979

Katsouyanni K, Pantazopoulu A, Touloumi G, Tselepidaki I, Moustris K, Asimakopoulos D, Poulopoulou G, Trichopoulos D (1993) Evidence of interaction between air pollution and high temperatures in the causation of excess mortality. Archives of Environmental Health 48:235–242

Katsouyanni K, Trichopoulos D, Zavitsanos X, Touloumi G (1988) The 1987 Athens heatwave [letter]. Lancet ii:573

Katz RW, Brown BG (1992) Extreme events in a changing climate: variability is more important than averages. Climatic Change 21:289–302

Kiernan V (1996) If you can't stand the heat, go shopping. New Scientist 20:10

Kilbourne EM (1992) Illness due to thermal extremes. In: Last JM, Wallace RB (eds) Public Health and Preventative Medicine. Appleton Lang, Norwalk, Conneticut, pp 491–501

Kilbourne EM, Choi K, Jones TS, Thacker SB (1982) Risk factors for heatstroke. A case-control study. JAMA 247:3332–3336

King K, Negus K, Vance JC (1981) Heat stress in motor vehicles: a problem in infancy. Pediatrics 68:579–582

Klein Tank AMG, Konnen GP (2004) Trends in Indices of Daily Temperature and Precipitation Extremes in Europe, 1946–99. Journal of Climate22:3665–3680

Klein Tank AMG, Wijngaard J, van Engelen A (2003) Climate of Europe: Assessment of observed daily temperature and precipitation extremes. European Climate Assessment 2002. De Bilt, KNMI

Koppe C (2005) Gesundheitsrelevante Bewertung von thermischer Belastung unter Berücksichtigung der kurzfristigen Anpassung der Bevölkerung an die lokalen Witterungsverhältnisse. PhD, University of Freiburg

Koppe C, Jendritzky G (2005) Inclusion of short term adaptation to thermal stresss in a heat load warning procedure. Meteorologische Zeitschrift 14:271–278

Koppe C, Jendritzky G, Kovats RS, Menne B (2003) Heatwaves: impacts and responses. World Health Organization, Copenhagen

Kovats RS, Hajat S, Wilkinson P (2004) Contrasting patterns of mortality and hospital admissions during heatwaves in London, UK. Occup Environ Med 61:893–898

Kovats RS, West C (2005) Heat-waves: health impacts and acute responses. Chemical Hazards and Poisons Report, May 20–22

Kunst AE, Looman CW, Mackenbach JP (1993) Outdoor air temperature and mortality in the Netherlands: a time series analysis. American Journal of Epidemiology 137:331–341

Kysely J (2005) Mortality and displaced mortality during heat waves in the Czech Republic. International Journal of Biometeorology 49:91

Kysely J, Hutt R (2003) Heat-related mortality in the Czech Republic examined through synoptic and "traditional" approaches. Climate Research 25:265–274

Langford IH, Bentham G (1995) The potential effects of climate change on winter mortality in England and Wales. International Journal of Biometeorology 38:141–147

Laschewski G, Jendritzky G (2002) The effects of thermal environment on human health: an investigation of 30 years daily mortality data from SW Germany. Climate Research 21:91–103

Lecomte D, de Pennaster D (2004) [People living in Paris, dead during the August 2003 heatwave and examined in Medico-legal Institute]. Bull Acad Natl Med 188:459–469

Livada I, Santamouris M, Niachou K, Papanikoloau N, Mihalakakou G (2002) Determination of places in the great Athens area where the heat island effect is observed. Theor Appl Climatol 71:2719–2730

LoVecchio F, Stapczynski JS, Hill J, Haffer AF (2005) Heat-related mortality – Arizona, 1993–2002, and the United States, 1979–2002. Mortality and Morbidity Weekly Report 54:628–630

Luterbacher J, Dietrich D, Xoplaki E, Grosjean M, Wanner H (2004) European seasonal and annual temperature variability, trends and extremes since 1500. Science 303:1503

Lye M, Kamal A (1977) Effects of a heatwave on mortality-rates in elderly inpatients. Lancet 1:529–531

Marmor M (1975) Heat wave mortality in New York City, 1949 to 1970. Archives of Environmental Health 30:130–136

Marmor M (1978) Heat wave mortality in nursing homes. Environmental Research 17:102–115

Martens WJ (1998) Climate change, thermal stress and mortality changes. Soc Sci Med 46:331–334

Martinez-Navarro F, Simon-Soria F, Lopez-Abente G (2004) Valoracion del impacto de la ola de calor del verano de 2003 sobre la mortalidad. [Evaluation of the impact of the heat-wave in the summer of 2003 on mortality]. Gac-Sanit 18 (Suppl 1):250–258

McMichael AJ et al (2004) Climate change. In: Ezzati M, Lopez AD, Rodgers A, Murray CJ (eds) Comparative Quantification of Health Risks: Global and Regional Burden of Disease due to Selected Major Risk Factors, Vol 2.World Health Organization, Geneva, pp 1543–1649

McMichael AJ, Kovats RS (1998) Assessment of the Impact on Mortality in England and Wales of the Heatwave and Associated Air Pollution Episode of 1976. LSHTM, London

McMichael AJ, Kovats RS, Martens P, Nijhof S, Livermore MT, Cawthorne A, deVries P (2000) Climate change and human health. Final Report to the Department of Environment, Transport and the Regions. London School of Hygiene and Tropical Medicine/ICIS, London Maastricht

McMichael AJ, Woodruff RE, Whetton P, Hennessy K, Nicholls N, Hales S, Woodward A, Kjellstrom T (2003) Human Health and Climate Change in Oceania: Risk Assessment 2002. Commonwealth of Australia, Department of Health and Ageing, Canberra

Meehl GA et al. (2000) An introduction to trends in extreme weather and climate events: observations, socio-economic impacts, terrestrial ecological impacts, and model projections. Bulletin of the American Meteorological Society 81: 413–416

Michelozzi P, de Donato F, et al. (2005) Heat waves in Italy: cause specific mortality and the role of educational level and socio-economic conditions, in Kirch W, Menne B: Extreme Weather Events and Public Health Response. Springer, Heidelberg

Michelozzi P, de Donato F, Accetta G, Forastiere F, D'Ovido M, Kalkstein LS (2004) Impact of heat waves on mortality – Rome, Italy, June–August 2003. JAMA 291: 2537–2538

Mirchandani HG, McDonald G, Hood IC, Fonseca C (1996) Heat-related deaths in Philadelphia–1993. Am J Forensic Med Pathol 17:106–108

Nafstad P, Skrondal A, Bjertness E (2001) Mortality and temperature in Oslo, Norway, 1990–1995. European Journal of Epidemiology 17:621–627

Naughton MP, Henderson A, Mirabelli M, Kaiser R, Wilhelm JL, Kieszak SM, Rubin CH, McGeehin MA (2002) Heat related mortality during a 1999 heatwave in Chicago. Am J Prev Med 22:221–227

Nogueira PJ, Dias CM (1999) Associacao entre morbilidade e clime em Portugal Continental. Observacoes ONSA 6

O'Neill M (2003) Air conditioning and heat-related health effects. Applied Environmental Science and Public Health, 1

O'Neill M, Zanobetti A, Schwartz J (2003) Modifiers of the temperature and mortality association. American Journal of Epidemiology 157:1074–1082

Oke TR (1997) Urban climates and global environmental change. In: Thompson RD, Perry AH (eds) Applied climatology: principles and practice. Routledge, London, pp 273–287

Paldy A, Bobvos J, Vamos A, Kovats RS, Hajat S (2005) The effect of temperature and heat-waves on daily mortality in Budapest, Hungary 1970–2000. In; Kirch W, Menne B, Bertollini R (eds) Extreme weather events and Public Health Responses. Springer, pp 99–108

Panagiotakos DB, Chrysohou C, Pitsavos C (2004) Climatological variations in daily hospital admissions for acute coronary syndromes. International Journal of Cardiology 94:229–233

Pattenden S, Nikiforov B, Armstrong B (2003) Mortality and temperature in Sofia and London. Journal of Epidemiology and Community Health 57:628–633

Roaf S, Crighton D, Nicol F (2004) Adapting buildings and cities to climate change, London. Architectural Press, London

Roaf S (2005) Adapting buildings and cities to climate change. London Architectural Press, London

Rogot E, Sorlie PD, Backlund E (1992) Air-conditioning and mortality in hot weather. American Journal of Epidemiology 136:106–116

Rooney C, McMichael AJ, Kovats RS, Coleman M (1998) Excess mortality in England and Wales, and in Greater London, during the 1995 heatwave. Journal of Epidemiology and Community Health 52:482–486

Rozzini R, Zanetti E, Trabucchi M (2004) Elevated temperature and nursing home mortality during 2003 European heat wave. J Am Med Dir Assoc 5:138–139

Saez M, Sunyer J, Tobias A, Ballester F, Anto JM (2000) Ischaemic heart disease mortality and weather temperature in Barcelona, Spain. European Journal of Public Health 10:58–63

Sakamoto-Momiyama M (1977) Seasonality in human mortality. University of Tokyo press, Tokyo

Sartor F (2004) Surmortalité en Belgique au cours de l'été 2003. Institut Scientifique de Santé Publique Section d'Epidemiologie, Brussels

Sartor F, Snacken R, Demuth C, Walckiers D (1995) Temperature, ambient ozone levels and mortality during summer, 1994, in Belgium. Environmental Research 70:105–113

SBW (Sozialministerium Baden-Wuerttemberg) (2004) Gesundheitliche Auswirkungen der Hitzewelle im August 2003. SBW, Freiburg

Schar C, Vidale PL, Luthi D, Frei C, Haberli C, Liniger MA, Appenzeller C (2004) The role of increasing temperature in European summer heatwaves. Nature 472:332–336

Schuman SH (1972) Patterns of urban heatwave deaths and implications for prevention: data from New York and St Louis during July 1966. Environmental Research 5:59–75

Schwartz J, Samet J, Patz JA (2004) The effects of temperature and humidity on hospital admissions for heart disease. Epidemiology 15:755–761

Schwartz J, Spix C, Touloumi G, Bacharova L, Barumamdzadeh T, Tertre A, Piekarksi T (1996) Methodological issues in studies of air pollution and daily counts of death or hospital admissions. Journal of Epidemiology and Community Health 50 (Suppl 1):S3–S11

Semenza JC, McCullough JE, Flanders WD, McGeehin MA, Lumpkin JR (1999) Excess hospital admissions during July 1995 heat wave in Chicago. Am J Prev Med 16:269–277

Semenza JC, Rubin CH, Falter KH, Selanikio JD, Flanders WD, Howe HL, Wilhelm JL (1996) Heat-related deaths during the July 1995 heat wave in Chicago. N Engl J Med 335:84–90

Senat (2004) La France et les Francais face a la canicule: les lecons d'une crise. Rapport d'information no. 195 (2003–2004) de Mme Letard, MM Flandre, S Lepeltier, fait au nom de la mission commune d'information du Senat, depose le 3 Fevrier 2004. Rapport d'information no. 195. Paris, France

Shanks NJ, Papworth G (2001) Environmental factors and heatstroke. Occupational Medicine 51:45–49

Sheridan S (2003) Heat, mortality and level of urbanisation. Climate Research 24:255–265

Shumway RH, Azari AS, Pawitan Y (1988) Modelling mortality fluctuations in Los Angeles as functions of pollution and weather effects. Environmental Research 45:224–241

Sierra Pajares Ortiz M, Diaz Jemenez J, Montero Robin JC, Alberdi JC, Miron Perez IJ (1997) Daily mortality in the Madrid community during 1986–1991 for the group between 45 and 64 years of age: its relationship to air temperature. Rev Esp Salud Publica 71:149–160

Smoyer KE (1998b) Putting risk in its place: methodological considerations for investigating extreme event health risks. Soc Sci Med 47:1809–1824

Smoyer KE (1998a) A comparative analysis of heat-waves and associated mortality in St Louis, Missouri – 1980 and 1995. International Journal of Biometerology 42:44–50

Stedman JR (2004) The predicted number of air pollution related deaths in the UK during the August 2003 heatwave. Atmospheric Environment 38:1087–1090

Stott PA, Stone DA, Allen MR (2004) Human contribution to the European heatwave of 2003. Nature 432:610–614

Todisco G (1987) Indagine biometeorologica sui colpi di calore verificatisi a Roma nel'estate del 1983 [Biometeorological study of heat stroke in Rome during summer of 1983]. Rivista di Meteorologica Aeronautica, XLVII:189–197

Vandentorren S, Suzan F, Medina S, Pascal M, Maulpoix A, Cohen J-C, Ledrans M (2004) Mortality in 13 French cities during the August 2003 heat wave. American Journal of Public Health 94:1518–1520

Vanhems P, Gambotti L, Fabry J (2003) Excess rate of in-hospital death in Lyons, France, during the August 2003 heat-wave. New England Journal of Medicine 348:2077–2078

Whitman S, Good G, Donoghue ER, Benbow N, Shou W, Mou S (1997) Mortality in Chicago attributed to the July 1995 heat wave. American Journal of Public Health, 87:1515–1518

WHO (1987) WHO Health impact of low indoor temperatures. WHO Regional Office for Europe, Copenhagen

WHO (1990) WHO Indoor Environment: Health aspects of air quality, thermal environment, light and noise. World Health Organization/United Nations Environment Programme, Geneva

WMO (2003) According to the World Meteorological Organization, extreme weather events might increase. World Meteorological Organization, Geneva (http://www.wmo.ch/web/Press/Press695.doc, accessed 29 October 2003)

WRI (2000) Urban environment and human health, World Resources 1996–97.World Resources Institute, New York, pp 31–55

Yan Z et al. (2002) Trends of extreme temperatures in Europe and China based on daily observations. Climatic Change 53:355–392

5 Floods and Human Health

Kristie L. Ebi

Contributing Authors: Shakoor Hajat,
Edmund Penning-Rowsell, R. Sari Kovats, Theresa Wilson

Summary

Floods are the most common natural disaster in Europe. Flooding events in the Ukraine, Poland, the Czech Republic, Germany, the Netherlands, France, Switzerland, Spain, the United Kingdom, and others during the 1990s and early 2000s highlighted European vulnerability to flooding. Throughout the 20th century, flood-related deaths have been either stable or decreasing. Limited data are available on other adverse health outcomes due to flooding; these health outcomes can occur during or immediately after the event (such as injuries and an increase in communicable diseases) or can occur later as a consequence of the event or its clean-up (such as mental health disorders). Anxiety and depression may last for months or years after a flood event. As a result, the full health burden of a flooding event may be much larger than the number of deaths and injuries initially estimated.

Extreme precipitation events in Europe are likely to increase in intensity and frequency with climate change, with wetter winters over most of Europe, wetter summers in northern Europe, and drier summers in southern Europe. Changes in overall flood frequency will depend on the generating mechanisms, with floods due to heavy rainfall events possibly increasing and floods due to spring snowmelt and ice-jams possibly decreasing.

The extent to which a community is vulnerable to flood-related health impacts depends on the degree of awareness of the flood hazard, the degree to which the hazard may be avoided and the availability of effective measures for post-flood clean-up. In countries where flooding risk is likely to increase, a comprehensive vulnerability-based emergency management program of preparedness, response and recovery has the potential to reduce the adverse health impacts of floods. These plans need to include activities designed to decrease vulnerability before, during and after a flood. Furthermore, these plans should enable a process to find effective and efficient approaches to coping with an uncertain future.

As part of the EU-funded project "Climate Change and Adaptation Strategies for Human Health," the health impacts of floods and strategies to increase preparedness and post-flood interventions are evaluated.

5.1   Introduction

Unusually severe floods in Europe during the 1990s and early 2000s increased public awareness of the potential human health and economic consequences of flooding. In 2002, the summer flooding of the Elbe and Danube rivers resulted in some of the worst floods seen in Europe for more than a century. The flooding in central Europe was of unprecedented proportions, with dozens of people losing their lives, extensive damage to the socioeconomic infrastructure, and destruction of the natural and cultural heritage (Commission of the European Communities 2002). Germany, the Czech Republic, and Austria were the three countries most severely affected. Estimates of the economic and insured losses were € 11.0 billion in Germany, € 3.9 billion in the Czech Republic, and € 3.4 billion in Austria (Munich Re, 2003). The proportion of the losses that was insured was relatively low (about 20% in Germany and 30% in the Czech Republic). The majority of losses were public facilities such as roads, railway lines, dykes, riverbeds, bridges, and other infrastructure.

Particularly for the transition countries of central and Eastern Europe, floods pose a significant risk to economic development, and can be devastating to the usually uninsured victims and to governments that are often ill-prepared to provide flood relief and recovery (Linnerooth-Bayer and Amendola, 2003). For example, economic damages from the 1997 floods in Poland and the Czech Republic were 2.9 and 2.5% of their respective GDPs.

Europe experiences three types of floods: flash, riverine, and storm surges. Flash and riverine floods result from two main groups of meteorological events (Green et al., 1994; Estrela et al., 2001). In large- and medium-sized river basins in north and central Europe, flooding usually results either from wide-ranging and continuous precipitation or from snowmelt in connection with high antecedent soil saturation. Peak discharges last up to several days. The downstream characteristics of the flood differ from the upstream characteristics because of lag, effects of the route, scale effects, and changes in geology, physiography and climate from headwaters to the outlet. Because water levels rise relatively slowly, forecasting and emergency response measures (such as evacuation) can reduce impacts.

Flash floods are usually associated with isolated and localized, very intense rainfall events that occur in small- and medium-sized river basins; these events usually take place at the end of summer and in autumn. Peak discharges last for minutes to hours (commonly less than 24 hours). Although flash floods can occur under a broad range of climatological and geographical conditions, they are most common in Mediterranean rivers originating in mountains close to the sea. Their rapid onset, high velocity flow, and associated debris load limits early warning and response time. Flash flooding and the circumstances that surround the event cause most flooding deaths, with drowning the leading cause of death (Malilay, 1997).

The main areas of Europe prone to frequent flooding episodes are the Mediterranean coast, the dyked areas of the Netherlands, the Shannon callows in central Ireland, the north German coastal plains, the Rhine, Seine and Loire valleys, some coastal areas of Portugal, the Alpine valleys, the Po valley in Italy, and the Danube and Tisza valleys in Hungary (Estrela et al., 2001). Overall, the Netherlands and Hungary have the highest flood risks.

5.2 Risk and Vulnerability

It is useful to distinguish between the risk of a flood, which describes the probability of occurrence, from vulnerability, which describes the inherent characteristics of a system that create the potential for harm (Sarewitz and Pielke Jr., 2002). The tolerated levels of flood risk vary across European countries, and across time within a country. For example, the Netherlands requires that river dykes and other water works be built such that they fail less than once every 1250 years, while the United Kingdom determines the standard of protection on a case-by-case basis according to cost-benefit and other criteria (Handmer et al., 1998). Development of effective and efficient flooding responses requires that both vulnerability and risk be understood because the impact of an extreme event is determined by both the physical phenomena and the interaction of those characteristics with social and other systems.

It is largely human actions, decisions, and choices that result in flooding vulnerability. For example, shifts in the location of industries and homes due to changing economic factors and lifestyle choices are significant factors affecting current European vulnerability (Mitchell, 2003). The combination of economic and non-economic forces that determine vulnerability affects different countries and regions differently. For example, the 2001 floods in the Upper Tisza river basin in northeastern Hungary led to the evacuation of 17000 people (out of a population of 200000), with 1000 homes destroyed and another 2000 damaged (Vari et al., 2003). The consequences of this flood were due not only to the flood event (flood waves originated in upstream Ukraine and arrived at very high speed, with little time for warning and preparation), but also to the characteristics of this region, which is one of the poorest regions in Europe, with most settlements far from cities and with poor road connections and limited access to the railway network.

Because the probability of extreme floods cannot be determined with full accuracy, when risk management focuses on risks rather than vulnerabilities, interventions can result in outcomes different than those intended. This is further complicated by factors such as climate change that have non-stationary relationships with flooding risk.

Vulnerability to flooding depends on both natural and human factors. Flood vulnerability can be analysed based on hydrologic and hydraulic factors (referred to as hazard; this is measured by maximum water levels and discharges, and by flood duration) or based on the land use and socioeconomic perception of risk. Humans can influence flooding either by affecting runoff patterns (such as urbanization) or by increasing the possible impacts of flooding (such as increasing settlements on floodplains). Floods can and do affect multiple sectors, including agriculture, industry, urban settlements and tourism.

The main driving forces that induce or intensify floods and their impacts include climate change, changes in terrestrial systems (hydrological systems and ecosystems), and economic and social systems (Estrela et al., 2001). Land-use changes, which induce land-cover changes, control the rainfall-runoff relationship. Deforestation, urbanization, and reduction of wetlands decrease the available water storage capacity and increase the volume and speed of runoff coefficient, leading to growth in flood amplitude and reduction of the time-to-peak. Land sealing from urbanization adversely influences flood hazard by increasing the amount of impervious areas. Current population growth and urbanization trends are expected to continue over the coming decades. The social and community dimensions of flooding can significantly affect individuals and households (Tapsell et al., 2002). Community activity may break down following a serious flood, and it may be months or longer before normal functioning is achieved.

Therefore, determination of the potential impacts of climate change on flood risk should include an integrated analysis of the cause-effect chain of precipitation – runoff generation – runoff concentration – flood wave propagation (routing) – inundation – flood damage (Bronstert, 2003).

5.3   Past Trends and Future Projections of Climate Change in Europe

The European Climate Assessment found that the average climate in Europe has been warming in recent decades (Klein Tank et al., 2002). Trends in extreme precipitation events leading to flooding are less consistent, with some studies suggesting significant increases in climatic extremes during the twentieth century (Klein Tank et al., 2002; Frich et al., 2002; Milly et al., 2002) and others finding no evidence of a long-term flooding trend over the past 80 to 150 years (Mudelsee et al., 2003). In the United Kingdom, one study reported an increase in heavy rainfall in winter but not in summer (Osborn and Hulme, 2002). Another study found that changes in protracted high flows over the last 30–50 years could be accounted for by climate variability (Robson, 2002). Although these results suggest that flooding trends do not provide sufficient evidence that climate change has already influenced flooding frequency in Europe, these results do not disprove projections of what could occur with a changing climate.

The Third Assessment Report of the Intergovernmental Panel on Climate Change (IPCC) projected an increase in global mean temperature in the range of +1.4 °C to +5.8 °C by 2100 for the full set of Standardized Reference Emission Scenarios (Albritton and Meira Filho, 2001). As temperatures warm, the atmosphere can hold more water vapour, resulting in increasing precipitation in many areas. The IPCC concluded that by 2100 the general pattern of changes in annual precipitation over Europe is for widespread increases in northern Europe (between +1% and +2% per decade), smaller decreases across southern Europe (maximum –1% per decade), and small or ambiguous changes in central Europe (i.e. France, Germany, Hungary) (Kundzewicz and Parry, 2001). The climate change projections suggest a marked contrast between winter and summer patterns of precipitation change. Most of Europe is projected to become wetter during the winter season (+1% to +4% per decade), with the exception of the Balkans and Turkey where winters are projected to become drier. In summer, there is a projected strong gradient of change between northern Europe (increasing precipitation of as much as 2% per decade) and southern Europe (drying as much as 5% per decade). These projected changes underscore the need to increase the development and implementation of measures to prevent adverse health impacts from flooding (Baxter et al., 2001).

The European project PRUDENCE used a high-resolution climate model to quantify the influence of anthropogenic climate change on heavy or extended summer precipitation events lasting for one to five days; these types of events historically have inflicted catastrophic flooding (Christensen and Christensen, 2003). During the months of July to September, increases in the amount of precipitation that exceed the 95th percentile are projected to be very likely in many areas of Europe, despite a possible reduction in average summer precipitation over a substantial part of the continent. Consequently, the episodes of severe flooding may become more frequent even with generally drier summer conditions. Changes in overall flood frequency will depend on the generating mechanisms – floods that are the result of heavy rainfall may increase while those generated by spring snowmelt and ice-jams may decrease.

Climate change could affect the shape of the flood frequency-magnitude distribution in several ways. One possibility is that the more rare or extreme events will be affected disproportionately such that only these events become more frequent. In this situation, the probable maximum flood would increase, perhaps substantially. Another possibility is that the whole frequency-magnitude distribution might shift such that all events become more frequent. A third possibility is that only smaller events become more frequent. These possible shifts in the distribution would have different consequences for the annual average damages that result from floods.

Another possible impact of climate change that could affect flooding vulnerability is the affect of increasing temperatures and changes in the hydrologic cycle on vegetation and soil conditions in a catchment area (Bronstert, 2003). Climate change may alter water retention and evaporation processes, thus, changing natural vegetation cover, which could have a feedback on flood development and severity.

5.4   Health Impacts of Flooding in Europe

The intensity and duration of rainfall in a river catchment area, the amount of rain during the preceding weeks and months, the topography and the preparedness of the population determine the health burden from a flooding event (Green et al., 1994). Despite the causal association, there is limited research on intensity of precipitation, the likelihood of a declared disaster and the magnitude of the health impacts experienced by a population. On average, the higher the water depth and the greater the flow velocity of a flood, the greater the damage to property (Estrela et al., 2001). Rapid speeds of onset or long flood duration increase the possibility of impacts.

The adverse human health consequences of flooding can be complex, far-reaching and difficult to attribute to the flood event itself (Hajat et al., 2003). Floods can cause major infrastructure damage, including disruption to roads, rail lines, airports, electricity supply systems, water supplies and sewage disposal systems. The economic consequences are often greater than indicated by the physical effects of floodwater coming into contact with buildings and their contents. Economic damage may reach beyond the flooded area and last longer than the event.

Adverse health impacts of flooding can arise from a combination of some or all of the following factors: characteristics of the flood event itself (depth, velocity, duration, timing, etc.); amount and type of property damage and loss; whether flood warnings were received and acted upon; the victims' previous flood experience and awareness of risk; whether or not flood victims need to relocate to temporary housing; the clean-up and recovery process, and associated household disruption; degree of difficulty in dealing with builders, insurance companies, etc.; pre-existing health conditions and susceptibility to the physical and mental health consequences of a flooding event; degree of concern over a flood recurrence; degree of financial concern; degree of loss of security in the home; and degree of disruption of community life.

Adverse health impacts are broadly categorized into:

- ░ Physical health effects sustained during the flood event itself or during the clean-up process, or from effects brought about by damage to infrastructure including displacement of populations. These physical effects largely manifest themselves within weeks or months following flooding, and are largely related to the shock of the flood, damp or dusty living conditions, and the recovery process (Tapsell et al., 2003).
- ░ Mental health effects directly attributable to the experience of being flooded or indirectly during the recovery process. These psychological

effects tend to be much longer lasting and can be worse than the physical effects of being flooded.

The physical health effects can be further categorized into direct effects caused by the floodwaters (such as drowning and injuries) and indirect effects caused by other systems damaged by the flood (such as water- and vector-borne diseases, acute or chronic effects of exposure to chemical pollutants released into floodwaters, food shortages etc.). There is a common perception that the problems associated with a flooding event end once the floodwaters have receded. However, for many victims, this is when most of their problems begin. Table 1 lists the health effects of flooding in Europe, with selected examples.

There is no common database for flood events in Europe, and the various national authorities do not collect data on floods using common criteria (Estrela et al., 2001). Swiss Re, Munich Re, WMO, and other national agencies record data on flood events. The worldwide occurrence and deaths from reported disasters, including flood events, have been recorded in the EM-DAT database (the OFDA/CRED International Disaster Database) from 1900 to the present. The database is compiled from various sources, including UN agencies, non-governmental organizations, insurance companies, research institutes, and press agencies. In order for a disaster to be entered into the database, at least one of the following criteria must be fulfilled: (1) ten or more people killed; (2) 100 or more people reported affected; (3) a call for international assistance; and (4) a declaration of a state of emergency.

Throughout the 20th century, there have not been clear trends in flood-related deaths while economic burdens of flooding and social disruption have become worse as humans move into unsafe and flood-endangered areas, thus, increasing the damage potential (Mitchell 2003).

5.4.1 Flooding-related Deaths and Injuries

Between 1980 and 1999, an annual rate of 1.3 deaths and 5.7 injuries occurred per 10 000 000 population due to inland floods and landslides in western Europe (McMichael et al., 2002). Table 2 lists floods in Europe in which more than 10 people died during the period 1995–2004, and Figure 1 graphs the number of flood events from 1970 to 2004, along with the number of deaths. The number of deaths is one measure of the severity of a flood; another is the economic damage. Some European floods caused limited numbers of deaths and large economic damage, such as the October 2000 flood in Kent, Sussex, and Hampshire in the United Kingdom that resulted in US $ 5 836 859 000 in damage and no deaths, or the August 2002 flood in Austria that caused 7 deaths and US $, 2 046 707 000 in damage (EM-DAT).

Comprehensive surveillance of morbidity following floods is limited. One survey of emergency rooms conducted in the United States following a

Table 1. The health outcomes of floods in Europe, with examples

Outcome	Comment	Examples
▌ **Deaths**	Most flood-related deaths can be attributed to high floodwater velocities; rapid speed of flood onset; deep floodwaters, where floodwater is in excess of 1 metre depth; long duration floods; debris load of flood-waters; characteristics of accompanying weather and clean up activities in the aftermath of floods	– A storm surge in February 1953 caused 307 deaths in the United Kingdom and 1795 deaths in the Netherlands. – In the United Kingdom, Bennet (1970) conducted a retrospective study of the 1968 Bristol floods, and found a 50% increase in the number of deaths among those whose homes had been flooded, with the most pronounced rise in those 45–64 years of age. – A flash flood in October 1988 in Nimes, France caused 9 deaths. – In 1996, 86 people died in the town of Biescas, Spain because of the water and mud that suddenly covered a campsite. – In 1997, river floods in central Europe caused more that 100 fatalities. – In the 1998 flood in Sarno, Italy, 147 people were killed by a river of mud that rapidly destroyed an urban area.
▌ **Injuries**	Surveillance of morbidity following floods is limited and little information is available	A community survey conducted following the 1988 floods in Nimes, France found that 6% of households reported mild injuries (contusions, cuts, and sprains) related to the flood.
▌ **Infectious disease outbreaks**	Small risk of communicable disease following flooding, although severe occurrences are rare due to the public health infrastructure (including water treatment and effective sewage pumping)	– Leptospirosis outbreak occurred after the flooding in the Czech Republic in 1997. – No increase in infectious disease was observed following the 1988 flash flood in Nimes, the 1995 river floods in eastern Norway or the 2002 floods in the United Kingdom – Finland reported 13 waterborne disease outbreaks with an estimated 7300 cases during 1998–1999, associated with untreated groundwater from mostly flooded areas.
▌ **Respiratory disease**	Very little information is available	Following the floods in the northeastern Republic of Sakha (Yakutia) in July 1998, a high incidence of respiratory diseases was observed by the International Federation of Red Cross (IFRC, personal communication)

Source: Hajat et al. 2003; Few et al. 2004

Table 2. European floods in which more than 10 people died, 1995 –2004

Year	Country	Location	Victims (and Damage in Million US $)
March 2004	Turkey	Erzurum, Batman, Bitlis, Konya, Silifke	15 Deaths
July 2002	Turkey	Rize province	39 Deaths
August 2002	Germany	Lower Saxony, Saxony-Anhalt, Saxony, Bavaria, Baden-Wurttemberg, Thuringa	27 Deaths 108 Injured 330 000 Affected (9129)
September 2002	France	Gard, Hérault, Vaucluse, Rhone, Provence departments	23 Deaths 2500 Affected (1190)
August 2002	Czech Republic	Prague, Central Bohemia, Southern Bohemia, Pilsner, Carlsbad, Usti districts	18 Deaths 200 000 Affected (2000)
July 2001	Poland	Malopolskie, Swietokrzyskie, Donoslaskie, Oploskie, Slaskie, Warminsko-Mazurkie, Podlaskie, Gdansk, Slupsk regions	27 Deaths 15 000 Affected (700)
January 2001	Greece	Athens, Corinth, Cape Sounion, Zakynthos	11 Deaths 450 Affected
October 2000	Italy	Piémont, Val d'Aoste, Ligurie	29 Deaths 43 000 Affected (434)
September 2000	Italy	Soverato (Near Catanzaro, Calabria)	16 Deaths 22 Injured
June 2000	Spain	North-East	16 Deaths 500 Affected
November 1999	France	Aude, Tarn, Herault, Pyrenees-Orientales	36 Deaths 3000 Affected (2.9)
June 1999	Romania		19 Deaths 3 Injuired 4578 Affected
July 1999	Romania	Northern & Western parts of Romania	15 Deaths 22 Injured 3840 Affected
July 1999	Serbia and Montenegro	Belgrade, Podunavlje, Sumadija, Morava, Pomoravlje and Bor districts	11 Deaths 60 339 Affected
May 1998	Italy	Campania region	147 Deaths 100 Affected (28.7)
August 1998	Turkey	Beskoy (Trabzon province)	60 Deaths 1000 Affected

Table 2 (continued)

Year	Country	Location	Victims (and Damage in Million US $)
July 1998	Slovakia	Sabinov, Presov districts	54 Deaths 61 Injured 10 850 Affected (24.5)
June 1998	Romania	Bacau, Vaslui, Vrancea (Northern Moldavia), Salaj, Mures, Neamt, Cluj, Alba, Sibiu, Hundoara (Transilvania)	23 Deaths 12 000 Affected (150)
June 1998	Turkey	Diyarbajir	22 Deaths
May 1998	Turkey	Zonguldak, Karabul, Bartin	10 Deaths 47 Injured 1 200 000 Affected (1000)
June 1997	Poland	Katowice, Opole, Walbrzych provinces	55 Deaths 162 500 Affected (4300)
July 1997	Czech Republik	Moravia, Bohemia regions	29 Deaths 2 409 Injured 87 725 Affected (150.3)
July 1997	Romania	Alba, Arad, Bihor, Bistrita-Nasaud, Botosani, Braila, Buzau, Dimbovita, Galati, Hundedoara, Maramures, Mures, Sibiu, Timis, Tulcea, Vaskui, Vrancea, Prahova, Bacau, Iasi, Suceava, Teleorman, Olt, Dolj, Caras-Severin	20 Deaths (110)
November 1997	Portugal	Southern part of the country	11 Deaths
June 1996	Italy	Tuscany, Lucca, Massa, Carrara, Udine, Florence, Veneto, Emilia Romagna, Lombardia	26 Deaths (1000)
January 1996	Portugal	Central and north regions	10 Deaths 1050 Affected (13)
July 1995	Turkey	Ankara, Istanbul, Senirkent	74 Deaths 10 000 Affected (65)
November 1995	Turkey	Izmir, Antalaya, Isparta	63 Deaths 300 000 Affected (1000)
January 1995	France	Basse-Normandie, Champagne-Ardennes, Bretagne, Pays de Loire, Ile-de-France	16 Deaths 3000 Affected (570)

Source: "EM-DAT: The OFDA/CRED International Disaster Database – www.em-dat.net – Université Catholique de Louvain – Brussels – Belgium"

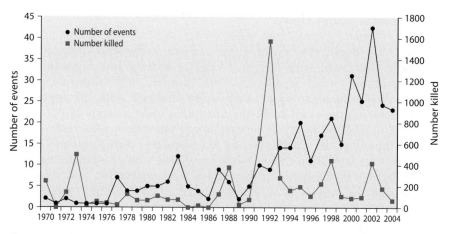

Fig. 1. Annual number of European floods 1970–2004, with number of deaths (EM-DAT v05.05)

1993 flood found that the most common reported injuries were sprains/strains, lacerations, other injuries, and abrasions/contusions (CDC, 1993). Chronic health effects also have been reported, including respiratory problems and high blood pressure (Tapsell et al., 2002). Exposure to toxic molds and fungi can occur because of flooding or its clean-up.

5.4.2 Infectious Diseases

There is a small risk of communicable disease following flooding, although outbreaks are rare in Europe due to the public health infrastructure, including water treatment and sanitation. Fever and waterborne disease have been reported following flood events. One example of an outbreak occurred when cases of leptospirosis were reported in the Czech Republic following flooding in 1997; however, the quality of data appears to be poor (Kriz et al., 1998). Analysis of waterborne disease outbreaks in Finland over the period 1998–1999 found that thirteen of fourteen outbreaks were associated with groundwater that was not disinfected, mostly related to flooding (Miettinen et al., 2001).

Evidence from Sweden demonstrates that along meandering flood-stricken rivers, floodwater mosquitoes (i.e. *Aedes rossicus*, *Ae. sticticus* and *Ae. vexans*) can appear in great numbers when floods occur during the warm season when the water temperature is favourable for mosquito development (Lindgren and Jaenson, personal communication, 2003). These mosquitoes can harbour tularaemia and arboviruses, although apparently no human cases have been reported. A future increased frequency of floods in combination with higher water temperatures, particularly in Central Sweden, during the summer and early autumn would increase the likelihood of pest occurrence of floodwater mosquitoes.

5.4.3 Mental Health Effects

There is no doubt that flooding, in common with other traumatic life events, is associated with increased rates of anxiety and depression, the most common mental health disorders (Bennet, 1970; Sartorius, 1990). Floods should be regarded as multi-strike stressors, with the sources of stress including the event itself; the disruption and problems of the recovery period; and the worry or anxiety about the risk of recurrence of the event (Tapsell et al., 2003). During the recovery period, mental health problems can arise from the problems associated with geographic displacement, damage to the home or loss of familiar possessions, and stress involved with the process of repairing (Fullilove, 1996; Keene, 1998; Tapsell and Tunstall, 2001). The full impact of a flood often is not appreciated until after people's homes have been put back in order. The lack of insurance may exacerbate the impacts of floods (Ketteridge and Fordham, 1995).

The stress associated with a perceived risk of the reoccurrence of flooding could include a perceived failure on the part of relevant institutions to alleviate flood vulnerability or to provide adequate warnings (Tapsell et al., 2003). Loss of confidence that the relevant authorities will warn about or protect against future events, with the attendant loss of security in the home, could exacerbate the existing stress and anxiety experienced by flood victims. Research shows that this additional stress and anxiety, along with the stress of the event itself, the stress associated with recovery, and pre-existing health conditions, can have significant impacts on the overall health and well-being of flood victims. Factors mediating between stress and health may include flood warnings, coping strategies, and social support. However, where flooding is unexpected or sudden, these meditating factors may be weakly developed or non-existent in their effect.

The health impacts of flooding may last long past the event itself. A study in the United States of elderly adults interviewed before and after flood events found that the persistence of health effects was directly related to flood intensity (Phifer et al., 1988). Sociodemographic status did not moderate the impacts of flood exposure on physical health. Men, those with lower occupational status and individuals 55–64 years of age were at significantly greater risk for increases in psychological symptoms. A study in the Netherlands on the health and well-being of exposed subjects six months after a flood event suggested that 15–20% of children were having moderate to severe stress symptoms and 15% of adults experienced very severe symptoms of stress (Becht et al., 1998). A United Kingdom study found a consistent pattern of increased psychological problems among flood victims in the five years following a flood (Green et al., 1985).

5.4.4 Vulnerable Subgroups

Certain groups within affected populations have been reported to be more vulnerable to the effects of flooding, including the elderly, women, children, minorities, individuals with disabilities, and those with low incomes (Hajat et al., 2003; Tapsell and Tunstall, 2001). People with lower incomes may be more vulnerable to the effects of a flood if they do not have adequate insurance coverage or the financial resources for flood clean up and repair (Ketteridge and Fordham, 1995). Various racial and ethnic groups may be differentially affected, both physically and psychologically, during the periods of emergency response, recovery and reconstruction (Fothergill et al., 1999). Factors such as language, housing patterns, building construction, community isolation, and cultural insensitivity of the majority population may affect vulnerability.

5.5 Current Approaches for Preparedness and Post-flood Interventions

A common typology for categorizing flood adjustments is to distinguish those measures that modify the flood event; modify human vulnerability to floods; and distribute the flood losses, with insurance being the most common measure. Generally, measures that modify a flood event are structural and those that control the impacts of floods are non-structural.

The main structural measures are usually designed to reduce flood vulnerability by reducing the exposure of properties and people to floods; they include flood control reservoirs, areas for controlled flooding, soil protection and reforestation, river channelization, protection dykes, the protection and cleaning of riverbeds, road and railway culverts, bridges, and the re-establishment of meanders and riparian zones (Estrela et al., 2001). None of these measures provides absolute protection. In fact, an important factor influencing the flood hazard is an unjustified belief in the absolute safety of structural defences. When a dyke breaks, the damage may be higher than it would have been in a levee-free area. The main categories of structural measures include measures that reduce the peak runoff, such as flood control reservoirs, areas for controlled flooding, soil protection and afforestation; measures that reduce the level of flooding for a given runoff, such as river channelization, protection dykes, and the protection and cleaning of riverbeds; and measures that reduce the duration of flooding, such as road and railway culverts, and bridges.

The main categories of non-structural measures include measures that reduce possible flood impacts on existing structures, such as reinforcement of buildings; measures that reduce possible impacts through changes in land use planning, such as restriction in uncontrolled building; and development of early warning systems and flood management measures, including forecasting, flood management rules and development of flood evacuation plans (Estrela et al., 2001).

Past structural and non-structural measures may not be as effective in the future with the increased heavy precipitation events projected under climate change. Under certain circumstances, they may even increase vulnerability such as when a dam designed to hold back a certain volume of river flow is overwhelmed, putting an unprepared population at risk.

5.6   Convention on the Protection and Use of Transboundary Watercourses and International Lakes

The catchment areas for many of the major rivers in Europe cross multiple national boundaries, requiring multilateral coordination and collaboration to harmonize strategies and policies to reduce flood-related impacts. Guidelines on Sustainable Flood Protection (UN/ECE, 2000) and Guidelines on Monitoring and Assessment of Transboundary Rivers (UN/ECE Task Force on Monitoring and Assessment, 2000) were developed under the Convention on the Protection and Use of Transboundary Watercourses and International Lakes. The Guidelines on Sustainable Flood Protection include recommendations on measures and best practices to prevent, control, and reduce the adverse impacts of flood events on human health and safety, while the Guidelines on Monitoring and Assessment of Transboundary Rivers focus on water quality. Both recommend that when considering flood protection measures, the human dimensions should be assessed in addition to technical and economic issues.

Primary and secondary preventive measures are needed for the protection of human health and safety. Primary prevention measures include building codes, legislation to relocate structures away from flood-prone areas, planning appropriate land use; adequately designed floodplains and flood-control structures; and early warning systems. Risk managers should map potential flood risks, including the location of chemical and nuclear plants and other sources of potentially hazardous materials; analyse vulnerability of communities potentially affected by a flood; develop an inventory of resources that could be mobilized during and following a flood; and establish a regional or national coordination mechanism that includes the health sector to deal with floods. A risk communication program should work to increase awareness of the dangers posed by floodwaters, including the characteristics of stream-flow velocity and the risks posed to motorists and others.

Secondary prevention measures are actions taken in response to early evidence of human health impacts, which includes flood response activities, guidelines on how populations should act during and after a flood event, and disease surveillance activities, including diseases related to water quality.

5.7 Early Warnings of Flooding Risk

The ability to predict the weather conditions that cause floods is improving, resulting in increasing interest in the development of flood forecasting and early warning systems that communicate information about impending risks to vulnerable communities and populations. Such a system should include identification of weather situations that could lead to floods that could adversely impact human health, monitoring of meteorological forecasts, mechanisms to issue flood warnings to vulnerable communities and populations, public health and other interventions to reduce or prevent flood-related illnesses and death, and evaluation of the effectiveness of communication and interventions (Hajat et al., 2003).

Early warning of flooding risk, with appropriate response by citizens and emergency responders, has been effective in reducing flood-related deaths (Malilay, 1997). Planning for response to flooding during the inter-flood phase aims to enable communities to effectively respond to the health consequences of floods, and local and central authorities to organize and effectively coordinate relief activities, including making the best use of local resources and properly managing national and international relief assistance (Hajat et al., 2003). The design of effective measures to prompt desired behaviour outcomes is important. Specific warnings should be designed to target the needs of vulnerable groups. Recent studies on the potential mental health consequences of being flooded suggest that medium- to long-term interventions also should be considered when planning responses. Further research is needed to determine if the mental health impacts of being flooded respond better to psychological and/or pharmacological interventions, or whether the interventions would best be targeted at providing financial or other assistance with recovery.

An example of the effectiveness of flood protection is the differences between the 1993–1994 flooding along the Rhine and Meuse in Germany and the 1995 flooding along the same rivers (Estrela et al., 2001). The two floods had similar characteristics, although the 1993–1994 flood had a second peak discharge. Persistent high precipitation caused both events; in December 1993, the accumulated precipitation was more than double the amount of the long-term average for that month. Ten people lost their lives in the 1993–1994 flood and total damage was estimated at US$ 900 million for Belgium, Germany, France, and the Netherlands (EM-DAT; Estrela et al., 2001). The total cost of flood damage in Germany in 1995 was half of this amount, presumably because people were aware of the risks and were better prepared.

Emergency response activities should include public health surveillance of morbidity and mortality associated with the flood (Malilay, 1997). Mortality surveillance should determine the nature and circumstances surrounding deaths so that appropriate preventive actions to prevent future mortality can be enacted. Morbidity surveillance should determine whether endemic diseases are increasing, whether outbreaks of infectious diseases that need to be controlled through vaccination or other means are occurring, and whether public advisories are needed to prevent further injuries, particularly during clean-up activities. Surveillance should determine whether drinking water is safe and whether there are any increases in vector populations. Surveillance also should measure mental well-being to identify opportunities to help victims and responders cope with the flood.

The potential complexity of health outcomes following a flooding event means that effective vulnerability reduction requires the involvement of local, regional, and national level actors across multiple sectors. Local knowledge can be valuable in the design of flood strategies and communication regarding flood warnings (Parker and Handmer, 1998; Correia et al., 1994).

Measures to reduce the potential health impacts of a flood can be undertaken by the population at risk, emergency responders, and policy makers before, during, and after a flood event (Malilay, 1997). The extent to which a community is vulnerable to flooding-related health impacts depends on the degree of awareness of the flood hazard, the degree to which the hazard may be avoided and the availability of effective measures for post-flood clean-up.

Physical, technological, economic and social factors affect the range of choice for adjustment to flood vulnerability (White, 1961). There are theoretical and practical ranges of choice. The practical range of choice is always narrower than the theoretical range of choice because any particular activity may not be feasible if the technology has not been demonstrated to be effective or economically viable, or because the activity is not socially and legally acceptable.

Table 3 a. The theoretical range of response options for flooding in the United Kingdom

Theoretical range of choice	Technically feasible?	Effective to address health outcome?	Environmentally acceptable?	Economically feasible?	Socially and legally acceptable?	Closed/ Open (Practical range of choice)
Land use planning to reduce risk of exposure	Yes at county and district levels only	Yes	Depends on situation	Yes	Yes	Open
Engineering works to reduce the risk of exposure	Yes	Yes	Depends on situation	Yes	Yes	Open
Insurance	Limited availability					Closed
Emergency relief	Yes	Yes	Yes	Yes	Yes	Open

Table 3 b. The practical range of response options for flooding in the United Kingdom

Practical range of choice	Magnitude of event/ exposure intensity	Techni- cally viable?	Financial capacity (including needed infra- structure)?	Human skills and insti- tutional capacity?	Compati- ble with current policies?	Other?
▮ Land use planning to reduce risk of ex- posure	Cure dominates prevention	Yes		Over 400 local planning authorities; little central coordina- tion	Variable	
▮ Engineering works to reduce the risk of exposure	Yes		Grant aid to supple- ment local resources for flood defense is provided only for capital schemes	Through environ- ment agency & county councils	Variable	
▮ Emergency relief	Yes			County and district councils; emergency services; local and regional health authorities	Yes	

Appendix 1 compares the roles and responsibilities for floodplain man- agement in France and in England and Wales (Penning-Rowsell and Wil- son, 2003). Tables 3a and 3b use this information for the United Kingdom as an example to illustrate the ranges of theoretical and practical choice to address flooding vulnerability; this table would need to be completed for each river basin to identify specific options. This information can be used by policy makers to determine specific options available for reducing flood vulnerability, and for identifying policy changes that would be required to increase the range of available options, such as making insurance available.

Table 4 lists some health-specific interventions that, when taken, could reduce the impacts of flooding. Recommendations from the WHO/EEA

Table 4. Health-specific interventions to reduce the potential impacts of floods

Health outcome	Intervention
▌ Mental health outcomes (anxiety and depression etc.)	– Post-flood counselling – Medical assistance (drugs etc.) – Visits by health workers or social workers to vulnerable people (elderly, disabled etc.)
▌ Infectious diseases and other physical heath effects, etc.	– Treatment of respiratory problems and skin rashes – Treatment for mold and other exposures – Treatment for strains and other effects of physical exertion – Vaccination (i.e. Hepatitis A) of general population – Boil water notices and general hygiene advice – Outbreak investigations where appropriate – Enhanced surveillance
▌ Pre-flood activities	– Pre-flood awareness raising campaigns, with messages targeted to different groups – Emergency planning – Inter-institutional co-ordination activities

Source: Penning-Rowsell and Wilson, 2003

Workshop on Extreme Weather and Climate Events and Public Health Responses (WHO, 2002) include providing better information to those especially vulnerable to suffer health impacts in floods; this includes the elderly, those with prior-event health problems, the poor and those with dependents (particularly children). Lack of awareness of appropriate behaviour during and following a flood can cause injuries and deaths when people underestimate the risks of their actions, such as returning home to rescue a pet. Approaches need to be developed to locate and target the vulnerable populations. Improved flood warnings are needed, including the establishment of more effective vulnerability management programs; this includes longer warning lead times, warnings that are more accurate, more advisory warning messages and better warnings for agencies. Better post-event social care is needed for those most affected, even those who initially appear not to be affected; this includes visits to identify problems, assistance with recovery work phases, financial assistance and advice, and medical/social advice. Self-help measures to reduce stress and damage to property can be encouraged, such as flood-proofing of property, development of a family flood plan etc.

In many areas, policies to restrict development in floodplains may be needed. Much of Europe is likely to need to adapt to an increased vulnerability to flooding, and so will need to develop national coping strategies. Part of this strategy must include increasing public awareness of flooding vulnerability. For example, the U.K. Environment Agency has improved its flood forecasting and warning systems, and increased public awareness through annual campaigns and 'Flood Awareness' weeks.

Improved protection against floods requires the integrated application of a package of measures, including natural water retention, structural flood protection, implementation of preventive actions against risks, raising the awareness of the remaining flood risk and individual preventive measures (Estrela et al., 2001). A review of the floodplain management plans in European countries found that many countries, regions, and states have emergency plans (Penning-Rowsell and Wilson, 2003). However, few plans refer explicitly to health impacts. Conclusions of this review include that flood responses need to focus on local capacity building, that health status is a major contributor to flood vulnerability and that it is possible to predict flood vulnerability from secondary sources. The health sector should be included in national and international emergency planning for floods and in flood vulnerability mapping. The health sector should be more pro-active in flood planning and in providing post-flood event assistance.

5.8   Conclusions

Despite floods being the most common natural disaster in Europe, surprisingly little is known about the health impacts of floods. What is known is based on studies conducted following large flooding events; much less is known about smaller events. Studies to date suggest that the biggest disease burden of floods is psychological distress, and the likelihood of infectious disease outbreaks is small in temperate, industrialized countries. More and better quantitative data are needed on the physical and mental health impacts associated with floods, including the long-term impacts. Disease surveillance needs to be increased during and after flooding, including surveillance for longer-term psychological impacts with its associated need for medical and social support during the recovery period.

In addition, there is a need for statistical indicators of vulnerability to assess the health risks. For example, results of epidemiologic studies of floods help medical care providers match resources to needs and facilitate better contingency planning (Binder and Sanderson, 1987). Data from vulnerability indicators will assist in identifying communities at risk, monitoring the impacts of future events, and evaluating the effectiveness of interventions. Effective vulnerability reduction will require the involvement of a range of sectors at the local, regional, and national level.

It is important for policy makers and the public to recognize that flooding vulnerability can never be wholly eliminated. The emphasis in disaster management should shift from post-disaster response to pre-disaster planning. A comprehensive, vulnerability-based emergency management program of preparedness, response and recovery has the potential to reduce the adverse health impacts of floods. Different programs may be needed for different types of floods. Coordination and collaboration are needed among all the actors involved, including meteorological agencies emergency response agencies and civil society. National plans of action should include potential transboundary impacts.

References

Albritton DL, Meira Filho LG, Coordinating Lead Authors (2001) Technical Summary. Contribution of Working Group I to the Third Assessment Report of the Intergovernmental Panel on Climate Change. Cambridge University Press, Cambridge

Baxter PJ, Moller I, Spencer T, Spence RJ, Tapsell S (2001) Flooding and Climate Change. In: Health Effects of Climate Change in the U.K. Department of Health (ed), pp 152–192

Becht MC, van Tilburg M-AL, Vingerhoets A-JJM et al (1998) Flood: a pilot study on the consequences for well-being and health of adults and children. Uitgeverij Boom, The Netherlands

Bennet G (1970) Bristol floods 1968: controlled survey of effects on health of local community disaster. BMJ 3:454–458

Binder S, Sanderson LM (1987) The role of the epidemiologist in natural disasters. Ann Emerg Med 16:1081–1084

Bronstert A (2003) Floods and climate change: interactions and impacts. Risk Analysis 23:545–557

Centers for Disease Control and Prevention (1993) Morbidity surveillance following the Midwest flood – Missouri. Morb Mortal Wkly Rep 42:797–798

Christensen JH, Christensen OB (2003) Severe summertime flooding in Europe. Nature 421:805

Commission of the European Communities (2002) Communication from the Commission to the European Parliament and the Council. The European Community Response to the Flooding in Austria, Germany and Several Applicant Countries. Brussels, 28. 8. 2002

Correia FN, da Graca Saraiva M, Rocha J, Fordham M, Bernardo F, Ramos I, Marques Z. Soczka L (1994) The planning of flood alleviation measures: interface with the public. In: Penning-Rowsell EC and Fordham M (eds) Floods Across Europe. Middlesex University Press, London, pp 167–193

EM-DAT: The OFDA/CRED International Disaster Database. Université Catholique de Louvain – Brussels – Belgium (www.em-dat.net, accessed 05-05-2005)

Estrela T, Menendez M, Dimas M, Marcuello C et al (2001) Sustainable water use in Europe. Part 3: Extreme hydrological events: floods and droughts. Environment issue report No 21. European Environment Agency, Copenhagen

Few R, Ahearn M, Matthies F, Kovats S (2004) Floods, health and climate change: a strategic review. Tyndall Centre for Climate Change Research Working Paper 63

Fothergill A, Maestas EG, Darlington JD (1999) Race, ethnicity and disasters in the United States: a review of the literature. Disasters 23:156–173

Frich P, Alexander LV, Della-Marta P, Gleason B, Haylock M, Klein Tank AMG, Peterson T (2002) Observed coherent changes in climatic extremes during the second half of the twentieth century. Clim Res 19:193–212

Fullilove MT (1996) Psychiatric implications of displacement: contributions from the psychology of place. Am J Psychiatry 153:1516–1523

Green CH, Emery PJ, Penning-Rowsell EC et al (1985) The health effects of flooding: a case study of Uphill. Middlesex University Flood Hazard Research Centre, Enfield

Green C, van der Veen A, Wierstra E, Penning-Rowsell (1994) Vulnerability refined: analyzing full flood impacts. In: Penning-Roswell EC, Fordham M (eds) Floods across Europe: hazard assessment, modeling and management. Middlesex University Press, London

Hajat S, Ebi KL, Kovats S, Menne B, Edwards S, Haines A (2003) The human health consequences of flooding in Europe and the implications for public health: a review of the literature. Applied Environmental Science Public Health 1:13–21

Handmer J, Penning-Rowsell E, Tapsell S (1998) Flooding in a warmer world: The view from Europe. In: Downing TE, Olsthoorn AA, Tol RSJ (eds) Climate, change and risk. Routedge, London, pp 125–161

Keene EP (1998) Phenomenological study of the North Dakota flood experience and its impact on survivors' health. Int J Trauma Nurs 4:79–84

Ketteridge AM, Fordham M (1995) Flood warning and the local community context. In: Enfield HJ (ed) Flood warning: issues and practice in total system design. University Flood Hazard Research Centre, Middlesex, pp 189–199

Klein Tank A, Wijngaard J, van Engelen A (2002) Climate of Europe: assessment of observed daily temperature and precipitation extremes. KNML, De Bilt, the Netherlands, p 36

Kriz B, Benes C, Castkova J et al (1998) Monitoring the epidemiological situation in flooded areas of the Czech Republic in 1997). In: Davidova P, Rupes V (eds) Proceedings of the Conference DDD '98) 11–12 May 1998, Prodebrady, the Czech Republic. National Institute of Public Health, Prague

Kundzewicz ZW, Parry M, Coordinating Lead Authors (2001) Europe. In: Climate Change 2001: Impacts, Adaptation, and Vulnerability – The Contribution of Working Group II to the Third Scientific Assessment of the Intergovernmental Panel on Climate Change. Cambridge University Press, Cambridge, UK, pp 641–692

Linnerooth-Bayer J, Amendola A (2003) Introduction to special issue on flood risks in Europe. Risk Analysis 23:537–543

Malilay J (1997) Floods. In: Noji E (ed) The public health consequences of disasters. Oxford University Press, New York, pp 287–301

McMichael AJ, Campbell-Lendrum D, Kovats S, Edwards S, Wilkinson P (2002) Comparative risk assessment: climate change. Centre on Global Change and Health, London School of Hygiene and Tropical Medicine, London

Mitchell JK (2003) European river floods in a changing world. Risk Analysis 23:567–574

Miettinen IT, Zacheus O, von Bonsdorff CH et al (2001) Waterborne epidemics in Finland in 1998–1999. Water Sci Technol 43:67–71

Milly PC, Wetherald RT, Dunne KA, Delworth TL (2002) Increasing risk of great floods in a changing climate. Nature 415:514–517

Mudelsee M, Borngen M, Tetzladd F, Grunewald U (2003) No upward trend in the occurrence of extreme floods in central Europe. Nature 425:166–169

Munich R (2003) Topics. Natural Catastrophes. Annual Review 2002

Osborn TJ, Hulme M (2002) Evidence for trends in heavy rainfall events over the UK. Phil Transactions of the Royal Society 360(A):1313–1325

Parker DJ and Handmer JW (1998) The role of informal flood warnings. J Contingencies Crisis Management 6:45–60

Penning-Rowsell DC, Wilson T (2003) Flooding and health impacts in Europe: An exploratory overview of potential responses and emergency planning. A report for the cCASHh project

Phifer JF, Kaniasty KZ, Norris FH (1988) The impact of natural disaster on the health of older adults: a multiwave prospective study. J Health Soc Behav 29:65–78

Robson AJ (2002) Evidence for trends in UK flooding. Philos Transact Ser A Math Phys Eng Sci 360:1327–1343

Sarewitz D, Pielke Jr R (2002) Vulnerability and risk: some thoughts from a political and policy perspective. A discussion paper prepared for the Columbia-Wharton/Penn Roundtable on "Risk Management Strategies in an Uncertain World"

Sartorius N (1990) Coping with disasters: the mental health component. Int J Ment Health 19:3–4

Tapsell SM, Penning-Rowsell EC, Tunstall SM, Wilson T (2002) Vulnerability to flooding: health and social dimensions. Phil Trans R Soc Lond 360(A):1511–1525

Tapsell SM and Tunstall SM (2001) The health and social effects of the June 2000 flooding in the North East region. Report to the Environment Agency, Thames region. Middlesex University Flood Hazard Research Centre, Enfield

Tapsell SM, Tunstall SM, Wilson T, Penning-Rowsell EC (2003) The long-term health effects of flooding: Banbury and Kidlington four years after the 1998 flood. A report for the cCASHh project

United Nations Economic Commission for Europe (2000) Meetings of the Parties to the Convention on the Protection and Use of Transboundary Watercourses and International Lakes. Guidelines on Sustainable Flood Protection. United Nations, Geneva

United Nations Economic Commission for Europe Task Force on Monitoring and Assessment under the Parties to the Convention on the Protection and Use of Transboundary Watercourses and International Lakes (2000) Guidelines on Monitoring and Assessment of Transboundary Rivers. United Nations, Geneva

Vari A, Linnerooth-Bayer J, Ferencz Z (2003) Stakeholder views on flood risk management in Hungary's Upper Tisza basin. Risk Analysis 23:585–600

White GF (1961) The Choice of Use in Resource Management. Natural Resources Journal 1:23–40. In: Kates RW, Burton I (eds) Geography, Resources, and Environment. Vol. 1. Selected Writings of Gilbert F. White (1986) University of Chicago Press, Chicago, pp 143–165

WHO Regional Office for Europe (2002) Floods: Climate Change and Adaptation Strategies for Human Health. Report of a WHO Meeting. London, UK

Appendix: The Roles and Responsibilities for Floodplain Management: Contrasting Situations in England and Wales, and in France

Introduction: Who is Responsible for What and Why?

The first lines of defence in mitigating the health impacts of floods are the reduction of flood risk[1] and of primary and secondary flood effects. In most countries, these are accomplished through risk management and control of the use of floodplain land, with the goal of reducing the number and extent of the population at risk. The responsibilities for controlling the interface between land and water are generally split – somewhat uncomfortably – between central and local governments, as illustrated by describing the situations in France and in England and Wales, which have very different strategies and policies for risk management.

France: Centralized Preventative Actions Against Flood Hazards

French floodplain policy has only been developed actively since the middle of the 1980s (Commissariat Général du Plan, 1997), although it has rudimentary 19th century roots. Legally, France's national policy for controlling development in flood prone areas has been buried in scattered legal provisions. Until 1982, only two specific tools were extant: the 1935 Submersible Surface Plan (PSS), with a hydraulic emphasis to facilitate the disposal of flood flows, and the Risk Perimeter system derived from the 1955 Planning Code. This system was specifically designed to protect people and properties by delimiting potential risks (Gendreau, 1998).

Flood prevention policy really started with the 1982 Compensation Law for victims of natural disasters. Beginning in 1984, the State developed Exposure to Predictable Natural Hazards Plans (PER) that combined risk zoning and associated prevention measures with land use planning and flood insurance. In a 1994 Circular, the central government then asked its representatives in Departments (Prefects) to strengthen the floodplain policy.

The 1995 law for the Protection of the Environment has reinforced these arrangements and simplified the legal system by substituting a single legal planning document for the Prevention of Predictable Natural Hazards Plan (PPR), which retains many of the PER aims and principles. Additional catastrophic events led to a further legislative review, and the Senate adopted a new law on "Hazards, technological risks and damage compensation" in May 2003. The aim is to enhance levels of compensation and encourage greater public involvement in risk reduction decisions.

Four interrelated components now appear to underlie the thinking that the formal arrangements embody, as discussed in the following sections.

---

[1] 'Flood risk' is usually defined as 'probability' or 'magnitude'

▌ **A Hazard/Risk Management Approach:** France has devised a specific planning system to control the existing and future use of land and buildings in areas liable to floods and other natural hazards. This PER system (1984–1995), and the PPR developments since 1995, is a centrally directed set of arrangements applied at the local level of the Communes (the lowest level of government). Its measures are aimed at considering risk in the town planning documents developed by local government.

Central government requires the preparation of maps delimiting hazard zones – for floods, avalanches earthquakes, etc. – and the imposition of corresponding land use zoning and control measures (Hubert and Reliant, 2003). Flood risk is delimited on the zoning maps by showing both floodplain storage areas (champs d'expansion des crues) and developed areas. Both the maps and the measures are formalized and included in the PPR (MATE, 1997; MATE, 1999; MATE, 2002).

▌ **The All-pervading Power of Central Government:** The French system involves a 'coercive' approach (May and Handmer, 1992; Handmer, 1996; Johnson and Handmer, 2003) within the strong national focus implemented by central government. The system of land use and building regulation in the hazard zones is initiated by the MEDD (Ministry for Ecology and Sustainable Development – ex MATE) and implemented by its decentralized services provided by Departments (the 2nd level of government).

This is done under the authority of the Prefects of Departments who are directly nominated representatives of the central government operating locally, and must be approved and applied by the local authorities at the commune scale headed by the Mayor (Maire). The relevant circulars and decrees are directives (not advice) to those communes that are chosen by central government to be subject to this system (i.e. those deemed to have significant hazard exposure, whether its Mayor has requested the arrangement or not).

Therefore, while Mayors are in charge of land use planning and development in their communes, their land use planning strategies, such as the Local Town Planning Plan (Plan Local d'Urbanisme, PLU since 2001), must comply with the State's specific risk regulations and, therefore, with the PPR system directed by the central government. This division of responsibility and some sharing of power between the centre and the periphery are quite deliberate. Within this approach, there is no financial dimension, and there are no incentives for landowners, builders, or developers to keep development out of flood hazard zones.

Communes and property owners in PPR-regulated floodplains do not receive compensation in exchange for accepting the constraints imposed on them in applying the regulations. The system is not one of encouragement or exhortation, but one of regulation and compulsion. However, the State stresses the need for the local authorities' agreement before it acts: representatives of central government must be in contact with the local government officials concerned from the beginning to the end of the PPR-elaborating process.

▮ **A Non-structural Approach: Moving away from Engineering:** The overall philosophy has recently given explicit priority to integrated non-structural flood protection measures over structural flood control. The two main non-structural measures are tougher land use regulation and compulsory flood insurance.

Engineering strategies are not ignored, but recent flood alleviation measures have generally taken non-traditional forms (e.g. small-scale works and greater ecological consideration). The national insurance and compensation system under the law of 13 July 1982 combines insurance with 'solidarity' (Gilbert and Gouy, 1998), in which everyone who purchases building insurance is required to pay, through a common and additional premium, for the costs of potential flood losses, whether the insured are in a hazard zone or not (Dubois-Maury, 2002).

▮ **Risk Catchments: Moving away from Fixed Institutional Boundaries:** The fourth key element in the French system is the "risk catchment" approach. Instead of focusing on administrative areas, the system of PPRs has progressed towards a 'territorial' system for the management of risk, within naturally delimited areas (e.g. river catchments) rather than political jurisdictions.

The 'territorial management' is designed at the geographical scale appropriate to the risk phenomenon (floods, avalanches etc.), within a sustainable development perspective. This approach is not new. In theory it was initiated with the implementation of PERs in 1984 (MATE, 1999), but in reality it was somewhat neglected until the 1995 law bringing in the PPRs.

The concept of "risk catchment" derives from the principles of 'hydrosystem' planning and management (which echoes the concept of ecosystem management). France developed this approach starting in the middle of the 1960s and further reinforced it through the Water Law of 1992. The idea is to include hazard management in a planning and development approach.

The effect is that the system of floodplain risk zones and their land use regulations can be implemented across more than one commune – and more than one Department – whereas previously this was administratively possible but not systematically practised.

Seeking Flood Mitigation: The Counter-hazard Arrangements in England and Wales

In England and Wales, the emphasis has been on flood prevention through floodwater control, known as flood defence. As in all countries, responsibilities for flood hazard mitigation vary at the different levels of the state and of government, as illustrated in the following table.

Flood Hazard Mitigation and Health Impacts:
Responsibilities of Central and Local Government

Mitigation strategy	Typical local or regional responsibilities	Typical central government or national-level responsibilities
Land use planning to reduce risk exposure	Yes: County and district councils	Generally not: some reserve powers
Engineering works to reduce risk exposure	Yes: Environment agency and county councils	Generally only in providing grant aid finance (e.g. DEFRA)
Insurance	Generally none	None although there may be some national regulation (e.g. ABI); insurance by private companies
Emergency relief	Yes: County and district councils (housing; social services); emergency services (police etc.); local and regional health authorities	Generally only central co-ordination in major flood events (e.g. Autumn 2000 floods in the United Kingdom)

The principles underlying flood hazard mitigation and, hence, the interface between land use and flood protection in England and Wales are mainly a product of the decentralized land use planning system, focussed on democratically elected local authorities. This approach was first properly codified in the 1947 Town and Country Planning Act (coinciding, by chance, with major floods in the winter of that year). Four issues are important, as discussed in the following sections.

Land Use Planning: The Main 'Line of Defence' Against Flood Hazards: The normal land use planning system underpins all attempts in England and Wales to control floodplain encroachment. Over 400 local planning authorities of various kinds (county, district, metropolitan and unitary councils) have the responsibility for land use planning, serving populations of between 50 000 and several hundred thousand. There are no special arrangements for the designation and 'protection' of flood risk zones that are akin to the PER/PPR system in France or, for example, to the system of Conservation Areas for historic towns in England and Wales and their role in safeguarding historical buildings.

The reason for this arrangement is that the British land use planning system is intended to encompass any relevant issue, including hazards such as floods or any other factor relevant to decisions about a local authority's area and its land use. It does this through a system comprising forward planning through local authority development plans and development control decisions relating to specific applications and sites. Indeed, it is one of the system's supposed strengths that it should be able to react to any new situation or evolving local circumstances rather that having to adopt fixed solutions.

The Dominance of Local Interests: Countering the 'Localness' of Floods: There is a dominance of local interests over regional or national concerns in decision-making over all floodplain use. Central government admittedly scrutinizes the local authority development plans to gauge whether they are in accordance with its national policy as reflected in Planning Policy Guidance (PPG) and Regional Planning Guidance. The Minister even has powers to 'call in', modify, and reject local authorities' development plans. On appeal, the Minister can also call in, modify, or revoke a development control planning application decision if it is of more than local importance.

However, these central government powers are used sparingly. Legislation and planning guidance in the 1990s sought to enhance consistency and efficiency by changing the relationship between development plans and development control to give pre-eminence to the former in a 'plan-led' planning system. However, significant floodplain decisions are still made at the development control stage for individual land use change initiatives. Moreover, development plans as well as development control overwhelmingly still have a local focus because all concerned believe that the best balance between competing uses for land can generally only be judged in the local context; the reason is that the democracy-based local authorities are deemed by central government to be the prime movers in land use affairs. Land use affects both the local economy and the local environment, which are acknowledged as relatively undisputed issues of local self-determination. Without these responsibilities, local authorities would lose much of their power and *raison* d'tre, so are jealously guarded. The principal focus is the optimal use of all land, determined from a local standpoint, rather than a concern for hazard reduction in designated risk zones. As the main flood defence operating authority, the Environment Agency advises on flood risk and seeks to influence the local authority development plans and decisions, but it cannot veto them.

Guidance, Not Coercion: The role for central government, is therefore, one of guidance through policy documents, circulars and PPGs. There is no compulsion from central government in seeking the control of development in flood risk areas. Nevertheless, this guidance has become firmer and more detailed over the years (Department of the Environment (DoE)/Ministry of Agriculture, Fisheries and Food (MAFF), 1982; DoE/MAFF, 1992; Department of the Environment, Transport, Local Government and the Regions (DTLR), 2001).

Both the Department of the Environment, Food and Rural Affairs (DEFRA; formerly MAFF), the central government department with responsibility for flood defence policy, and the Environment Agency have always accepted the primacy of the local authorities in land use planning.

However, they have become increasingly concerned that the net effect of the devolution of this responsibility is increased 'encroachment' and increasing flood damage potential, hence the firmer guidance.

▮ **Land and Water: Flood Mitigation in Action:** These arrangements show the primacy of land use planning in the form of development plans for local authority administrative areas over water planning for catchments and coastal cells.

The Environment Agency, as the relevant non-departmental governmental body, has a general supervisory and regulatory duty over all matters relating to flood defense and functions relating to other aspects of the water environment such as water resources, water quality and conservation. It also has responsibility for integrated and functional planning for the water environment through various forms of catchment planning, such as Catchment Flood Management Plans.

Other bodies and groupings produce plans for other aspects of the water environment, such as Shoreline and Estuary Management plans. However, all these are non-statutory plans, and although local authorities are encouraged in PPG 25 to consider them, they are all subordinate to their land use development plans. Furthermore, the administrative boundaries of the bodies dealing with land use planning and planning for the water environment are different, making integration of the two forms of planning more difficult.

As in France, the formal arrangements, the actual practices, and the evolution of events affect what happens on the ground. The way that resources are allocated by central government to flood defence in England and Wales means that 'cure' dominates 'prevention' because DEFRA grant aid to supplement locally raised resources for flood defence is provided only for capital schemes (not annual revenue support), as dictated in the relevant legislation. The outcome is that engineering approaches have dominated nonstructural alternatives such as catchment and floodplain land use management or source control for many decades. Expenditure on 'solutions' has tended to follow the creation of problems – such as by encroachment or upstream urbanization increasing runoff – rather than it seeking their prevention: a policy of protection rather than prevention (White and Howe, 2002).

However, the situation is not static and policies constantly change in an incremental way even without new legislation or new institutions. In the context of sustainable development, a new Environment Agency approach involving broadly based flood risk management rather than flood defence is evolving. This, for example, is embodied in the Agency's Thames Estuary Flood Risk Management Project to develop a long-term strategy for development and flood risk for the Thames Estuary area. This strategy will encourage or at least allow development in flood risk areas so long as a proactive strategy for flood defence investment in structural works is implemented in advance of the changes in land use that will result.

References

Commissariat General du Plan (1997) La prevention des risques naturels, rapport d'evaluation. Comite interministeriel de l'evaluation des politiques publiques. Premier Ministre, La Documentation Francaise, Paris, France

Department of the Environment (DoE)/MAFF/Welsh Office (1982) Circular 17/82. Development and Flood Risk. London

Department of the Environment (DoE)/MAFF/Welsh Office (1992) Circular 30/92. Development and Flood Risk. London

Department of the Environment, Transport, Local Government and the Regions (DTLR) (2001) Policy and Planning Guidance 25 (PPG 25). London

Dubois-Maury J (2002) Les risques naturels en France, entre reglementation spatiale et solidarite de l'indemnisation. Annales de geographie, 111th annee, no. 627/628, pp 637–651

Gendreau N (1998) La gestion du risqué d'inondation et l'amenagement des cours d'eau. Anneles des Ponts-et-Chaussees 87:53–59

Gilbert C, Gouy C (1998) Flood management in France. In: Rosenthal U, Hart P (eds) Flood response and crisis management in Western Europe. Springer, Berlin, pp 15–56

Handmer JW (1996) Policy design and local attributes for flood hazard management. Contingencies and Crisis Management 4:189–197

Hubert G, Reliant C (2003) Cartographie reglementaire du risqué d'inondation: decision autoritaire ou negociee. Annales des Ponts-et-Chaussees 105:24–31

Johnson CL, Handmer JW (2003) Coercive and cooperative policy designs: moving beyond the irrigation system. Irrigation and drainage 52:193–202

May P, Hander JW (1992) Regulatory policy design: cooperative versus deterrent mandates. Australian Journal of Public Administration 51:43–53

MATE (Ministere de l'Amenagement du territoire et de l'Environnement), Ministere de l'Equipement, des Transports et du Tourisme (1997) Plans de prevention des risques naturels previsibles (PPR), Guide general. La Documentation Francaise, Paris, France

MATE (Ministere de l'Amenagement du territoire et de l'Environnement), Ministere de l'Equipement, des Transports et du Tourisme (1999) Plans de prevention des risques naturels previsibles (PPR), Risques d'inondation, guide methodologique. La Documentation Francaise, Paris, France

MATE (Ministere de l'Amenagement du territoire et de l'Environnement), Ministere de l'Equipement, des Transports et du Tourisme (2002) Plans de prevention des risques naturels previsibles (PPR), Risques d'inondation, mesures de prevention. La Documentation Francaise, Paris, France

White I, Howe J (2002) Flooding and the role of planning in England and Wales: a critical review. Journal of Environmental Planning and Management 45:735–745

6 Vector- and Rodent-borne Diseases

Bettina Menne, Kristie L. Ebi

Vector-borne diseases (VBD) are infections transmitted by the bite of infected arthropod species, such as mosquitoes, ticks, bugs and flies. VBDs are important health outcomes to be associated with climatic changes due to their widespread occurrence and sensitivity to climatic factors. Climate change can affect vector-borne disease in several ways, namely

▌ the survival and reproduction rates of vectors, in turn determining their distribution and abundance
▌ the intensity and temporal pattern of vector activity (particularly biting rates) throughout the year and
▌ the rates of development, survival and reproduction of pathogens within vectors.

A wide variety of ecological trends are associated with the long-term warming trend that has been occurring over the past century. The range of possible ecological changes varies from the ecosystem level to the community, species and population levels. It is expected that ecosystem responses will be one mediator of the potential effects of changes in climate on infectious diseases, especially vector-borne diseases. The extent and magnitude of climate-driven changes in human and animal infectious diseases mediated by ecological systems will be determined by two major factors:

▌ the nature (functional, structural), type, extent, distribution and magnitude of the ecosystem responses to climate change
▌ the degree of association between the components of the cycle of the disease (infectious agents, arthropod vectors, invertebrate intermediate hosts and vertebrate animal reservoirs) to the natural biological systems.

Despite the ecological changes, the distribution and seasonality of diseases that are transmitted by cold-blooded insects or ticks are likely to be affected by climate change.

Increases in temperatures in Europe might allow the establishment of tropical and semitropical vector species, permitting transmission of diseases in areas where low temperatures have hitherto prevented their over-wintering. A change in the distribution of important vector species may be among the first signs of the effect of global climate change on human health.

Global warming may have a more immediate effect on mosquito populations and only later on mosquito-borne diseases. With increased average temperatures, the length of the breeding seasons for mosquito populations would be extended and population densities would increase. Warmer northern climates would cause an expansion of the distribution of species now found only in warmer climates. Should these changes occur, it seems likely that the distribution of mosquito-borne diseases would also subsequently expand.

The Chapters 6.1 to 6.6 describe the results of
- an in-depth literature review, expert workshops, mapping of the geographical distribution
- ecological studies, to present the current situation and geographical distribution in Europe of common endemic vector-borne diseases
- review of the occurrence and transmission of vector- and rodent-borne diseases in Europe and their climatic dependencies
- identification of vulnerable regions and current preventive measures and
- case studies on tick-borne encephalitis and its main host, reservoir and vector species and habitat vegetation in the Czech Republic.

6.1 Leishmaniasis: Influences of Climate and Climate Change Epidemiology, Ecology and Adaptation Measures

ELISABET LINDGREN, TORSTEN NAUCKE

Contributing authors: CLIVE DAVIES, PHILIPPE DESJEUX, PIERRE MARTY and BETTINA MENNE

6.1.1 Introduction

Leishmaniasis is a protozoan parasitic infection that is transmitted to humans through the bite of an infected female sandfly. The *Leishmania* parasite can also be transmitted directly from person-to-person through the sharing of infected intravenous needles, or through blood transfusions (Kubar et al., 1997). The disease is endemic in 88 countries. Overall prevalence is 12 million people and the population at risk is 350 million. The global burden of leishmaniasis has remained stable for some years, causing 2.4 million Disability Adjusted Life Years lost and 59 000 deaths in 2001 (Davies et al., 2003). Human infections with *Leishmania* can be classified into four main forms: visceral (VL), cutaneous (CL), mucocutaneous leishmaniasis (MCL), and the rare diffuse cutaneous leishmaniasis (DCL). CL is the most common form with about 1 to 1.5 million new cases worldwide each year. VL is the most severe form with about half a million new cases each year. Ninety percent of VL cases occur in five countries: Bangladesh, India, Nepal, Sudan and Brazil. Leishmaniasis can be either zoonotic, or anthroponotic, i.e. humans are the sole reservoirs and sole sources of infection for the vector. Anthroponotic leishmaniasis is the source of severe epidemic outbreaks and is common in Africa and south-western parts of Asia. However, in the Mediterranean basin (as well as in Latin America) leishmaniasis is zoonotic with dogs, foxes and rodents being the main animal reservoirs. Recent studies have shown that cats (Sanchez et al., 2000; Pennisi, 2002; Del Guidice and Marty, 2003) and horses (Köhler et al., 2002; Solano Gallego et al., 2003) can be carriers of the parasites as well.

CL produces skin lesions, and is usually self-healing but can cause disfiguring scars. In contrast, VL may be lethal if untreated. Death is mainly due to secondary infection, and often occurs within two to three years of being infected. This Chapter will focus on the VL form of leishmaniasis in Europe because of its serious public health implications. VL affects internal organs and the symptoms include irregular fever, weight loss, swelling of the liver and spleen, and anaemia. The incubation period can last from a week to more than a year. Relatively few people will develop symptoms after being bitten by an infected sandfly, but dormant infections may emerge if the immune response is lowered. People with defected cell-mediated immune systems, such as patients infected with human immunodeficiency virus (HIV) are more vulnerable to *Leishmania* infection, with a risk 100–1000 greater of developing symptoms. In Europe, approximately 95–99% of *Leishmania*/HIV co-infections involve the visceral form of the disease.

Leishmaniasis is distributed mainly in arid and semi-arid areas of the world. The distribution of the vector sandflies is closely linked to climatic conditions, with the different vector species having different climatic preferences (see Geographical Distribution Section). In Europe, transmission foci are usually south of latitude 45 °N and less than 800 m above sea level, but recent findings have detected sandflies (*Phlebotomus perniciosus*) in Germany as high as 49 °N (Naucke, 2002; Naucke and Schmitt, 2004). The biting activity of the European sandflies is strongly seasonal and in most areas is restricted to the summer months.

In Europe VL is considered to be a rare disease. However, the incidence of VL increased significantly in the region during the 1990s. This was due, in part to the large proportion of *Leishmania*/HIV co-infections (approximately 25–70% of adult VL cases are co-infected with HIV) and to better reporting after the establishment of a WHO surveillance network in 1994 (Desjeux et al., 2000). Non-HIV VL incidence rates have increased in Italy and France (including a four-fold increase in Alpes-Maritimes), and new endemic areas have been detected where no previous autochtonous cases had been reported, e.g. in northern Italy, North Croatia, Switzerland and Germany.

6.1.2 Geographical Distribution

The geographical distribution of leishmaniasis worldwide is limited by the distribution ranges of the different sandfly species (which in turn are limited by their susceptibility to cold climates, by conditions affecting the development of the different species of *Leishmania* and whether animals or humans are the principal pathogen reservoir hosts (WHO, 2000b). Populations of *Phlebotomus perniciosus*, one of the known vectors of Mediterranean VL, have been found as far north as Paris, France (Parrot, 1922), and recently also in Gerweiler, Germany (Naucke and Schmitt, 2004). Table 1 lists the distribution of the different vector species that may carry *L. infantum*, the pathogen that causes VL in Europe.

Since the mid-1990s, the reported worldwide geographical distribution of endemic leishmaniasis has expanded (WHO, 2000a). This spread is probably due to a combination of factors, among them increased monitoring, more intensive research, demographic changes, land use/land cover changes that create new habitats and/or changes in microclimate, and changes in seasonal climate. The overlap between geographical areas with high risk of both leishmaniasis and HIV is increasing with the spread of leishmaniasis (typically a rural disease) into urban areas and the increased spread of HIV into rural areas (WHO, 2000a).

Table 1. Distribution of different sandfly species in Europe that may carry *L. infantum*, the cause of European visceral Leishmaniasis

Vector	Countries
▌ *P. ariasi*	France, Italy, Portugal, Spain
▌ *P. neglectus*	Albania, Greece, Italy[a], Malta, Romania, Serbia and Montenegro[b], southern Austria[c]
▌ *P. perfiliewi*	Albania, Greece, Italy, Serbia and Montenegro[b]
▌ *P. perniciosus*	France (incl. Corsica), Germany, Italy, Malta, Portugal, Spain, southern Austria[c]
▌ *P. tobbi*	Albania, Cyprus, Greece, Italy (Sicily), Serbia and Montenegro[b]
▌ *P. mascittii*[d]	Belgium, Germany, France (incl. Corsica), Italy, Monaco, Serbia and Montenegro[b], Spain, Switzerland

[a] Maroli et al., 2002; [b] Bosnia and Herzegovina, Croatia, The former Yugoslav Republic of Macedonia, Serbia and Montenegro; [c] Suspected geographical distribution because of autochthonous visceral cases (Beyreder, 1965; Kollaritsch et al., 1989; Dornbusch et al., 1999); [d] Suspected vector species because of autochthonous visceral cases in human and animal in Germany (Gothe, 1991; Naucke and Pesson, 2000; Bogdan et al., 2001; Köhler et al., 2002; Deplazes and Mettler, 2003; Naucke and Schmitt, 2004)

Within the European Region, cases of VL have been reported from Albania, Bosnia and Herzegovina, Bulgaria, Croatia, France, Greece, Hungary, Italy, The former Yugoslav Republic of Macedonia, Malta, Monaco, Portugal, Romania, Spain and Serbia and Montenegro (Fig. 1, Table 1). Leishmaniasis is also transmitted in the adjoining countries of Azerbaijan, Cyprus, Georgia, Kazakhstan, Tajikistan, Turkey, Turkmenistan and Uzbekistan. Figure 1 shows the regional distribution of VL, compared to CL.

Most cases of co-infection in Europe are reported from the most densely populated areas and provinces and, as shown in Figure 2, there is a predominance of cases in coastal areas (75%). In south-western Europe, 80% of co-infection cases are from urban areas, the main cities being Lisbon and Porto in Portugal; Barcelona, Granada, Madrid and Seville in Spain; Marseille and Nice in France; and Genoa, Milan and Catania (Sicily) in Italy. This distribution pattern, however, is probably partly related to the current location of the WHO surveillance centres (which specifically monitor *Leishmania*/HIV co-infections) rather than to other factors. Most of the centres are located near cities like Rome and Catania in Italy; Madrid, Barcelona, Seville, Bilbao, Granada and Palma de Mallorca in Spain; Paris, Marseille, Montpellier and Nice in France; and Lisbon in Portugal.

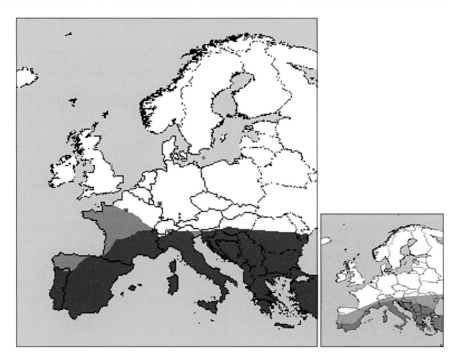

Fig. 1. Approximate distribution of visceral leishmaniasis and its vectors in the European Region compared to cutaneous leishmaniasis and its vectors

6.1.3 Population at Risk in Europe

Populations at risk include people living in rural and peri-urban areas where both sandflies and reservoir animals are prevalent. The sandfly vector is mainly active during the night, and the highest risk for contracting the disease from sandfly bites is therefore between dusk and dawn.

VL used to be found predominately in children but during recent years an increasing proportion of adult cases (non-HIV) have been reported. This change in age distribution is probably caused by several factors, such as changes in human exposure patterns, environmental changes, and improvements in case diagnosis and notification. Better nutrition and general health status in European children have probably played a role in reducing the characteristic high susceptibility of children to this disease.

Urban areas have become high-risk locations of late. This is mainly due to other transmission pathways (i.e. blood-to-blood transfer) becoming more common, as well as to increased susceptibility to the pathogens from co-infection with HIV. New high-risk groups are, thus, intravenous drug users who share syringes (Amela et al., 1996; Cruz et al., 2002; Pineda et al., 2002). In Europe 77% of *Leishmania*/HIV co-infected patients are aged between 31 and 50 years and 83% of them are men (WHO, 1997). The

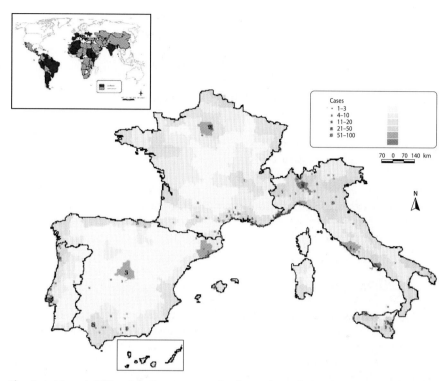

Fig. 2. Leishmania/HIV co-infection cases per locality and population density, 1990–1998. The inserted world map shows the global distribution of Leishmaniasis (light blue) and *Leishmania/* HIV co-infection (dark blue) (Desjeux, 2000)

highest risk for VL is found among HIV-infected intravenous drug users, who account for 71% of co-infection cases (Desjeux et al., 2000).

In addition, *Leishmania* can be transferred through blood-transfusion (Kubar et al., 1997), and transplacental route of infection is also possible, but rare (Meinecke et al., 1999).

Leishmania/HIV co-infections accelerate and aggravate symptoms of both leishmaniasis and acquired immunodeficiency syndrome (AIDS) (Desjeux et al., 2000). Relapses of leishmaniasis after treatment are common among persons with HIV co-infection. The mean survival of co-infected patients is only 13 months (WHO, 2000b). The current number of HIV-infected persons in western and eastern Europe (not including the Russian Federation) is approximately 580000 (WHO, 2001). In the Mediterranean basin, 1.5–9% of AIDS patients develop VL (WHO, 2000a).

Most of the co-infection cases have been reported from France, Italy, Portugal and Spain, but co-infection is prevalent in Albania, Croatia and Greece as well. On average, the number of reported cases of *Leishmania/* HIV co-infection increased during the late 1990s relative to the number of reported HIV cases in the same period (Desjeux et al., 2000). However, the

Table 2. Number of reported cases of visceral leishmaniasis/HIV co-infection in southern Europe, 1990–2001

Country	1990–1995	1996–1998	1999–2001
▌ France	127	132	59
▌ Italy	144	85	106
▌ Portugal	29	88	42
▌ Spain	473	412	214

Source: P. Desjeux, personal communication

numbers of cases of co-infection have recently started to fall in southern Europe, owing to the use of new therapeutic methods, i.e. highly active antiretroviral therapies (HAART) (Table 2).

6.1.4 Pathogen Transmission Cycle

All types of leishmaniasis are caused by parasites belonging to the genus *Leishmania*. The life-cycle of *Leishmania* protozoa comprises two developmental stages. The first, so-called amastigote stage is found in vertebrate reservoir hosts such as dogs and rodents, and is transmitted to other vertebrate hosts (such as humans) by sandflies. While feeding, the sandfly ingests the amastigotes. Over a period of 4 to 25 days the amastigotes continue their development inside the sandfly, where they undergo major transformation into flagellated promastigotes. When the now infected sandfly feeds on a fresh source of blood, it passes the promastigotes into the new host (Stage 2). The promastigotes are incorporated and differentiate into the amastigote form in the new vertebrate host, and hence the life-cycle is completed (Cunningham, 2002) (Fig. 3).

Dogs are not only passive parasite reservoirs but may become sick themselves. Canine VL has proven to be of importance in Europe, and dogs with active disease have until now usually been destroyed. New preventive methods, such as the use of insecticide-impregnated dog collars are likely to take the place of such radical methods (Maroli et al., 2001).

Sandfly Infection Rate

Sandfly infection rates vary across Europe (Table 3). The highest infection rates have been generally found in *P. perniciosus* (about 3.3% on average), and the lowest in *P. perfiliewi* (average of 0.1%). Infection rates detected in Spain and France (about 4% on average) have typically been higher than those reported from Italy (Table 3). However, the data published to date are too scarce to draw any firm conclusions, as reported rates will be greatly dependent on a range of potential biases, so direct comparisons

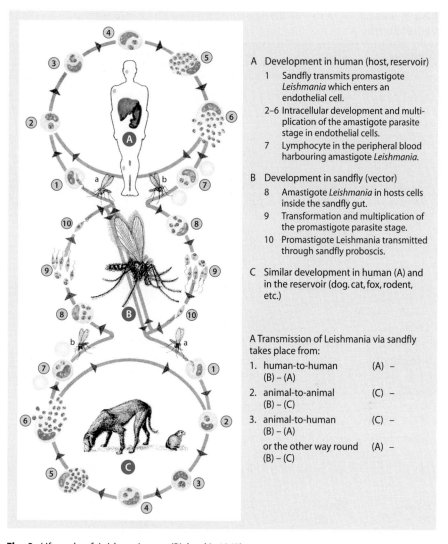

A Development in human (host, reservoir)

  1 Sandfly transmits promastigote *Leishmania* which enters an endothelial cell.

  2–6 Intracellular development and multiplication of the amastigote parasite stage in endothelial cells.

  7 Lymphocyte in the peripheral blood harbouring amastigote *Leishmania*.

B Development in sandfly (vector)

  8 Amastigote *Leishmania* in hosts cells inside the sandfly gut.

  9 Transformation and multiplication of the promastigote parasite stage.

  10 Promastigote Leishmania transmitted through sandfly proboscis.

C Similar development in human (A) and in the reservoir (dog. cat, fox, rodent, etc.)

A Transmission of Leishmania via sandfly takes place from:

1. human-to-human   (A) – (B) – (A)

2. animal-to-animal   (C) – (B) – (C)

3. animal-to-human   (C) – (B) – (A)

   or the other way round   (A) – (B) – (C)

Fig. 3. Life-cycle of *Leishmania* spp. (Piekarski, 1962)

with few data will be unreliable. Furthermore, the annual inoculation rate will depend not only on infection rates, but also seasonal variation in sandfly activity.

Table 3. Summary of studies on infection rates of *L. infantum* in different sandfly species in southern Europe

Country	Area	Sandfly species	Sandfly infection rate *L. infantum* (%)	References
France	Cevennes	*P. ariasi*	1.6	Rioux et al., 1984
	Nice	*P. ariasi*	4	Izri et al., 1992a
		P. perniciosus	4	
Greece	Corfu	*P. neglectus*	0.12	Léger et al., 1988
	Cyprus	*P. tobbi*	0.14	Léger et al., 2000a
Italy	Abruzzi	*P. perfiliewi*	0.5	Maroli et al., 1987
	Apulia	*P. perniciosus*	1.7	Maroli et al., 1988
	Campania (Ercolano)	*P. perniciosus*	2.8	Maroli et al., 1994
	Campania (Maddaloni)	*P. perniciosus*	6.2	Maroli et al., 1994
	Sardinia	*P. perniciosus*	10.5	Bettini et al., 1986
Malta	Gozo island	*P. perniciosus*	0.25	Léger et al., 1991
Monaco		*P. ariasi*	28.6	Izri et al., 1996
		P. perniciosus	2.5	
Portugal	Lisbon (Arrabida)	*P. ariasi*	4.9	Alves Pires et al., 1984
		P. perniciosus	2	
	Algarve	*P. perniciosus*	5	Schrey et al., 1989
		P. perniciosus	0.4	Alves Pires et al., 2001
	Alto Douro	*P. ariasi*	1.3	Alves Pires et al., 1991
Spain	Almeria	*P. perniciosus*	5.4	Martin Sanchez et al., 1993
	Catalonia	*P. ariasi*	0.52	Guilvard et al., 1996
		P. perniciosus	0.15	
	Malaga	*P. perniciosus*	3.7	Morillas Marquez et al., 1996
	Zaragoza	*P. ariasi*	4.2	Lucientes Curdi et al., 1988
		P. perniciosus	1.1	

6.1.5 Vector Ecology

There are more than 700 species of sandflies worldwide, of which at least 30–40 are thought to be involved in the transmission of *Leishmania* (Killick-Kendrick, 1990). European leishmaniasis vectors are sandflies of the genus *Phlebotomus*. In the Mediterranean area, five species have been shown to transmit VL, namely *P. ariasi, P. neglectus, P. perfiliewi, P. tobbi* and *P. perniciosus* (Minter, 1989; WHO, 1996; Leger et al., 2000a) (Table 3). In addition, *P. langeroni*, present in central Spain (Martinez Ortega et al., 1996) has been found to carry *L. infantum* in Egypt (Doha and Shehata, 1992).

In contrast to most mosquito species, which use stagnant water for breeding, the female sandfly generally lays her eggs on, or in the ground of specific environments where there is sufficient organic matter, heat and humidity to complete larval development, i.e. in the bark of old trees, in rodent burrows, in water-wells, along rifts, ruins of old buildings, inside barns, in cracks of house walls and household rubbish (Naucke, 2002). Female adults search for blood during the early and late evening and may cover a radius varying from a few metres to a kilometre. The flight range of sandflies differs according to the species and habitat, but is often around a kilometre. *P. papatasi* has been observed to fly as far as 1 500 m, but that distance is not achieved by engorged or gravid females (Doha et al., 1991). In France, *P. ariasi* has frequently been shown to fly a kilometre and occasionally as far as 2 200 m (Killick-Kendrick, 1990). A female sandfly lays between 50 and 100 eggs at each oviposition. Depending on temperature and larval diet, the normal time from egg laying to the emergence of adults ranges from 35 to 60 days. (WHO, 1984). In the Mediterranean, sandflies develop between one and three generations per year (Naucke, 1998).

Influence of Environmental and Climatic Factors on Disease Risk

The distribution of VL in Europe is significantly less than the distribution of the sandfly vectors (Fig. 1). The occurrence of disease transmission within the range of the vectors will depend on vector abundance, vector survival, vector biting rates (i.e. the gonotrophic cycle), the extrinsic incubation period and the length of the season. All of these parameters are climate dependent, but the precise relationships with climate need to be further studied and evaluated. Caution is also required in the interpretation of laboratory experiments, as sandflies are able to evade extreme weather conditions in the field by their choice of resting site (Fig. 4) and time of activity.

Temperature and humidity are the two most important climatic factors for sandfly survival, development and activity. *Phlebotomus* sandflies can survive cold temperatures in diapause (overwintering). Diapause is initiated by a combination of low temperatures and reduced daylight and can last between four and eight months, depending on the location. In Europe,

Fig. 4. Two female *P. perfiliewi* sandflies resting during the day in poroton-stones in Halkidiki, Greece. Photo and copyright Torsten Naucke

the biting activity of sandflies is strongly seasonal, and restricted to summer months in most areas. Adult activity as well as larval development slows down considerably when temperatures drop below 20 °C. However, some species such as *P. neglectus* and *P. mascittii* have been shown to be active temperatures of 13 °C and 13.5 °C, respectively (Naucke, 1998; Schmitt, 2002).

Temperature also affects the parasite itself; evidence indicates that *L. infantum* is prevalent within the 5 to 10 °C January and 20 to 30 °C July isotherms, usually located below latitude 45 °N and below 400–600 m above sea level (Kuhn, 1999). The worldwide distribution of sandflies is considered to be confined to areas that have at least one month with a mean temperature of 20 °C (WHO, 1984). However, it has been suggested that the limit of the European distribution of vectors and parasites better corresponds to the 10 °C annual isotherm (Maier et al., 2003). This is in accordance with recent findings of vectors and isolated autochthonous cases of leishmaniasis in southern Germany where the July isotherm is between 16 °C and 18 °C. The highest leishmaniasis-focus in Europe is the Guadix focus (Andalusia, Spain) at an altitude of 900 m to 950 m above sea level.

Sandflies are sensitive to sudden temperature changes and usually prefer regions with small differences between maximum and minimum temperatures. Sandfly survival can be reduced if the climate gets too hot and dry, even though they can rest in cold, humid places during the daytime (McCarthy et al., 2001). Resting sites of adults are known for a few species of sylvatic sandflies; they include tree-holes and trunks. Peridomestic species rest on walls and, at hot times of the day, retreat into cracks and crevices (WHO, 1984). Poroton-stone buildings, for example, have the ability to store humidity during the night and evaporate it during the day, which

Table 4. Important *Phlebotomus* species in Europe and their minimum and maximum temperature thresholds, their optimum temperatures and humidity, and their preferred altitudes

Vector	Minimum temperature (larvae)	Optimum temperature (larvae)	Maximum temperature (larvae)	Preferred humidity (adults)	Preferred altitude	Causative organism
P. perniciosus	0 °C	25 °C	Unknown	Unknown	< 900 m	*L. infantum*
P. perfiliewi	–4 °C	25 °C	33 °C	60–80%	< 450 m	*L. infantum*
P. ariasi	5 °C	Unknown	30 °C	Unknown	300–500 m	*L. infantum*
P. neglectus	–4 °C	25 °C	30 °C	60–80%	300–1000 m	*L. infantum*
P. papatasi	5 °C –4 °C (Greece)	28–34 °C	35 °C	36–45%	0–1000 m	*L. infantum*
P. sergenti	Unknown	31–33 °C	Unknown	0–45%	Unknown	*L. tropica*
P. mascittii [d]	–4 °C	19 °C	28 °C	60–80%	0–400 m	Unknown

Sources: Killick-Kendrick, 1999; Rioux et al., 1985; Singh, 1999; Naucke and Schmitt, 2004

are favourable conditions for adult sandflies to survive the hot and dry summer days (Fig. 4).

In addition to direct associations between climate and leishmaniasis transmission, climate will have indirect impacts by influencing the distribution of hosts, the local vegetation (important as resting sites and sugar sources), and patterns of human exposure to sandfly vectors.

Table 4 summarizes some characteristics of important *Phlebotomus* species in Europe.

6.1.6 Possible Effects of Climate Change in Europe

With climate change, the distribution ranges of both the sandfly vectors and the pathogens may extend northwards and into higher altitudes (WHO, 1999). In currently endemic areas, higher seasonal temperatures would lead to a prolonged activity period and a shorter diapause period. This could result in an increased number of sandfly generations per year. In addition, higher temperatures are likely to accelerate the maturation of the protozoan parasite, thereby increasing the risk of infection (Rioux et al., 1985; McCarthy et al., 2001). However, if the climate becomes too hot and dry for vector survival, the disease may disappear from some localities, although the vectors may adapt by resting in cool, humid places during daytime (Fig. 4).

Imported infected dogs may contribute to the emergence of leishmaniasis in new locations. They are a potential source of the pathogen if the vectors expand their geographical distribution owing to a change in climate

(WHO, 1999). Several imported cases of canine leishmaniasis are reported from Germany (Gothe et al., 1997) and the Netherlands (Slappendel, 1988) every year, for example.

In order to develop more sophisticated climate-vector-disease scenario models, increased knowledge is needed on the short- and long-term effects of climate variations on leishmaniasis risk. This can be achieved in several ways, such as strengthening

▌ experimental evidence measuring the influence of temperature or humidity (for example) on sandfly or *Leishmania* biology;

▌ field evidence that sandfly biological parameters vary with climate;

▌ field evidence that geographic variation in sandfly abundance (including presence vs. absence) or *Leishmania* infection rates correlate with climatic variables; and

▌ field evidence that temporal variation in sandfly abundance or *Leishmania* infection rates (including seasonal and long-term trends) correlate with climatic variables.

As different sandfly species are responsible for *Leishmania* transmission across Europe, climate change will impact VL risk differently in various parts of the region (see Tables 1 and 4). Regional climate models for leishmaniasis should, therefore, be constructed according to the local vector species.

6.1.7 Adaptation and Preventive Measures

WHO (2002) identified the following objectives to reduce the incidence of leishmaniasis worldwide:

▌ to provide early diagnosis and prompt treatment

▌ to control the sandfly population through residual insecticide spraying of houses and through the use of insecticide-impregnated bednets

▌ to provide health education and produce training materials

▌ to detect and contain epidemics in the early stages [1] and

▌ to provide early diagnosis and effective management for *Leishmania/ HIV* co-infections.

Adaptation and prevention of VL in Europe should, in addition to the above, also include methods concerning zoonotic hosts, such as surveillance of movements of dogs and changes in infectious rates, and updates on preventive measures and treatments.

---

[1] In Europe epidemic outbreaks of VL are rare compared to other parts of the world where humans instead of animals are the parasite reservoirs. The last VL epidemic in Europe occurred in the Emilia Romagna region in northern Italy between 1971 and 1973, and affected 60 individuals in total (42 adults and 18 children) and caused 13 deaths (Pampiglione et al., 1974).

Diagnosis and Treatment

The minimal incubation period for VL is between two and six months, but incubation periods as long as 16 years have been recorded (Grant et al., 1994). If left untreated, VL has a mortality rate of almost 100% (WHO, 1996). Death is mainly due to secondary infection, and often occurs within two to three years. In more acute forms, death may occur within 6 to 12 months. For patients with a compromised immune system, the mean survival even with optimal treatment is only 13 months.

Differential diagnoses for VL include other diseases with massive splenomegaly, such as malaria, portal hypertension, leukaemia, lymphoma, typhoid, miliary tuberculosis and haemolytic anaemia. The clinical diagnosis has to be verified by laboratory methods. Direct evidence of infection is found when the parasite can be detected in tissue smears, with splenic aspirate being more sensitive than bone marrow or lymph node aspirates.

However, these methods are complicated and time consuming, and it can be several weeks before results are known (Vidyashankar et al., 2001). Therefore, methods showing indirect evidence of infection are often used, such as the direct agglutination test (DAT) and rK39-antigen-based dipstick test, which are currently the best available serological tests in terms of sensitivity and specificity for use in primary health care services (Le Ray et al., 1999; Chappuis et al., 2003). Nevertheless, positive results are still uncertain as cross-reaction can occur with other tropical diseases, including malaria and schistosomiasis (Vidyashankar et al., 2001). New promising methods, such as detection of leishmanial antigen in urine through a latex agglutination test (Katex) are being tested (Attar et al., 2001).

Leishmaniasis is difficult to diagnose serologically in HIV patients. This is due both to the impaired immune system of HIV patients and to the frequent presence of other opportunistic diseases. Bone marrow aspiration is often necessary. Studies from Spain have demonstrated the use of laboratory-colonized sandflies (*P. perniciosus*) as xenodiagnostic tool for HIV patients (Molina et al., 1999).

Several drugs are currently available for the treatment of VL. Pentavalent antimony compounds have been in use for over six decades, but the treatment is time and resource consuming. The drug should, if possible, be administered in hospital in the form of daily intramuscular injections over a period of 20 to 30 days. Relapses after treatment are common for *Leishmania*/HIV co-infected patients, and other types of drug sometimes have to be used.

HAART has some effect on the relapse rate (De La Rosa et al., 2002), and the use of HAART is believed to have contributed to the recent decline in the number of *Leishmania*/HIV co-infected cases in Europe.

Therapeutic drug resistance to pentavalent antimony is becoming a growing problem, and the severe side-effects of the antimony compounds have for some time been considered to be a main concern in the treatment of leishmaniasis. Another available drug is liposomal amphotericine B (AmBisome®), which is expensive but only requires five days of therapy. The drug has been tested in parts of Southern Europe for the treatment of adults and children (Castagnola et al., 1996; Minodier et al., 1998; Minodier and Garnier, 2000; Kafetzis and Maltezou, 2002; Marty and Rosenthal, 2002; Cascio and Colomba, 2003; Syriopoulou et al., 2003).

However, from a worldwide perspective it is alkylphosphocholine miltefosine (Miltefosine®), first developed as an anticancer drug and which can be taken orally, that offers the most hope (Davies et al., 2003). Miltefosine® has been in use in India since 2002 (Sundar et al., 2002). Because of its teratogenic potential, this drug should be used with caution in women of childbearing age.

Due to the threat of drug resistance, new drugs should be available under controlled therapeutic forms for humans only, and not for veterinary use.

Vaccination

First- and second-generation (e.g. recombinant antigen) *Leishmania* vaccines have been under development for several years (Dumonteil et al., 2001). A recent major paradigm shift that reflects the role of the cell-mediated resistance may encourage the development of vaccines with more long-term immunity (Davies et al., 2003). So far, the situation for VL is less promising than for CL.

Vaccination of dogs is another alternative, aimed at reducing parasite transmission. A vaccine against a component of sandfly saliva that prevents *L. major* (CL) in mice, is currently being tested on dogs in the United States. Recent reports suggest some progress is also being made in vaccines for canine VL (Gradoni, 2001).

Control Targeted at the Vector

Sandfly vectors are not actively controlled in Europe (WHO, 1999). One way to control sandflies is by the complete destruction of their habitat, followed by development of the land to ensure that conditions suitable for the flies do not return (Killick-Kendrick, 1999). Such drastic methods may be effective in some cases but can be both expensive and environmentally degrading. Removal of resting and breeding sites of the sandflies is a more moderate preventive method. An alternative might be the use of larvicides on breeding places (Naucke, 1998).

The use of insecticides indoors is another method aimed at controlling sandfly populations. However, most female sandflies rest outdoors during the maturation of eggs and also bite outdoors, which reduces the efficiency of this method (Killick-Kendrick, 1999). The same short-coming applies to the use of pyrethroid-impregnated bednets and impregnated curtains. However, the use of insect repellents on skin is effective in preventing sandfly bites (Desjeux, 1996) as long as the repellent remains active.

Control Targeted at the Reservoir Host

In Europe, canine leishmaniasis is a major veterinary problem (WHO, 1999). In much of the Mediterranean area, most dogs become infected during their lifetime and between 20 and 30% of the population is seropositive at any time. It is, however, extremely difficult to cure dogs, and in almost all cases the infection relapses after each attempt at treatment. It is recommended that drugs used to treat human infections should not be used to treat dogs, in order to avoid the development of resistance (WHO, 1996).

At present, the most effective preventive method for canine leishmaniasis as well as for the control of parasite reservoirs in endemic areas is deltamethrin-impregnated dog collars, which have shown good results if used for more than a week before exposure to the insects (Killick-Kendrick et al., 1997; Maroli et al., 2001; Reithinger et al., 2001). Dogs have been found to be protected from the majority of sandfly (*P. perniciosus*) bites and the collar remains effective for a complete sandfly season. Community-wide application of deltamethrin-impregnated dog collars not only protects domestic dogs from *L. infantum* infections, but has also been shown to reduce the risk of *L. infantum* infection in children (Mazloumi Gavgani et al., 2002). Such dog collars could have a role in control of VL and replace controversial dog culling programmes in some countries. However, the effectiveness of dog collars on disease prevention in risk areas will depend on the importance of wild animal species versus domestic dogs as reservoir hosts of *L. infantum* (Mazloumi Gavgani et al., 2002).

Health Education

Health education must be strongly supported in order to increase the level of awareness among the exposed population, and to promote community control measures so as to bring both the vectors and the reservoirs under control (Desjeux, 1996).

Surveillance and Monitoring

The WHO surveillance network that was set up in 1994 to monitor *Leishmania*/HIV co-infection still has incomplete coverage. It would be of interest to expand this existing network as well as start monitoring and registering VL cases that are not co-infected with HIV. This is of interest not only for preventive and adaptive purposes but also for early detection of changes in disease incidence over time, and to provide better data for studies of locations with different endemicity.

Movements of domestic dogs between endemic and non-endemic countries in Europe should be monitored. This is particularly important since the increased risk from climate change will allow vector species to become established in new locations over the coming decades. Infected dogs could, thus, contribute to the establishment of VL transmission in new areas.

Pathogens other than *Leishmania* spp. could become a future problem. *P. papatasi* is, for example, well known as a good virus vector, and is both an anthropophilic and autogenious sandfly species (El Kammah, 1972; Miscevic, 1980).

Autogeny has been described in a number of sandfly species. This phenomenon has an obvious survival advantage, since it ensures maintenance of the species in the event that a suitable vertebrate host is not available. This is especially important for an insect like a sandfly, which has a limited flight range and a relatively short life. Autogeny may also have important implications for the survival of Phleboviruses, since a number of these agents are transovarially (vertically) transmitted in their sandfly vectors (Tesh, 1988). *P. mascittii* is an anthropophilic, zoophilic and autogenious sandfly species (Toumanoff and Chassignet, 1954; Ready and Ready, 1981), and thus fulfills the criteria as a virus vector. Regional monitoring of changes in sandfly distribution of both proven and non-proven *Leishmania* vector species would, therefore, be of interest.

Future Research Needs

There is a need for further research on VL and its vectors species in Europe. Better understanding of the current situation will allow more specific risk evaluations, and form the basis for predictions of changes in distribution and endemicity due to environmental and climatic changes in different parts of Europe. Some main issues to address have been summarized below:

- collate available data on sandfly and parasite distributions, and sandfly seasonality, and relate to climate and environmental variables
- collect new field data on biological parameters of sandflies (including activity patterns, seasonality, survival and infection rates) in relation to defined climatic and environmental variables, including latitude and altitude transects over full seasons. Develop network of fieldworkers across Europe able to use standard methodology for measuring these parameters (especially infection rate)
- collect laboratory data on vectorial competence, including extrinsic incubation period, and diapause determinants of different European vector species
- collect new field data on dog prevalence, and set up a network of vets across Europe to use standardized methodology
- analyse and register changes in human infection rates – with and without HIV co-infection.

6.1.8 Conclusions

While the evidence is not compelling that either sandfly or VL distributions in Europe have altered in recent times as a result of climate change, this hypothesis cannot be discounted. The potential for future climate-driven changes remains considerable. Sandfly vectors already have a wider range than that of *L. infantum*, and imported cases of dogs infected with *L. infantum* are common in central and northern Europe (Gothe et al., 1997). Once conditions make transmission suitable in northern latitudes, these imported cases could act as a plentiful source of infections to permit the development of new endemic foci.

In order to prevent or adapt to future changes in VL endemicity in Europe continuous monitoring and surveillance of sandfly populations, infections in dogs, and human cases are recommended, as well as targeted information to populations at risk, including public health personnel. Local control of sandfly populations and the use of insecticide-impregnated dog collars are recommended.

Although only a small percentage of human *L. infantum* infections in Europe currently cause VL symptoms, the large population in Europe who have sub-clinical infections, as well as HIV, remain a concern as they are at risk from immunosuppression. An additional public health threat is the ability of several of the European sandfly species to be potent vectors of viruses. Climate-induced changes in sandfly abundance may, thus, increase the risk for the emergence of new diseases in the region.

References

Adler S, Theodor O, Lourie EM (1930) On sandflies from Persia and Palestine. Bulletin of Entomological Research 21:529–539

Adler S, Theodor O (1932) Vectors of Mediterranean kala-azar. Nature 130:507

Adler S, Theodor O, Witenberg GG (1938) Investigations on Mediterranean kala azar. XI. A study of leishmaniasis in Canea (Crete). Proceedings of the Royal Society, B, 125:491–516

Adler S (1946) The sandflies of Cyprus (Diptera). Bulletin of Entomological Research 36:497–511

Al Zahrani MA et al (1988) Phlebotomus sergenti, a vector of Leishmania tropica in Saudi Arabia. Transactions of the Royal Society of Tropical Medicine and Hygiene, 82:416

Alptekin D et al (1999) Sandflies (Diptera: Psychodidae) associated with epidemic cutaneous leishmaniasis in Sanliurfa, Turkey. Journal of Medical Entomology 36: 277–281

Alves Pires C et al (1984) Les phlébotomes du Portugal. I. Infestation naturelle de Phlebotomus ariasi Tonnoir, 1921 et Phlebotomus perniciosus Newstead, 1911, par Leishmania dans le foyer zoonotique de Arrabida (Portugal). Annales de Parasitologie Humaine et Comparée 59:521–524

Alves Pires C et al (1991) Phlebotomes du Portugal. IV. Infestation naturelle de Phlebotomus ariasi par Leishmania infantum MON-24 dans le foyer de. [Phlebotomus in Portugal. Natural infestation of Phlebotomus ariasi by Leishmania infantum in the foci of l'Alto Douro]. Annales de Parasitologie Humaine et Comparée 66:47–48

Alves Pires C et al (2001) The phlebotomines of Portugal. X. Natural infestation of Phlebotomus perniciosus by Leishmania infantum MON-1 in Algarve. Parasite 8: 374–375 (in French)

Amela C et al (1996) Injecting drug use as risk factor for visceral leishmaniasis in AIDS patients. European Journal of Epidemiology 12:91–92

Attar ZJ et al (2001) Latex agglutination test for detection of urinary antigens in visceral leishmaniasis. Acta Tropica 78:11–16

Balducci M (1988) Virus trasmessi all'uomo da flebotomi: Ruolo del virus Toscana (Bunyaviridae, Phlebovirus) nell'eziologia di infezione del sistema nervoso centrale [Virus transmitted to men: The role of the Toskana virus]. Parassitologia 30:179–185 (in Italian)

Ben Ismail R et al (1987) Isolation of Leishmania major from Phlebotomus papatasi in Tunisia. Transactions of the Royal Society of Tropical Medicine and Hygiene 81:749

Bettini S et al (1986) Leishmaniasis in Sardinia. 2. Natural infection of Phlebotomus perniciosus Newstead, 1911, by Leishmania infantum Nicolle, 1908, in the Province of Cagliari. Transactions of the Royal Society of Tropical Medicine and Hygiene 80:458–459

Beyreder J (1965) Ein Fall von Leishmaniose in Niederösterreich. [A case of Leishmania in Niederösterreich] Wiener Medizinische Wochenschrift 115:900–901

Bogdan C et al (2001) Visceral leishmaniasis in a German child who had never entered a known endemic area: Case report and review of the literature. Clinical Infectious Diseases 32:302–306

Borcic B, Punda V (1987) Sandfly fever epidemiology in Croatia. Acta Medica Iugoslavica 41:89–97

Cascio A, Colomba C (2003) Childhood Mediterranean visceral leishmaniasis. Le Infezioni in Medicina 11:5–10 (in Italian)

Castagnola E et al (1996) Early efficacy of liposomal amphotericin B in the treatment of visceral leishmaniasis. Transactions of the Royal Society of Tropical Medicine and Hygiene 90:317–318

Chaniotis B, Gozalo Garcia G, Tselentis Y (1994) Leishmaniasis in greater Athens, Greece. Entomological studies. Annals of Tropical Medicine and Parasitology, 88:659–663

Chappuis F et al (2003) Prospective evaluation and comparison of the direct agglutination test and an rK39-antigen-based dipstick test for the diagnosis of suspected kala-azar in Nepal. Tropical Medicine and International Health 8:277–285

Croset H, Abonnenc E, Rioux JA (1970) Phlebotomus (Paraphlebotomus) chabaudi n.sp. (Diptera-Psychodidae). Annales de Parasitologie Humaine et Comparée 45: 863–873

Cross ER, Hyams KC (1996) The potential of global warming on the geographic and seasonal distribution of Phlebotomus papatasi in Southwest Asia. Environmental Health Perspectives 104:724–727

Cruz I et al (2002) Leishmania in discarded syringes from intravenous drug users. Lancet 359:1124–1125

Cunningham AC (2002) Parasitic adaptive mechanisms in infection by Leishmania. Experimental and Molecular Pathology 72:132–141

Davies CR et al (2003) Leishmaniasis: new approaches to disease control. British Medical Journal 326:377–382

De La Rosa R et al (2002) Incidence of and risk factors for symptomatic visceral leishmaniasis among human immunodeficiency virus type 1-infected patients from

Spain in the era of highly active antiretroviral therapy. Journal of Clinical Microbiology 40:762–767

Dedet JP et al (1979) Isolation of Leishmania major from Mastomys erythroleucus and Tatera gambiana in Senegal (West Africa). Annals of Tropical Medicine and Parasitology 73:433–437

Del Giudice P, Marty P (2003) Cat-associated zoonosis: Don't forget rabies and leishmaniasis. Archives of Internal Medicine 163:1238

Depaquit J, Léger N, Ferté H (1998) The taxonomic status of Phlebotomus sergenti Parrot, 1917, vector of Leishmania tropica (Wright, 1903) and Phlebotomus similis Perfiliev, 1963 (Diptera-Psychodidae). Morphologic and morphometric approaches. Biogeographical and epidemiological corollaries. Bulletin de la Société de Pathologie Exotique 91:346–352 (in French)

Deplazes P, Mettler M (2003) Epidemiologische und klinische Aspekte der caninen Leishmaniose in Zentraleuropa [Epidemiology and clinical aspects of dog leishmania in Central Europe]. Tagung der DVG-Fachgruppe „Parasitologie und Parasitäre Krankheiten", Leipzig

Desjeux P (1996) Leishmaniasis: public health aspects and control. Clinics in Dermatology 14:417–423

Desjeux P et al (2000) Leishmania/HIV co-infection in south-western Europe 1990–1998. Retrospective analysis of 965 cases. World Health Organization, Geneva (document WHO/LEISH/2000.42)

Doha S et al (1991) Dispersal of Phlebotomus papatasi (Scopoli) and P. langeroni Nitzulescu in El Hammam, Matrouh Governorate, Egypt. Annales de Parasitologie Humaine et Comparée 66:69–76

Doha S, Shehata MG (1992) Leishmania infantum MON-98 isolated from naturally infected Phlebotomus langeroni (Diptera: Psychodidae) in El Agamy, Egypt. Journal of Medical Entomology 29:891–893

Dornbusch HJ et al (1999) Viszerale Leishmaniose bei einem 10 Monate alten österreichischen Mädchen: XXXIII. Tagung der Österreichischen Gesellschaft für Tropenmedizin und Parasitologie [Visceral Leishmania in a 10 month old girl in Austria. XXXIII. Meeting of the Austrian Society of Tropical Medicine and Parasitology], Innsbruck

Dumonteil E et al (2001) Report of the Fourth TDR/IDRI Meeting on Second-generation Vaccines against Leishmaniasis, 13 May 2001, Universidad Autónoma de Yucatán, Mexico. World Health Organization, Geneva (document TDR/PRD/LEISH/VAC/01.1)

Eitrem R (1991) Sandfly fever: Epidemiological, clinical and virological studies. [Dissertation]. Karolinska Institute, Stockholm

El Kammah KM (1972) Frequency of autogeny in wild-caught egyptian Phlebotomus papatasi (Scopoli) (Diptera: Psychodidae). Journal of Medical Entomology 9:294

George JE (1970) Isolation of Phlebotomus fever virus from Phlebotomus papatasi and determination of the host ranges of sandflies (Diptera: Psychodidae) in West Pakistan. Journal of Medical Entomology 7:670–676

Gligic A et al (1982) First isolation of Naples sandfly fever virus in Yugoslavia. Acta Biologica Jugoslavica (B) 19:167–175

Gligic A et al (1983) Jug Bogdanovac virus – a new member of the vesicular stomatitis serogroup (Rhabdoviridae: Vesiculovirus) isolated from Phlebotomine sandflies in Yugoslavia. Mikrobiologija 20:97–105

Gothe R (1991) Leishmaniasis in dogs in Germany: Aethiology, biology, epidemiology, clinic pathogenesis, diagnosis, therapy and disease prevention. Kleintierpraxis 36:69–84 (in German)

Gothe R, Nolte I, Kraft W (1997) Leishmaniasis of dogs in Germany: epidemiological case analysis and alternative to conventional causal therapy. Tierarztiche Praxis 25:68–73

Gradoni L (2001) An update on antileishmanial vaccine candidates and prospects for a canine Leishmania vaccine. Veterinary Parasitology 100:87–103

Gramiccia M, Gradoni L, Pozio E (1985) Il genere Leishmania in Italia [The Leishmania genre in Italy]. Parassitologia 27:187–201

Grant A et al (1994) Laryngeal leishmaniasis. Journal of Laryngology and Otology 108:1086–1088

Grassi GB (1908) Intorno ad un nuevo flebotomo. Atti della Reale Accademia dei Lincei Rendiconti [Around a new flebotomo. Acts of the Reale Accademia dei Lincei] (5 Ser) 17:681–682

Guan L, Xu Y, Li B, Dong J (1986) The role of Phlebotomus alexandri Sinton, 1928 in the transmission of kala-azar. Bulletin of the World Health Organization 64:107–112

Guilvard E et al (1991) Leishmania tropica au Maroc. III. Rôle vecteur de Phlebotomus sergenti. A propos de 89 isolats [Tropical leishmania in Maroc. The role of Phlebotomus sergenti]. Annales de Parasitologie Humaine et Comparée 66:96–99

Guilvard E et al (1996) Infestation naturelle de Phlebotomus ariasi et Phlebotomus perniciosus (Diptera-Psychodidae) par Leishmania infantum [Natural infestation of Phlebotomus ariasi and Phlebotomus perniciosus (Diptera-Psychodidae) by Leishmania infantum] (Kinetoplastida-Trypanosomatidae) en Catalogne (Espagne). Parasite 3:191–192

Izri MA et al (1992a) Phlebotomus perniciosus Newstead, 1911 naturellement infeste par des promastigotes dans la region de Nice (France) [Phlebotomus perniciosus, Newstead, 1911, naturally infested by promatygotes in the region of Nice, France]. Bulletin de la Société de Pathologie Exotique 85:385–387

Izri MA et al (1992b) Isolement de Leishmania major chez Phlebotomus papatasi a Biskra (Algérie) fin d'une épopée écoépidémiologique [Isolation of Leishmania majeur in Biskra, end of a ecoepidemiological period]. Annales de Parasitologie Humaine et Comparée 67:31–32

Izri MA, Belazzoug S (1993) Phlebotomus (Larroussius) perfiliewi naturally infected with dermatropic Leishmania infantum in Tenes, Algeria. Transactions of the Royal Society of Tropical Medicine and Hygiene 87:399

Izri MA et al (1996) Presumed vectors of leishmaniasis in the Principality of Monaco. Transactions of the Royal Society of Tropical Medicine and Hygiene 90:114

Kafetzis DA, Maltezou HC (2002) Visceral leishmaniasis in paediatrics. Current Opinion in Infectious Diseases 15:289–294

Kaul SM (1991) Phlebotomine sandflies (Diptera: Psychodidae) from Khandwa and Hoshangabad Districts of Madhya Pradesh, India. Journal of Communicable Diseases 23:257–262

Killick-Kendrick R (1990) Phlebotomine vectors of the leishmaniases: A review. Medical and Veterinary Entomology 4:1–24

Killick-Kendrick R (1999) The biology and control of phlebotomine sandflies. Clinics in Dermatology 17:279–289

Killick-Kendrick R et al (1997) Protection of dogs from bites of phlebotomine sandflies by deltamethrin collars for control of canine leishmaniasis. Medical and Veterinary Entomology 11:105–111

Killick-Kendrick R, Rioux JA (2002) Mark-release-recapture of sandflies fed on leishmanial dogs: The natural life-cycle of Leishmania infantum in Phlebotomus ariasi. Parassitologia 44:67–71

Kollaritsch H et al (1989) Suspected autochthonous kala-azar in Austria. Lancet, 901–902

Kubar J et al (1997) Transmission of L. infantum by blood donors. Nature Medicine, 3:368–368

Kuhn KG (1997) Climatic predictors of the abundance of sandfly vectors and the incidence of leishmaniasis in Italy [Dissertation]. London School of Hygiene and Tropical Medicine, London

Kuhn KG (1999) Global warming and leishmaniasis in Italy. Bulletin of Tropical Medicine and International Health 7(2):1–2

Köhler K et al (2002) Cutaneous leishmaniosis in a horse in southern Germany caused by Leishmania infantum. Veterinary Parasitology 109:9–17

Léger N et al (1988) Isolation and typing of Leishmania infantum from Phlebotomus neglectus on the island of Corfu, Greece. Transactions of the Royal Society of Tropical Medicine and Hygiene 82:419–420

Léger N et al (1991) Les phlébotomes impliques dans la transmission des leishmanioses dans l'ile de Gozo (Malte) [The flebotome in the transmission of leishmania on the Gozo island in Malta]. Annales de Parasitologie Humaine et Comparée 66:33–41

Léger N et al (2000a) Phlebotomine sandflies (Diptera-Psychodidae) of the isle of Cyprus. II. Isolation and typing of Leishmania (Leishmania) infantum Nicolle, 1908 (zymodeme MON-1) from Phlebotomus (Larroussius) tobbi Adler and Theodor, 1930. Parasite 7:143–146 (in French)

Léger N, Depaquit J, Ferté H (2000b) Phlebotomine sandflies (Diptera-Psychodidae) of the isle of Cyprus. I. Description of Phlebotomus (Transphlebotomus) economidesi n. sp. Parasite 7:135–141 (in French)

Léger N, Pesson B (1987) Taxonomy and geographic distribution of Phlebotomus (Alderius) chinensis s.l. and P. (Larroussius) major s.l. (Diptera-Psychodidae). Status of species present in Greece. Bulletin de la Société de Pathologie Exotique Filiales 80(2):252–269 (in French)

Le Ray D, Desjeux P, Modabber F (1999) Multicentric evaluation of DAT (direct agglutination test) for diagnosis and treatment of visceral leishmaniasis. Journal of eukaryotic microbiology 46 (Suppl). Abstracts, Groupement des Protistologues de Langue Française, 37th Annual Meeting, 1316 May 1999, Lille, France (http://www.jeukmic.org/abstr/int/f2/f215.html, accessed 7 February 2003)

Lucientes Curdi J et al (1988) Sobre la infeccion natural por Leishmania en Phlebotomus perniciosus Newstead, 1911 y Phlebotomus ariasi Tonnoir, 1921 en el foco de Leishmaniosis de Zaragoza [Over the natural infection of Leishmania in Phlebotomus perniciosus]. Revista Iberica de Parasitologia 48:7–8

Maier WA et al (2003) Possible effects of climatic change on the distribution of arthropode (vector)-borne infectious diseases and human parasites in Germany. Umweltbundesamt, pp 1–386 (in German)

Maroli M, Gramiccia M, Gradoni L (1987) Natural infection of Phlebotomus perfiliewi with Leishmania infantum in a cutaneous leishmaniasis focus of the Abruzzi Region, Italy. Transactions of the Royal Society of Tropical Medicine and Hygiene 81:596–598

Maroli M et al (1988) Natural infections of phlebotomine sandflies with Trypanosomatidae in central and south Italy. Transactions of the Royal Society of Tropical Medicine and Hygiene 82:227–228

Maroli M et al (1994) Natural infection of Phlebotomus perniciosus with MON 72 zymodeme of Leishmania infantum in the Campania region of Italy. Acta Tropica 57:333–335

Maroli M, Bettini S (1997) Past and present prevalence of Phlebotomus papatasi (Diptera: Psychodidae) in Italy. Parasite 4:273–276

Maroli M et al (2001) Evidence for an impact of canine leishmaniasis by the mass use of deltamethrin-impregnated dog collars in southern Italy. Medical and Veterinary Entomology 15:358–363

Maroli M et al (2002) Recent findings of Phlebotomus neglectus Tonnoir, 1921 in Italy and its western limit of distribution. Parassitologia 44:103–109

Martin Sanchez J et al (1993) Infeccion natural de Phlebotomus perniciosus Newstead, 1911 (Diptera, Phlebotomidae) por dos zimodemos distintos del complejo Leishmania infantum en el sudeste de Espana [Natural infection of Phlebotomus perniciosus Newstead, 1911 (Diptera, Phlebotomidae)]. Boletim da Sociedade Portuguesa de Entomologia 3(Suppl):513–520

Martinez Ortega E, Conesa Gallego E, Romera Lozano H (1996) Phlebotomus (Larroussius) langeroni Nitzulescu, 1930 (Diptera, Psychodidae), espèce nouvelle pour l'Espagne [Phlebotomus (Larroussius) langeroni Nitzulescu, 1930 (Diptera, Psychodidae), a new species in Spain]. Parasite 3:77–80

Marty P, Rosenthal E (2002) Treatment of visceral leishmaniasis: a review of current treatment practices. Expert Opinion on Pharmacotherapy 3:1101–1108

Mazloumi Gavgani AS et al (2002) Effect of insecticide-impregnated dog collars on incidence of zoonotic visceral leishmaniasis in Iranian children: a matched-cluster randomised trial. Lancet 360:374–379

McCarthy JJ et al (2001) Climate change 2001. Impacts, adaptation, and vulnerability. Cambridge University Press, Cambridge

Meinecke CK et al (1999) Congenital transmission of visceral leishmaniasis (Kala Azar) from an asymptomatic mother to her child. Pediatrics 104:1–5

Merdan AI et al (1992) Two successive years studies on Phlebotomus papatasi in North Sinai Governorate, Egypt. Journal of the Egyptian Society of Parasitology 22:91–100

Minodier P et al (1998) Pediatric visceral leishmaniasis in southern France. Pediatric Infectious Disease Journal 17:701–704

Minodier P, Garnier JM (2000) Childhood visceral leishmaniasis in Provence. Archives de Pédiatrie 7:572s–577s (in French)

Minter DM (1989) The leishmaniases. In: Geographical distribution of arthropod-borne diseases and their principal vectors. World Health Organization, Geneva (document WHO/VBC/89.967)

Miscevic Z (1980) The origin of blood-meals in some sandflies (Diptera, Phlebotominae) in one of the endemic foci of visceral leishmaniasis in Yugoslavia. Acta Parasitologica Jugoslavica 11:25–37 (in Yugoslavian)

Molina R et al (1999) Infection of sandflies by humans coinfected with Leishmania infantum and human immunodeficiency virus. American Journal of Tropical Medicine and Hygiene 60:51–53

Morillas Marquez F et al (1996) Leishmaniosis in the focus of the Axarquia region, Malaga province, southern Spain: A survey of the human, dog, and vector. Parasitology Research 82:569–570

Moritz A, Prinzinger S, Bauer N (2001) Canine visceral leishmaniasis: Infectious agent, infection, clinical signs, diagnosis, therapy and prophylaxis – a review. Kleintierpraxis 46(9):533–547 (in German)

Naucke TJ (1998) Investigation on vector control of phlebotomine sandflies in northeastern Greece. S Roderer, Regensburg (in German)

Naucke TJ, Pesson B (2000) Presence of Phlebotomus (Transphlebotomus) mascittii Grassi, 1908 (Diptera: Psychodidae) in Germany. Parasitology Research 86:335–336

Naucke TJ (2002) Leishmaniosis, a tropical disease and its vectors (Diptera, Psychodidae, Phlebotominae) in Central Europe. Denisia 6:163–178 (in German)

Naucke TJ, Schmitt C (2004) Is leishmaniasis becoming endemic in Germany? International Journal of Medical Microbiology 293(Suppl 37):179–181

Newstead R (1911) The papataci flies (Phlebotomus) of the Maltese Islands. Annals of Tropical Medicine and Parasitology 5:139–186

Nitzulescu V (1930) Phlebotomus langeroni n. sp. et P. langeroni var. longicuspis n. var. de Douar-Shot (Tunisie). Annales de Parasitologie Humaine et Comparée 8:547–553

Nitzulescu G, Nitzulescu V (1931) Essai de table dichotomique pour la détermination des phlébotomes européens [Table for the determination of European Phlebotoms]. Annales de Parasitologie Humaine et Comparée 9:122–133

Pampiglione S, Placa M, Schlick G (1974) Studies on Mediterranean leishmaniasis. 1. An outbreak of visceral leishmaniasis in Northern Italy. Transactions of the Royal Society of Tropical Medicine and Hygiene 68:349–359

Parrot LM (1917) Sur un nouveau phlebotome algerien Phlebotomus sergenti, sp. nov. [On a new phlebotome in Algerie]. Bulletin de la Société de Pathologie Exotique 10:564–567

Parrot LM (1921) Sur une variété nouvelle de Phlebotomus minutus, Rondani. [On a new variety of Phlebotomus minutus, Rondani] Bulletin de la Société d'Histoire Naturelle de l'Afrique du Nord 12:3740

Parrot LM (1922) Présence de Phlebotomus perniciosus Newstead dans la région parisienne. [Presence of Présence de Phlebotomus perniciosus Newstead in the aris region]. Bulletin de la Société de Pathologie Exotique 15:694

Parrot LM (1930) Notes sur les Phlébotomes [Notes on Phlebotomes]. IV. Phlebotomus perfiliewi [sic] n. sp. Archives de l'Institut Pasteur d'Algérie 8:383–385

Pennisi MG (2002) A high prevalence of feline leishmaniasis in Southern Italy. 2nd International Canine Leishmaniasis Forum, Seville 6–8 February

Perfiljew PP (1963) Paraphlebotomus grimmi Porchinskyi, 1876 and related species. Trudy Voenno-Meditsinskoi Akademii RKKA 149:69–79 (in Russian)

Piekarski G (1962) Medizinische Parasitologie in Tafeln [Tables of Medical Parassitologie]. Bayer AG Leverkusen, Pharmazeutisch-wissenschaftliche Abteilung

Pineda JA et al (2002) Leishmania spp. infection in injecting drug users. Lancet 360: 950–951

Pratlong F et al (1991) Leishmania tropica in Morocco. IV. Enzymatic diversity within a focus. Annales de Parasitologie Humaine et Comparée 66:100–104

Ready PD, Ready PA (1981) Prevalence of Phlebotomus spp. in southern France: Sampling bias due to different man-biting habits and autogeny. Annals of Tropical Medicine and Parasitology 75:475–476

Reithinger R, Teodoro U, Davies CR (2001) Topical insecticide treatments to protect dogs from sand fly vectors of leishmaniasis. Emerging Infectious Diseases 7:872–876

Rioux J, Knoepfler LP, Martini A (1969) Presence in France of Leishmania tarentolae parasite of Tarentola mauritanica. Annales de Parasitologie Humaine et Comparée 44:115–116

Rioux JA et al (1980) Ecology of leishmaniasis in the south of France. 13. Middle slopes of hillsides as sites of maximum risk of transmission of visceral leishmaniasis in the Cevennes. Annales de Parasitologie 55:445–453 (in French)

Rioux JA et al (1982) Confirmation de l'existence en France continentale de Phlebotomus sergenti Parrot, 1917 [Confirmation of the existence in continental France of Phlebotomus sergenti Parrot, 1917]. Annales de Parasitologie Humaine et Comparée 57:647–648

Rioux JA et al (1984) Écologie des leishmanioses dans le sud de la France [Ecology of leishmania in the south of France]. 18. Identification enzymatique de Leishmania infantum Nicolle, 1908, isolé de Phlebotomus ariasi Tonnoir, 1921 spontanément infesté en Cévennes. Annales de Parasitologie Humaine et Comparée 59:331–333

Rioux JA et al (1985) Ecology of leishmaniasis in the south of France. 21. Influence of temperature on the development of Leishmania infantum Nicolle, 1908 in Phlebotomus ariasi Tonnoir, 1921. Experimental study. Annales de Parasitologie Humaine et Comparée 60:221–229 (in French)

Rodhain F et al (1985) Le virus Corfou: Un nouveau phlébovirus isolée de phlébotomes en Grèce [The Corfou virus: A new phlebotome isolated in Greece]. Annales de l'Institut Pasteur (E) Annales de Virologie 136:61–166

Rondani C (1843) Species italicae generis Hebotomi, Rndn., ex insectis dipteris: Fragmentum septimum ad inserviendam dipterologiam italicam. Annales de la Société Entomologique de France 1:263–267 (in Latin)

Sanchez MA et al (2000) The Evaluación del gato común (Felis catus domesticus) como reservorio de la leishmaniosis en la cuenca mediterránea [Evaluation of cats Felis catus domesticus as reservoir of leishmania in the mediterranean]. Revista Técnica Veterinaria, Pequeños Animales 24:46–54

Schlein Y et al (1982) Leishmaniasis in the Jordan Valley. II. Sandflies and transmission in the central epidemic area. Transactions of the Royal Society of Tropical Medicine and Hygiene 76:582–586

Schlein Y et al (1984) Leishmaniasis in Israel: Reservoir hosts, sandfly vectors and leishmanial strains in the Negev, Central Arava and along the Dead Sea. Transactions of the Royal Society of Tropical Medicine and Hygiene 78:480–484

Schmidt JR, Schmidt ML, McWilliams JG (1960) Isolation of Phlebotomus fever virus from Phlebotomus papatasi. American Journal of Tropical Medicine and Hygiene 9:450–454

Schmidt JR, Schmidt ML, Said MZ (1971) Phlebotomus fever in Egypt, isolation of papatasi fever virus from P. papatasi. American Journal of Tropical Medicine and Hygiene 20:483–490

Schmitt C (2002) [Investigations on the biology and distribution of Phlebotomos mascittii, Grassi 1908] Untersuchungen zu Biologie und Verbreitung von Phlebotomus (Transphlebotomus) mascittii, Grassi 1908 (Diptera: Psychodidae) in Deutschland [Dissertation]. Institut für Medizinische Parasitologie, Bonn

Schrey CF, Pires CA, Macvean DW (1989) Distribution of phlebotomine sandflies and the rate of infection with Leishmania promastigotes in the Algarve, Portugal. Medical and Veterinary Entomology 3:125–130

Scopoli GA (1786) Deliciae florae et faunae insubricae seu novae, aut minus cognitae species plantarum et animalum quas in Insubria Austrica tam spontaneas, quam exoticas vidit, descripsit, et aeri incide curavit. Pars I, p 86; Pars II, p 115; Pars III, p 87

Singh KV (1999) Studies on the role of climatological factors in the distribution of Phlebotomine sandflies (Diptera: Psycodidae) in semi-arid areas of Rajasthan, India. Journal of Arid Environments 42:43–48

Sinton JA (1928) The synonymy of the Asiatic species of Phlebotomus. Indian Journal of Medical Research 16:297–324

Sinton JA (1933) Notes on some Indian species of the genus Phlebotomus. Part XXXII. Phlebotomus dentatus n. sp. Indian Journal of Medical Research 20:869–872

Slappendel RJ (1988) Canine leishmaniasis: A review based on 95 cases in the Netherlands. Veterinary Quarterly 10:1–16

Solano Gallego L et al (2003) Cutaneous leishmaniosis in three horses in Spain. Equine Veterinary Journal, London 35:320–323

Stratigos JD et al (1980) Epidemiology of cutaneous leishmaniasis in Greece. International Journal of Dermatology 19:86–88

Strelkova MV (1996) Progress in studies on Central Asian foci of zoonotic cutaneous leishmaniasis. A review. Folia Parasitologica 43:1–6

Sundar S et al (2002) Oral miltefosine for Indian visceral leishmaniasis. New England Journal of Medicine 347:1739–1746

Syriopoulou V et al (2003) Two doses of a lipid formulation of amphotericin B for the treatment of Mediterranean visceral leishmaniasis. Clinical and Infectious Diseases 36:560–566

Tesh RB (1988) The genus Phlebovirus and its vectors. Annual Review of Entomology 33:169–181

Tesh RB, Lubroth J, Guzman H (1992) Simulation of arbovirus overwintering: survival of Toscana virus in its natural sand fly vector Phlebotomus perniciosus. American Journal of Tropical Medicine and Hygiene 47:574

Theodor O (1947) On some sandflies (Phlebotomus) of the sergenti group in Palestine. Bulletin of Entomological Research 38:91–98

Theodor O (1958) Psychodidae-Phlebotominae. Schweizerbart'sche Verlagsbuchhandlung, Stuttgart, pp 1–55

Tonnoir A (1921 a) Une nouvelle espèce européenne du genre Phlebotomus (Phlebotomus ariasi) Annales de la Société Entomologique de Belgique 61:53–56

Tonnoir A (1921 b) Une nouvelle espèce européenne du genre Phlebotomus (Ph. neglectus). Annales de la Société Entomologique de Belgique 61:333–336

Toumanoff C, Chassignet R (1954) Contribution a l'étude des phlébotomes en Corse. Bulletin de l'Institut Nationale d'Hygiene 9:664–687

Ubeda Ontiveros JM et al (1982) Flebotomos de las Islas Canarias (Espana). Revista Iberica de Parasitologia 42:197–206

Vanni V (1939) Osservazioni e ricerche sperimentali in una endemia di leishmaniosi cutanea. Memorie della Classe di Scienze Fisiche, Matematiche e Naturali 10:87–104 (in Italian)

Verani P et al (1981) Ruolo di flebotomi nella transmissione del virus toscana. Parassitologia 23:281–284 (in Italian)

Verani P, Nicoletti L, Ciufolini MG (1984) Antigenic and biological characterization of Toscana virus, a new Phlebotomus fever group virus isolated in Italy. Acta Virologica 28:39–47

Vidyashankar C, Agrawal R (2001) Leishmaniasis. EMedicine Journal 2(12):1–16 (http://www.emedicine.com/ped/topic1292.htm, accessed 7 February 2003)

WHO (1984) The leishmaniases: report of a WHO Expert Committee. World Health Organization, Geneva (WHO Technical Report Series No 701)

WHO (1996) Manual on visceral leishmaniasis control. World Health Organization, Geneva (document WHO/LEISH/96.40)

WHO (1997) Leishmania/HIV co-infection. Epidemiological analysis of 692 retrospective cases. Weekly Epidemiological Record 72:49–54

WHO (1999) Early human health effects of climate change and stratospheric ozone depletion in Europe. WHO Regional Office for Europe, Copenhagen (document EUR/ICP/EHCO 02 02 05/15)

WHO (2000 a) Leishmaniasis and Leishmania/HIV co-infection. In: WHO report on global surveillance of epidemic-prone infectious diseases. Geneva, World Health Organization (document WHO/CDS/CSR/ISR/2000.1)

WHO (2000 b) Leishmaniases and Leishmania/HIV co-infections. World Health Organization, Geneva (WHO Fact Sheet No 116; http://www.who.int/inf-fs/en/fact116.html, accessed 26 January 2003)

WHO (2001) WHO web information: Report on the global HIV/AIDS epidemic 2002; (http://www.who.int/pub/epidemiology/hiv_aids_2001.xls, accessed 23 January 2003)

WHO (2002) Programme for the surveillance and control of leishmaniasis. World Health Organization, Geneva

Yaghoobi Ershadi MR, Javadian E, Tahvildare Bidruni GH (1995) Leishmania major MON-26 isolated from naturally infected Phlebotomus papatasi (Diptera: Psychodidae) in Isfahan Province, Iran. Acta Tropica 59:279–282

Zivkovic V (1974) Répartition de Phlebotomus chinensis balcanicus Theodor, 1958 (Diptera, Psychodidae) en Yougoslavie [Repartition of Phlebotomus chinensis balcanicus Theodor, 1958 (Diptera, Psychodidae) in Yugoslavia]. Acta Parasitologica Jugoslavica 5:3–9

Zivkovic V (1975) Recherches récentes sur les phlébotomes (Diptera, Psychodidae) dans un foyer endémique de leishmaniose viscérale en Serbie (Yougoslavie) [Research on phlebotomes (Diptera, Psychodidae) in an endemic foci of visceral leishmania in Serbia]. Acta Parasitologica Jugoslavica 6:37–43

6.2 Lyme Borreliosis in Europe: Influences of Climate and Climate Change, Epidemiology, Ecology and Adaptation Measures

ELISABET LINDGREN, THOMAS G. T. JAENSON

6.2.1 Introduction

Lyme borreliosis (LB) is the most common vector-borne disease in temperate zones of the Northern Hemisphere. About 85 000 cases are reported annually in Europe (estimated from available national data). However, this number is largely underestimated as case reporting is highly inconsistent in Europe and many LB infections go undiagnosed. In the United States between 15 000 and 20 000 cases are registered each year and the disease is currently endemic in 15 states (Steere, 2001).

LB is transmitted to humans during the blood feeding of hard ticks of the genus *Ixodes*: in Europe mainly *Ixodes ricinus,* and to a lesser extent *I. persulcatus.* The symptoms of LB were described almost a century ago by the Swedish dermatologist Arvid Afzelius, but the disease was not identified until 1977, in the area of Lyme in the United States – hence the name Lyme disease. Following the discovery in 1982 of the spirochete (spiral-shaped bacterium) *B. burgdorferi* s.l. as the causative agent of LB, the disease emerged as the most prevalent arthropod-borne infection in northern temperate climate zones around the world. In Europe the disease is nowadays commonly called Lyme borreliosis. LB is a multisystem disorder that is treatable with antibiotics. Neither subclinical nor symptomatic infections provide immunity. If early disease manifestations are overlooked or misdiagnosed, LB may lead to severe complications of the neurological system, the heart and the joints. Spirochetes are maintained in nature in ticks and in the blood of certain animal species: in Europe particularly insectivores, small rodents, hares and birds. Humans as well as larger animals, like deer and cattle, do not act as reservoirs for the pathogen.

Current knowledge of the impact of different climatic factors on vector abundance and disease transmission is rather extensive. Climate sets the limit for latitudinal and altitudinal distribution of ticks. In addition, daily climatic conditions during several seasons (as ticks may live for more than three years) influence tick population density both directly and indirectly. The pathogen is not in itself sensitive to ambient climatic conditions, except for unusually high temperatures, but human exposures to the pathogen – through tick bites – may be influenced by weather conditions.

During the last decades, ticks have spread into higher latitudes (observed in Sweden) and altitudes (observed in the Czech Republic) in Europe and have become more abundant in many places (Tälleklint and

Jaenson, 1998; Daniel et al., 2003). These tick distribution and density changes have been shown to be related to changes in climate (Lindgren et al., 2000; Daniel et al., 2004). The incidences of LB and other tick-borne diseases have also increased in Europe during the same time period. In some places this may be an effect of better reporting over time. However, studies from localized areas that have reliable long-term surveillance data show that such incidence increases are real, and that they are related to the same climatic factors that have been shown to be linked to changes in tick abundance (Lindgren, 1998; Lindgren and Gustafson, 2001; Daniel et al., 2004).

6.2.2 Geographical Distribution

Distribution

The geographical distribution of LB worldwide correlates with the known distribution of the ixodid vectors (Fig. 1). In Europe, the distribution of *I. ricinus* overlaps with the distribution of *I. persulcatus* in the coastal regions east of the Baltic Sea and further south along that longitude into middle Europe, from where the range of *I. persulcatus* stretches to the Pacific Ocean. Where the two species overlap there are microclimatic conditions separating their distribution. *I. persulcatus* is more flexible and less sensitive to hydrothermal changes in the environment than *I. ricinus* (Korenberg, 1994). Studies of the Baltic regions of Russia showed for example that 11.5% of *I. ricinus* ticks (development stages not stated) were carriers of *B. burgdorferi* s.l. in contrast to 26.3% of *I. persulcatus* (Alekseev et al., 2001). In addition, a large number of other tick species have been reported as

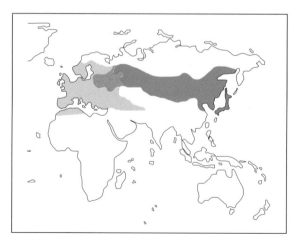

Fig. 1. Distribution of *I. ricinus* and *I. persulcatus* indicates areas where both tick species are present (the map is based on several sources)

carriers of *B. burgdorferi* s.l., but this does not necessarily mean that these ticks are effective in transmitting the disease.

Seasonal climatic conditions limit the latitude and altitude distribution of ticks in Europe (Daniel, 1993; Lindgren et al., 2000; Daniel et al., 2003, 2004). Both altitude and latitude distribution limits of *I. ricinus* have changed during recent years in Europe. Ticks are now found in abundance up to 1200 m above sea level (a.s.l.) in the Czech Republic (Daniel et al., 2003), up to 1300 m a.s.l. in the Italian Alps (Rizzoli et al., 2002), and along the Baltic Sea coastline up to latitude 65°N in Sweden (Jaenson et al., 1994; Tälleklint and Jaenson, 1998). At high northern latitudes, where the inland climate generally is too harsh for ticks to survive, small tick populations can be found in locations where the landscape characteristics help in modifying the climatic conditions. That is, close to large bodies of water, i.e. in river valleys, around inland lakes and along the coastlines (Lindgren et al., 2000).

Incidence

Surveillance in Europe varies and does not allow direct comparison between countries. In some regions the general public is not aware of the risk, and as the symptoms of LB is easily neglected – especially if the characteristic skin rash called erythema migrans does not occur initially – LB may go undetected. In addition, data obtained from various European laboratories are often not directly comparable because of different serological tests used to detect antibodies to *B. burgdorferi* s.l. (Santino et al., 2002). Even if LB is diagnosed, there is often a lack of reporting as few countries have made LB a compulsorily notifiable disease. Despite these caveats, it appears that both disease incidence and antibody prevalence are higher in the central and eastern parts of Europe than in the western parts (Table 1). A gradient of decreasing incidence from south to north in Scandinavia and from north to south in Italy, Spain and Greece has also been noted (e.g. Epinorth, 2003; EUCALB). The highest incidences of LB in Europe are found in the Baltic States and Sweden in the north, and in Austria, the Czech Republic, Germany, Slovenia and central Europe (Tables 1 and 2).

In much of Europe, the number of reported cases of LB has increased from the early 1990s (e.g. the Czech Republic, Estonia, Lithuania; see Fig. 2), and the geographic distribution of cases has also expanded. This is partly due to an increased level of awareness in the general population and among medical personnel, and to better reporting. However, studies from the Czech Republic and Sweden show changes in vector abundance as well as changes in latitudinal or altitudinal distribution of ticks during the same time period (Tälleklint and Jaenson 1998; Daniel et al., 2003). The possible factors underlying these reported changes will be discussed in the sections below.

Reliable epidemiologic data are sparse for most of the endemic LB areas. However, there exist valid serological data from some extensive studies,

Table 1. Incidence and annual number of cases of LB, and seroprevalence of antibodies (in human blood) in different European countries

Country	Incidence per 100 000 population (annual average)	Annual number of cases (average)	Antibody prevalence (in human blood)*
Austria	300	14–24 000	1997: General 7.7%[1]
Belgium	No data	500	No data
Bosnia and Herzegovina	LB is prevalent		
Bulgaria	55	3500	No data
Croatia	>200 (LB absent in Southern parts)[2]		
Czech Republic	27–35 (Fig. 2a)	3500	No data
Denmark	0.8 (Table 2)	<50	No data
Estonia	30–40 (Fig. 2b)	<500	1997: Risk pop. 2.7%[1]
Finland	12.7 (Table 2)	<700	1995: Risk pop. 16.9%[3] 1998: High risk area 19.7%[4]
France	16.5 40 (Berry-Sud)	7–10 000	1997: Risk pop. 15.2%[5]
Germany	25 111 (Würzburg)[6]	15–20 000	1997: General 5.6%[1]
Greece	No data	No data	1997: General 1–3%[1] 2000: Young males 3.3%[7]
Hungary	Neuroborreliosis 2.9 (Baranya)[8]	No data	No data
Iceland	B. garinii present in I. uriae[9]		
Ireland	0.6	<50	1998: General 3.4%[10]
Italy	~17 (Liguria)	<20 (Central Italy)	1997: General: 1.5–10%[1] 1998: Risk pop. 27%[11]
Latvia	15.6 (Table 2)	<400	No data
Lithuania	25–35 (Fig. 2c)	<1300	1994: General 4–32%[12]
Luxemburg	LB is prevalent[13]		
Malta	Uncertainty whether the pathogen is endemic or not		

Table 1 (continued)

Country	Incidence per 100 000 population (annual average)	Annual number of cases (average)	Antibody prevalence (in human blood)*
▌ Netherlands	43	6500	1993: Risk pop. 28%; non-risk pop. 5%[14] 1997: General 9%[15] 2001: Risk pop. 15%[16]
▌ Norway	2.8 (Table 2)	124	No data
▌ Poland	4.8 32.2 (Podlasie Province)[17]	No data	1995: Risk area 49.7%[1] 1999: General 33%; Risk pop. 48%[17] Podlasie Province: Risk pop. 1995: 39%; 2000: 4%[17]
▌ Portugal	Borrelia prevalent in ticks[18]		
▌ Romania	No data	No data	1999: General 4–8%; Risk pop. 9.3–31.7%[19]
▌ Serbia and Montenegro	LB is prevalent	No data	No data
▌ Slovakia	No data	1000	2001: General 5.4%; Risk pop. 16.8%[20]
▌ Slovenia	155	>2000	Children 12.6%[21]
▌ Spain	9.8 (La Rioja)[22]	26 (La Rioja)[22]	No data
▌ Sweden	80 (South)	10000 (South)	1992: General 7%[23] 1995: General 34% (south)[24]
▌ Switzerland	30.4	2000	No data
▌ United Kingdom	Before 1992: 0.06 After 1996: 0.3[25]	>2000	1998: Risk pop. 0.2%[26] 2000: Risk areas 5–17%[27]

* *General* general population; *Risk pop.* risk population, like hunters; *Risk areas* people living in high risk areas, like forests, etc.

Ref.: [1]Santino et al., 1997; [2]Mulic et al., 2000; [3]Oksi and Viljanen, 1995; [4]Carlsson et al., 1998; [5]Zhioua et al., 1997; [6]Huppertz et al., 1999; [7]Stamouli et al., 2000; [8]Pal et al., 1998; [9]Olsen et al., 1993; [10]Robertson et al., 1998; [11]Ciceroni and Ciarrocchi, 1998; [12]Montiejunas et al., 1994; [13]Reiffers-Mettelock et al., 1986; [14]Kuiper et al., 1993; [15]De Mik et al., 1997; [16]Goossens et al., 2001; [17]Pancewicz et al., 2001; [18]De Michelis et al., 2000; [19]Hristea et al., 2001; [20]Štefančiková et al., 2001; [21]Cizman et al., 2000; [22]José A. Oteo, 2003, personal communication; [23]Gustafson et al., 1993; [24]Berglund et al., 1995; [25]Smith et al., 2000; [26]Thomas et al., 1998; [27]Robertson et al., 2000

Table 2. LB Northern Europe and neighbouring areas, 1999–2002. Reported annual number of LB cases and LB incidence per 100 000 inhabitants (Epinorth, 2003)

Area	1999		2000		2001		2002	
	No. of cases	Inci-dence	No. of cases	Inci-dence	No. of cases	Inci-dence	No. of cases	Inci-dence
Archangelsk Region	No data		33	2.3	32	2.3	39	2.7
Denmark	18	0.3	64	1.2	53	1	41	0.8
Estonia	321	22.2	601	43.8	342	25	319	23.3
Finland	404	7.8	895	17.2	691	13.3	884	17
Iceland	No data						0	0
Kaliningrad Region	No data		189	19.9	No data		68	7.2
Latvia	281	11.5	472	19.4	379	16	328	14
St. Petersburg Region			140	8.5	107	6.5	111	6.7
Lithuania	766	20.7	1713	46.3	1153	33	894	25.7
Murmansk Region	No data		1	0.1	3	0.3	2	0.2
Nenets area	No data		0	0	0	0	0	0
Norway	No data		138	3.1	124	2.8	109	2.4
Republic of Karelia	No data		44	5.8	21	2.8	25	3.3
St. Petersburg (city)	No data		541	11.3	323	6.7	398	8.5

such as those of Berglund et al. (1995, 1996), who conducted a prospective, population-based study over one year in parts of southern Sweden. National data from countries where LB is a notifiable disease are important; such as the Czech Republic where national surveillance of LB started in 1990 (Fig. 2 a). Possible causes of inter-annual variations in incidence will be discussed in the sections below.

Population at Risk in Europe

High-risk groups are people living and/or working in LB endemic locations such as forested areas (Smith et al., 1991; WHO, 1995; Santino et al., 1997; Carlsson et al., 1998; Robertson et al., 2000). Occupations such as forest workers, hunters, rangers, gamekeepers, farmers and military field personnel have an especially high risk of contracting LB. This has been shown in several studies on antibody prevalence in human blood and on disease incidence (Cristofolini et al., 1993; Kuiper et al., 1993; Nuti et al., 1993; Oksi

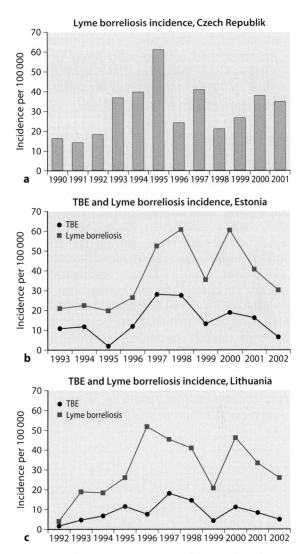

Fig. 2 a–c. Annual incidence of LB during the 1990s in the Czech Republic (Milan Daniel, 2003, personal communication), and TBE in Estonia and Lithuania (Kuulo Kutsar, 2003, personal communication)

and Viljanen, 1995; Santino et al., 1997; Zhioua et al., 1997; Ciceroni and Ciarrocchi, 1998; Stamouli et al., 2000; Goossens et al., 2001; Pancewicz et al., 2001; Štefančiková et al., 2001). Certain recreational habits are linked to increased risk of disease, such as orienteering, hunting, gardening and picnicking (EUCALB; Zeman, 1997; Gray et al., 1998 a). Higher seroprevalence of antibodies have been reported for men, probably due to higher exposure to ticks (Carlsson et al., 1998 a). In some studies LB is more common among children (EUCALB), whereas in others the highest incidences of LB are found in the working age group (Pancewicz et al., 2001).

Awareness of LB is generally lower in city dwellers than in long-term residents of endemic areas. Little knowledge of ticks and inaccurate perception of the disease increase the risk of acquiring infection and of not recognizing the condition (EUCALB).

High-risk Periods

Reports of cases often show distinct seasonality (Fig. 3). Symptoms of disease normally occur within 2–30 days of a tick bite; hence the seasonal pattern of LB cases lags slightly behind the seasonal pattern of tick activity. The highest risk periods from a public health point of view take place when peaks in tick activity occur simultaneously with peaks in human visits to tick-infested areas. During the tick activity season, there are no "safe" times as ticks may be active day and night (Mejlon, 1997).

In Sweden cases of LB are reported throughout the year but with the majority of affected people having been infected during July and August (Åsbrink et al., 1986). A study from southern Sweden showed that tick bites occur most frequently in July, with the highest number of cases with erythema migrans consequently being reported in August, and other LB manifestations peaking in September (Berglund et al., 1995). These seasonality patterns are explained by the fact that more people than usual are visiting *B. burgdorferi* s.l. infested areas during these periods since July is the main summer holiday month in Sweden, and the forest berry and mushroom picking season as well as the hunting season begins in July/August.

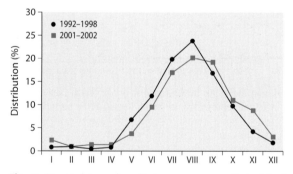

Fig. 3. Reported Lyme borreliosis cases per month in Estonia shown in percentage of annual occurrence (Kuulo Kutsar, 2003, personal communication)

6.2.3 Pathogen Transmission Cycle

The Pathogen

B. burgdorferi s.l., the causative agent of LB, is a gram-negative bacterium, which belongs to the family Spirochaetaceae. To date, *B. burgdorferi* s.l. can be divided into eleven genospecies, of which those with pathogenic significance are *B. afzelii*, *B. burgdorferi* sensu stricto, *B. garinii* and, possibly, *B. valaisiana* (Gylfe et al., 2000; Kurtenbach et al., 2002). The only pathogenic strain found in the United States is *B. burgdorferi* s.s. (Steere, 2001). In contrast, all four pathogenic genospecies of *B. burgdorferi* s.l. are present in Europe, although most of the LB cases are caused by *B. afzelii* and *B. garinii* (Fraenkel et al., 2002; Oehme et al., 2002; Ornstein et al., 2002; Santino et al., 2002). *B. afzelii* predominates in the northern, central and eastern parts of Europe, and *B. garinii* in the western parts.

Several genospecies may be present simultaneously both in infected ticks (Schaarschmidt et al., 2001; Oehme et al., 2002) and in patients diagnosed with LB in Europe (Tazelaar et al., 1997; Schaarschmidt et al., 2001). Different clinical manifestations are often associated with the different genospecies, as are different reservoir hosts (see Table 5) (Humair and Gern, 2000; Kurtenbach et al., 2002). However, this does not seem be the case in Russia and eastern parts of Europe where non-specific maintenance cycles involving small mammals and various borrelia species have been described (Gorelova et al., 1995; Richter et al., 1999).

The Tick

The European *I. ricinus* and *I. persulcatus,* as well as the closely related North American vectors of Lyme disease, *I. scapularis* and *I. pacificus,* belong to the hard tick family Ixodidae. Of about 850 described species of ticks worldwide, this family is the most important from a medical and veterinary point of view (Sonenshine, 1991). In addition to *Borrelia* spirochetes, *I. ricinus* is capable of transmitting a whole range of other pathogens including viruses, bacteria (e.g. rickettsiae), protozoa and nematodes (see Table 6).

I. ricinus is usually found in vegetation types that maintain high humidity. Woodlands are preferred to open land, with a preference of deciduous compared to coniferous woodlands (Adler, 1992; Glass et al., 1995; Tälleklint 1996; Gray et al., 1998a; Memeteau et al., 1998; Zeman and Januska, 1999). On-going studies show that not only are woodland habitats important but habitat configuration, i.e. forest patch structure and connectivity, appears to play a crucial role in determining tick abundance (Agustin Estrada-Peña, 2003, personal communication). On the British Isles ticks are more abundant in open grassy meadows (Gray et al., 1998a; Estrada-Peña, 1999). Ticks may also be found in suburban and urban environments (Spielman, 1994; Dister et al., 1997; Gray et al., 1998a; Junttila et al., 1999).

Preferred habitats often have thick undergrowth and ground litter, which provide cover against cold and drought and create microclimatic conditions with high humidity (Daniel et al., 1977; Mejlon 1997; Gray et al., 1998 a).

Ticks may live for more than three years, depending on climatic conditions (Balashov, 1972; Gray, 1991). Most of the tick's life is spent on the ground restoring water balance, undergoing metamorphosis (egg-larvae-nymph-adult), laying eggs, or hibernating. However, when the tick is in active search of a blood meal it climbs the vegetation. Larvae normally stay closer to the ground as they are more sensitive to ambient humidity than more mature stages (Gigon, 1985; Mejlon and Jaenson, 1997), whereas adults may be found on vegetation even as high as 1.5 metre (Mejlon and Jaenson, 1997). This is one reason why larvae are more often found to parasite smaller animals than nymphs and adults (e.g. Tälleklint and Jaenson, 1994).

Reservoir Hosts

Even though ticks feed on a large range of species, including mammals, birds and reptiles, only a few may act as reservoirs for the pathogen. The abundance of reservoir hosts in a particular habitat is the most important factor in the establishment of infected tick populations. Important competent reservoirs of *B. burgdorferi* s.l. in Europe are rodents, such as *Apodemus* mice and voles; insectivores, such as shrews and hedgehogs; hares; and several bird species, including migratory birds (Aeschlimann et al., 1986; Hovmark et al., 1988; Gern et al., 1991; Matuschka et al., 1992; De Boer et al., 1993; Olsen et al., 1993; Tälleklint and Jaenson, 1993, 1994; Gray et al., 1994; Ciceroni et al., 1996; Jaenson and Tälleklint, 1996; Craine et al., 1997; Gern et al., 1998; Kurtenbach et al., 1998 a, b; Zeman and Januska, 1999; Zore et al., 1999; Gylfe et al., 2000). Small mammals, which often are reservoir-competent hosts, are mainly infested by larval ticks, to a lesser extent by nymphs, but rarely by adult ticks. Medium-sized mammals, such as hares, and large mammals, such as game, cattle and horses, are infested by all tick stages. These latter mammals are reservoir-incompetent but are nevertheless important for pathogen transmission as they provide food for large numbers of adult females, thereby contributing to higher tick abundance (Jaenson and Tälleklint, 1999; Robertson et al., 2000). Studies have shown that in areas where game, like roe deer and cattle are present ticks are more abundant (Gray et al., 1992; Jensen et al., 2000; Jensen and Frandsen, 2000; Robertson et al., 2000), and the number of reported LB cases is higher (Zeman and Januska, 1999).

Dogs, horses and possibly cattle can suffer from manifestations of LB, particularly joint-associated symptoms (Stanek et al., 2002).

Pathogen Circulation in Nature

Once a tick has become infected with spirochetes, it will harbour the pathogens for the rest of its life. Transmission of pathogens often takes place between one and three days after an infected tick has attached itself to a host/human (Kahl et al., 1998). Thus, if a tick is detected and immediately removed after attachment the risk of infection in humans is reduced substantially. Little is known about the duration of borrelia infection in vertebrate hosts. In England and Wales pheasants are important reservoirs and may be infective for ticks for as long as three months (Kurtenbach et al., 1998 a). At present only small rodents – apart from the ticks themselves – are known to remain infected during the winter period (Humair et al., 1999). The transmission of *B. burgdorferi* s.l. in nature is less sensitive to seasonal variations in tick activity patterns than the TBE virus transmission cycle (Randolph, 2001) because of the longer period that the reservoir hosts stay infectious with *B. burgdorferi* s.l.

Borrelia spirochetes may be transferred directly from the female tick to its offspring, but such vertical transmission is rare (e.g. Mejlon and Jaenson, 1993; Gray et al., 1998 a; Humair and Gern, 2000). In general, less than 1% of host-seeking larvae are infected, compared with between 10% and 30% of the nymphs and between 15% and 40% of adults (Aeschlimann et al., 1986; Jaenson, 1991; Mejlon and Jaenson, 1993; Gray et al., 1998 a). The majority of the different tick stages become infected when feeding on blood from an infective reservoir animal. The skin of some host species (including some reservoir-incompetent hosts, like sheep) can constitute an interface for the transmission of spirochetes between infected and non-infected ticks that are co-feeding closely together (Randolph et al., 1996; Ogden et al., 1997). However, six times as many ticks acquire infection when feeding on infected mice than when co-feeding with infected ticks and only 1 out of 100 larvae appear to acquire spirochetal infection when co-feeding with infected nymphs (Richter et al., 2002).

The tick is attached to the host for several days during feeding. This allows the tick to be carried passively into new locations by the host it is attached to. Small mammals such as rodents have rather limited territories. Larger animals such as roe deer normally move around within a range of between 50 and 100 hectares (Cederlund and Liberg, 1995), while birds may deposit ticks far from their original habitat. Birds are often passive carriers of ticks infected with *B. burgdorferi* s.l. (Olsen et al., 1995). Therefore, migrating birds play a role in the introduction of pathogens into new locations along their migratory routes. In addition migratory birds have recently been found to be able to carry LB as a latent infection for several months, and that this infection can be reactivated and passed on to ticks as a result of the stress the birds experience during migration (Gylfe et al., 2000).

The dispersal of borrelia pathogens by birds into the European region may occur along the migrating routes northward/southward in Europe, from Africa in the south and along the west/eastern routes from Eurasia, depending on the bird species. Even though birds account for most of the long-range spread of ticks and pathogens this rarely leads to any major new colonization of ticks or to LB outbreaks in previously nonendemic areas (Jaenson et al., 1994), unless local ecological conditions have been altered, either by land use/land cover changes or climate change (Lindgren, forthcoming). However, transport of ticks and pathogens by migratory birds do have an impact on tick abundance and, in the case of *B. valaisiana* and *B. garinii* (Table 5), on the infectivity in areas that are already LB endemic and/or where suitable ecological conditions are present.

6.2.4 Influence of Environmental and Climatic Factors on Disease Risk

Invertebrate disease vectors, such as ticks, are highly sensitive to climatic conditions, but human infections are only the end product of a complex chain of environmental processes (Epstein, 1999). Some of the most important factors that may influence disease burden are listed in Table 3.

Climatic and environmental factors can affect LB risk in several ways by determining: the spatial tick distribution (at a range of scales from microhabitat to geographic region); daily variability in risk of infective tick bites;

Table 3. Factors influencing changes in the disease burden of LB (modified from Lindgren, forthcoming)

Factors influencing vector and reservoir animal abundance:
▌ Land use and land cover changes (the latter due to human activities or from natural causes) that affect tick habitats and host animal populations.
▌ Global changes, such as climate change, with direct effects on the survival and development of ticks, and indirect effects on tick abundance and pathogen transmission through impacts on the composition of plant and animal species, and on the on-set and length of the seasonal activity periods of the tick
Factors influencing human–tick encounters:
▌ Changes in human settlements and other demographic changes in relation to the proximity to risk areas
▌ Changes in human recreational behaviour, including changes caused by altered climatic conditions
▌ Changes in use of areas for commercial purpose, e.g. forestry, game-keeping, hunting and eco-tourism
▌ Effectiveness of information campaigns on LB disease risk, and the use of different self-protective methods.
Factors affecting society's adaptive capability to changes:
▌ Socioeconomic and technological level
▌ Presence of monitoring and surveillance centres and networks
▌ Capability of the public health sector and local communities to handle acute and long-term changes in disease outbreaks and risk
▌ Type of energy and transportation systems in use and other factors influencing the society's contribution to present and future greenhouse gas emissions

seasonal patterns in risk of infective tick bites; inter-annual variability in risk of infective tick bites; and long-term trends.

Climate and the Life-cycle Dynamics of the Tick

Both the length of each season as well as daily temperatures and humidity are important factors for the survival, development and activity of ticks. Ticks become active when the ambient temperature increases above 4 to 5 °C, below which they are in a chill coma (Balashov, 1972; Duffy and Campbell, 1994; Clark, 1995). Higher temperatures are needed for metamorphosis and egg hatching, i.e. between 8 °C and 10–11 °C respectively (Daniel, 1993).

I. ricinus activity has a prominent annual cycle. Depending on location, ticks start to search for blood meals in early or late spring: in the Czech Republic this usually occurs in March or April, whereas in Latvia the start is later (Fig. 4). Both unimodal and bimodal (with peaks in both spring and autumn) seasonal tick activity peaks have been reported and seasonal occurrence of the different tick stages may vary between years and regions (Nilsson, 1988; Mejlon and Jaenson, 1993; Tälleklint and Jaenson, 1997; Gray et al., 1998 a; Zakovska, 2000; Randolph et al., 2002). In habitats where desiccation is common, such as open areas, periods of activity will be shortened to only a few weeks, as opposed to several months in dense woodlands where moisture is higher (Gray, 1991).

I. persulcatus behaves similarly to I. ricinus, except that autumn activity rarely occurs (Piesman and Gray, 1994; Korenberg 2000).

The earlier the arrival of spring and the more extended the autumn season, the longer the period that allows ticks to be active and undergo metamorphosis. This may lead to a faster life-cycle with a reduction in the duration between tick bites (Balashov, 1972; Dobson and Carper, 1993).

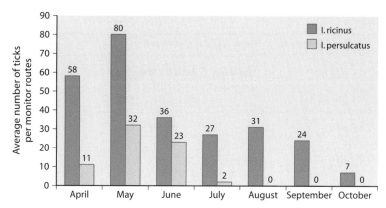

Fig. 4. Average seasonal activity of tick vectors in Latvia 1999–2002 (Kuulo Kutsar, personal communication, 2003)

I. ricinus larvae and nymphs that feed in the early parts of the season moult into the subsequent life stage in 1–3 months, whereas larvae and nymphs feeding in the latter part of the season enter diapause, hibernate, and moult the following year (Tälleklint, 1996).

There is always a risk that the tick will not survive during the winter; the survival rate of *I. ricinus* larvae is approximately 5% and around 20% for nymphs (Sonenshine, 1991). The longer the season of activity, the larger the proportion of the tick population that hibernates in a more advanced developmental stage. Winter survival depends on minimum temperatures, duration of exposure to cold, the tick's developmental stage and feeding status. Even if the tick survives the winter, further ability to undergo metamorphosis the following spring depends on the length and magnitude of exposure to the cold (Lindsay et al., 1995). Nymphs and adults may resist freezing temperatures well below –7 °C, whereas eggs and larvae, especially if fed, are slightly more sensitive to the cold (Table 4) (Balashov, 1972; Daniel et al., 1977; Gray, 1981). Laboratory studies have shown that ticks survive a couple of months at –5 °C (Fujimoto, 1994) and can resist air temperatures as low as –10 °C for up to one month if not in direct contact with ice (Dautel and Knülle, 1997). Ticks overwinter in ground cover vegetation. Deep snow conditions could be favourable for winter survival of the tick since deep snow may increase the ground temperature by several degrees (Berry, 1981). The effect of snow cover on ground temperatures depends on such factors as snow depth and duration, physical characteristics of soil and air temperatures (Berry, 1981).

During the host-seeking periods when ticks climb onto vegetation the tick is particularly vulnerable to low air humidity. Larvae are more sensitive than adults and nymphs to both temperature (Table 4) and desiccation (Daniel, 1993). The need for host-seeking ticks to maintain a stable water balance is an important factor in determining the location and duration of

Table 4. Climate factors linked to tick vector survival and activity

I. ricinus life-stages	Temperature thresholds			Humidity
	Minimum survival	Activity threshold	Optimum	
		Air	Soil		
∎ Larvae	–5 – –7 °C[*1]	No data		15–27 °C[2]	
∎ Nymph	No data	4–5 °C[4]		10–22 °C[2]	80–85%[5,6]
∎ Adult female	–20 °C[*1]	7 °C[3]	4–5 °C[3]	18–25 °C[2]	

* Ticks have been shown to resist very low temperatures; however the length of the cold exposure is important. Ref: [1]Dautel and Knülle, 1997; [2]Daniel and Dusabek, 1994; [3]Sonenshine, 1993; [4]Balashov, 1972; [5]Gray, 1991; [6]Kahl and Knülle, 1988

activity (Randolph and Storey, 1999). The non-parasitic (off-host) phases of *I. ricinus* require a humidity of at least 80–85% at the base of the vegetation (Kahl and Knülle, 1988; Gray, 1991). Vegetation characteristics are, thus, important for the maintenance of tick populations.

B. burgdorferi is not sensitive to ambient climatic conditions except for unusually high temperatures. Optimum temperatures for *B. burgdorferi* s.l. are between 33 °C and 37 °C (Barbour, 1984; Heroldova et al., 1998; Hubalek et al., 1998) and the maximum temperature threshold is 41 °C (Hubalek et al., 1998).

Indirect Effects of Climate

Tick density at a given time in a given place is the combined effect of climatic and environmental conditions that have occurred over several years. Long-term studies covering several decades have shown that tick density, as well as disease risk during a particular year, are linked to the number of days per season with temperatures favourable for tick activity, development and year-round survival during two successive years previous to the one studied (Lindgren, 1998; Lindgren et al., 2000; Lindgren and Gustafson, 2001). Such climatic conditions do not only have direct effects on the tick's survival and life-cycle dynamics but create indirect implications for tick prevalence and disease risk (Fig. 5).

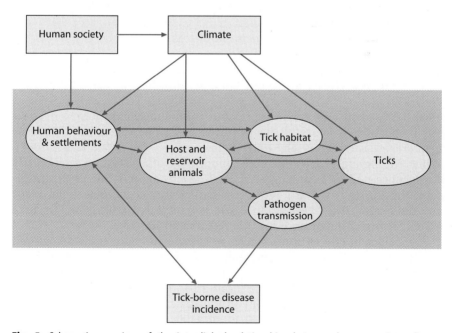

Fig. 5. Schematic overview of the inter-linked relationships between human society, climate change, ecological and demographic changes, and changes in tick-borne disease incidence (Source: Lindgren, 1998)

Weather conditions, such as temperatures and precipitation, affect the microclimate of the tick habitat, which in turn impacts the tick's immediate survival and activity. However, long-term effects of climate variability may affect the length of the vegetation period and cause changes in plant species composition, thus, affecting the spatial distribution and prevalence of both tick and host animal populations. The effect on vegetation affects the ecological dynamics between host animals and ticks through a complex chain of environmental processes, as described for North American conditions by Jones et al. (1998). Snow conditions, in turn, may impact the winter survival of both ticks and host animals. Deep snow conditions are favourable for small hibernating mammals as they create higher ground temperatures, whereas crusty deep snow may be lethal for larger hosts like the roe deer, which feed on sprigs and similar vegetation which are accessible to the animal if the snow cover is thin or easy to remove (Cederlund and Liberg, 1995).

The risk of human LB infection in a specific area depends both on the number of infective ticks in active search for a blood meal and on factors influencing human exposure to ticks. Variations in weather conditions influence human recreational behaviour and hence the risk of exposure to infected tick bites (Jaenson, 1991; Kaiser, 1995). Recreational surroundings such as forests and their marginal grasslands as well as parks are areas often preferred by ticks. Long-term climatic changes may affect vegetation zones and hence influence the commercial use of an area (Table 3) with, in some cases, increased exposures of humans to ticks and, in others, decreased encounters between humans and disease vectors.

6.2.5 Observed Effects of Recent Climate Variations in Europe

Since 1950, night-time temperatures (i.e. minimum temperatures) have risen proportionally more in the Northern Hemisphere than daytime temperatures (Easterling et al., 1997; Beniston and Tol, 1998; IPCC, 2001). Winter temperatures have increased more than other seasons, particularly at higher latitudes (Easterling et al., 1997; Beniston and Tol, 1998). In Europe, the spring now starts two weeks earlier than it did before the 1980s and the length of the vegetation season has increased (IPCC, 2001). These are all factors of importance for tick vectors and LB risk.

Early signs of effects from changes in climate are more easily recognized in areas located close to the geographical distribution limits (altitudinal or latitudinal) of an organism. Mountain studies on *I. ricinus* populations have been performed in the same locations in the Czech Republic in 1957, 1979–1980 and 2001–2002. A shift in the upper altitude boundary of permanent tick population from 700 m to 1100 m a.s.l. has been observed (Daniel et al., 2003). Specifically, tick surveys (on permanently resident dogs and by flagging) were carried out between 2001 and 2002 at altitudes between 700 m and 1200 m a.s.l. in the Sumava mountains. Ticks were

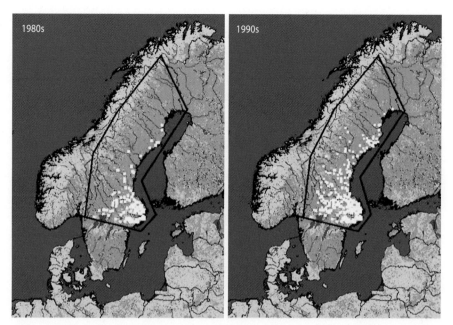

Fig. 6. Differences in tick prevalence in central and northern Sweden (southern parts not included). The left map illustrates conditions before 1980 and the right map illustrates tick distribution after the mid 1990s (from Lindgren et al., 2000)

found on all dogs up to 1100 m a.s.l. and similarly up to this altitude by flagging. These findings contrasted with analogous surveys carried out in the same region in 1957 at altitudes between 780 m and 1200 m a.s.l., when no ticks were detected above 800 m a.s.l., and few between 780 m and 800 m a.s.l. Similarly no ticks were detected in the same region during surveys at 760 m a.s.l. between 1979 and 1980. These tick distribution changes have been shown to be linked to changes in climate (Daniel et al., 2004).

Latitudinal changes in tick distribution between the early 1980s and mid 1990s have been reported from northern Sweden (Jaenson et al., 1994; Tälleklint and Jaenson, 1998) (Fig. 6). These distribution shifts were shown to be correlated to changes in daily seasonal climate (Lindgren et al., 2000). For example, less severe winter temperatures and increased number of days with temperatures vital for tick reproduction (i.e. $> 10\,^\circ\mathrm{C}$) were related to establishment of tick populations at the highest latitudes (Lindgren et al., 2000).

The abundance of *I. ricinus* vectors increased in central parts of Sweden during the early 1980s and mid 1990s (Tälleklint and Jaenson, 1998). Spatial and temporal analyses showed that such density increases were significantly correlated to "accumulated temperature days" that represented milder winters, earlier start of the tick activity period in spring, and prolonged autumn seasons over a consecutive number of years (Lindgren et al., 2000).

Studies on the effects of long-term climate variations on tick-borne disease prevalence are affected by several scientific problems. Data vary between places and over time due to differences in awareness, surveillance, and diagnostic methods, which so far have made long-term studies on LB difficult to perform. However, reliable data on TBE are available from some localized areas, for example from Stockholm County in Sweden where a surveillance programme ran for four decades after an outbreak in the late 1950s. Changes in incidence of TBE in the county during this time period have been shown to be related to changes in daily seasonal climate represented by "accumulated temperature days" (Lindgren, 1998; Lindgren and Gustafson, 2001). These TBE findings are of interest for the understanding of changes in LB prevalence as the incidences of both diseases show similar annual patterns, as shown in Fig. 2b,c. It has been suggested that TBE transmission should be negatively affected by a warmer climate, with the hypotheses being that rapid cooling in autumn will allow more larvae and nymphs to feed together in the following spring, and that this is more important for pathogen transmission for TBE than for LB due to shorter viraemia (Randolph, 2001), though these predictions were based on studies made on a pan-European level. However, new findings about the relationships between climate and the recent spread of TBE to higher altitudes in the mountains of the Czech Republic (Zeman and Benes, 2004) confirm the previous findings described above (Lindgren, 1998; Lindgren and Gustafson, 2001; Daniel et al., 2003, 2004).

6.2.6 Possible Future Effects of Climate Change in Europe

Global temperatures are predicted to continue to increase, and at a rate faster than at any time before in human history with the risk of causing greater instability in regional climates with changes in precipitation and wind patterns. In the latest report from United Nation's Intergovernmental Panel on Climate Change (IPCC, 2001) the following predictions for Europe are listed: in Northern Europe and the Alps the climate is likely to become generally milder and precipitation will increase over the next 50 to 100 years. Winter temperatures will rise proportionally more at high latitudes; the region as a whole will experience higher increases in night compared to day temperatures; the vegetation season is projected to increase in length (for example in Sweden this could be as much as 1–2 months) (Christensen et al., 2001); increased risk for flooding is expected to occur, particularly in the northern and north-western parts of Europe, whereas more summer droughts are expected in the southern parts. These are all conditions that could affect the distribution and population density of ticks and their host animals and alter the vegetation composition of tick habitats over the coming decades.

The possible impact of future climate change on LB risk in Europe can be surmised by investigating the impact of climate on the observed spatial and temporal heterogeneities described above. Soon it should be possible to generate biological "process-based" (i.e. taking into account the multi-year life cycles) models of *I. ricinus* abundance, seasonality, and distribution (Randolph et al., 2002). However, to date, attempts to predict *I. ricinus* or LB distributions have largely been limited to statistical "pattern-matching" models (e.g. Daniel and Kolar, 1991; Daniel and Dusabek, 1994; Estrada-Peña, 1997, 1998, 1999, 2002; Zeman, 1997; Daniel et al., 1999; Zeman and Januska, 1999; Randolph, 2001; Rizzoli et al., 2002). Such statistical models provide some insight into possible impacts of climate change, but conclusions should be drawn with caution (Randolph et al., 2002).

Taking current knowledge from different disciplines, it is possible to make theoretical projections of future changes in disease burden in Europe. Based on the results of the studies reviewed for this extensive report, it seems most probable that a future climate change in Europe will:
▌ facilitate a spread of LB into higher latitudes and higher altitudes
▌ contribute to an extended and more intense LB transmission season in some areas
▌ diminish the risk of LB, at least temporarily, in locations with repeated droughts or severe floods.

6.2.7 Adaptation and Preventive Measures

Diagnosis and Treatment

LB is a multisystem disorder, which can affect a complex range of tissues. The clinical presentations can generally be divided into three stages, but progress from an early to a later stage does not always occur. The early infection consists of localized erythema migrans (stage 1), which occurs in about 60–80% of cases within 2–30 days of a tick bite and consists of a red skin rash or lesion spreading from the site of the bite. If left untreated, a disseminated infection that affects the nervous system, joints and/or the heart (stage 2) may follow within days or weeks. Neuroborreliosis occurs in about 20% of LB cases, arthritis in 10%, while carditis is rare. Among children, *B. burgdorferi* s.l. is now the most common bacterial cause of encephalitis and facial palsy/paralyses. Chronic LB (stage 3) is nowadays uncommon. In the United States, chronic LB causes symptoms mainly in the joints, particularly the knee, whereas in Europe chronic symptoms are more diverse and include a rare degenerative skin condition called acrodermatitis chronica atrophicans, which mainly occurs in elderly women. It has been suggested that different genospecies of *B. burgdorferi* s.l. are associated with different clinical manifestations, as shown in Table 5. However the symptoms often overlap between genospecies (Schaarschmidt et al., 2001; Ornstein et al., 2002).

Laboratory evidence of infection, by demonstration of specific antibodies, is not required for the diagnosis of erythema migrans but for all other

Table 5. Distribution of the different pathogenic European genospecies of *B. burgdorferi* sensu lato, their main reservoir hosts, and their predominant LB symptoms

Genospecies	Predominant clinical manifestation	Main distribution	Predominant reservoir host
B. garinii	Neurological symptoms[1,2]	Western Europe[3]	Birds[4,8,12] (Rodents)[4]
B. afzelii	Acrodermatitis chronica atrophicans[1,2]	Central, eastern[3] and northern Europe and Scandinavia[5,6]	Rodents[7]
B. burgdorferi s.s.	Arthritis[7]	USA[9] Sparsely in Europe and Russia[10]	Rodents[4] Birds
B. valaisiana (suspected to be pathogenic)	Unknown	Mainly Ireland, but also the United Kingdom, Netherlands, Scandinavia, Switzerland, Italy[10,11]	Birds[4,12]

Ref.: [1] Van Dam et al., 1993; [2] Balmelli and Piffaretti, 1995; [3] Ruzic-Sabljic et al., 2002; [4] Kurtenbach et al., 1998b; [5] Jenkins et al., 2001; [6] Fraenkel et al., 2002; [7] Humair et al., 1999; [8] Gylfe et al., 2000; [9] Steere, 2001; [10] Alekseev, 2001; [11] Santino et al., 2002; [12] Humair et al., 1998

Table 6. Pathogens that may be co-transmitted with *B. burgdorferi* s.l. by ixodid ticks (EUCALB)

Organism	Disease
Babesia divergens	Babesiosis
Babesia microti	Babesiosis
Coxiella burnetii	Q fever
Anaplasma (previously named *Ehrlichia*)	Erlichiosis
Francisella tularensis	Tularaemia
Tick-borne encephalitis virus	Tick-borne encephalitis

clinical manifestations of LB; especially as *I. ricinus* may transmit several other zoonotic pathogenic organisms, some of which may interact with *B. burgdorferi* s.l. and affect LB diagnosis and epidemiology (Table 6). Co-infection can result in one of the diseases being overlooked, particularly when symptoms are similar, for example for TBE and early LB neuroborreliosis. Some other diseases transmitted by ticks, such as human erlichiosis and babesiosis, are immunosuppressive and can affect the severity of the infections and lead to treatment difficulties (EUCALB; Krause et al., 1996, 2002). Co-infection with human erlichiosis has, for example, been found in 11.4% of LB cases in a study from southern Germany (Fingerle et al., 1999).

Most infections with *B. burgdorferi* are asymptomatic and self-limiting, so no treatment is required if antibodies are found in individuals without clinical symptoms. However, patients showing symptoms with adequate supporting laboratory evidence for diagnosis should be treated to prevent possible progression of the disease. A range of antibiotics is available (penicillin, cephalosporin, tetracycline etc.) but their selection and use vary between countries. If treatment is initiated in a local or disseminated early stage, healing rates of more than 85% can be achieved (Hofmann, 2002).

Vaccination

There has been uncertainty whether a vaccine is cost-effective, due to the high cost of the vaccine, the low risk of LB in many areas and the largely curable nature of the disease. It has been shown that a vaccine is cost-effective only for people living or working in endemic areas who are frequently exposed to tick bites (Hsia et al., 2002). Since there appears to be much more heterogeneity among the different pathogenic European genospecies of *B. burgdorferi* s.l. (Ciceroni et al., 2001), it will probably be necessary to produce a "cocktail" of surface proteins for an effective vaccine in Europe. However, it is unlikely that adequate protection against all strains will be achieved.

Control Targeted at the Vector

Acaricides used on vegetation have been shown to reduce the density of ticks throughout the activity season if applied in late spring (Schulze and Jordan, 1995). However, most of the chemicals are costly, short-lived and nonspecific and may cause ecological disturbances. They should not be recommended unless severe, epidemic situations are prevalent (Jaenson et al., 1991). Acaricides, like permethrin, may be applied on animals that serve as tick hosts, and have been used successfully on livestock (Gray et al., 1998a). Permethrin should be used with caution, though, as it is known to be highly toxic to aquatic organisms including fish, and to honeybees (Begon et al., 1996).

The role of biological agents in regulating tick populations has so far been poorly investigated. Hill (1998) argues for the possibility of releasing certain nematode species that are pathogenic to engorged female ticks as a measure of tick control. The risk of unexpected negative ecological effects should be carefully considered before such methods are used.

Private gardens and city parks could be landscaped in such a way that it helps reduce tick and host animal abundance. Short-cut lawns hold few ticks for example (Duffy and Campbell, 1994). Choosing plants that are not attractive as forage for larger hosts such as roe deer and hares help reduce the risk of accidental tick introduction into gardens and parks (Jaenson et al., 1991). Repeated removal of undergrowth and leaf litter has been shown to reduce considerably the amount of *I. scapularis* (Schulze and Jordan, 1995).

Controlled burning of the vegetation affects ticks directly by exposure to lethal temperatures and indirectly by removing suitable vegetation for surviving ticks. Tick mortality will depend on the intensity of the fire, the time of the year that the burning takes place and the post-burn vegetation conditions (Schmidtmann, 1994). However, other species will be negatively affected, and extensive burning contributes to emissions of greenhouse gases.

Control Targeted at the Reservoir Host

Changes in the composition of host species in an area could change the risk of infection. However, to what degree depends on the new proportion between reservoir competent and incompetent host animals in relation to the density of the different tick stages (Jaenson and Tälleklint, 1999). Effective disease control would require massive population reduction of both reservoir species and larger host animals, such as roe deer (Tälleklint and Jaenson, 1994), and is therefore not feasible. A less radical method is to use deer fences. They significantly reduce the abundance of immature tick stages in the non-deer area, but it does not result in elimination of ticks. Medium-sized mammals play a role in introducing adult ticks into areas where deer have been excluded, so the density in the non-deer area will be related to the density of ticks on the other side of the deer fence (Daniels and Fish, 1995). The domestic cat can be used in residential areas for the control of mice populations, but this will only have limited effects on disease prevention.

Information and Health Education

The most effective preventive method available for LB is information to the general public (Gray et al., 1998b; O'Connell et al., 1998). Knowledge about LB has been shown to be notably higher in endemic areas compared to non-endemic ones – a difference mainly due to differences in media coverage (Gray et al., 1998b). Swedish media, for example broadcast or print information about risk areas, risk periods and personal preventive measures at the beginning of the tick activity period in spring each year. With increasing tourism within the European region, it is important to target visitors to tick endemic areas as well. Several informative web sites about LB risk and prevention are now available for the general public, for example, the National Public Health Service for Wales (http://www.phls.wales.nhs.uk/lyme.htm), and the Lyme Disease Network, Lymenet, in the United States (http://www.lymenet.org).

There are several effective methods available for personal protection against tick bites in addition to knowledge about risk areas, risk periods and the use of adequate gardening practices:

▌ Proper clothing, i.e. preferably wearing boots, when visiting tick-infested locations is important. Trousers and sweaters should be tightened or tucked in at ankles, wrists and waist (Tälleklint and Jaenson, 1995).

▌ Chemical repellents can be applied on the clothes. N-diethyl-toulamide (DEET), the same agent used in mosquito-repellents, has been shown to be effective in preventing *I. ricinus* from adhering to clothing (Jaenson et al., 2003).

▌ Routine control of body and clothes directly after activities in tick-infested vegetation can reduce tick attachment. Daily self-inspection and prompt tick removal is protective against LB.

▌ The fur of pet animals that are kept in tick-infested areas should be inspected regularly for ticks, particularly for attached female adults. Otherwise, such pets could contribute considerably to an increasing density of ticks in adjacent gardens and surroundings (Thomas GT Jaenson, unpublished observations).

In addition, in-depth education about clinical symptoms, diagnosis, treatment, risk factors, surveillance etc. should be provided to health personnel not only in endemic areas but also in potential new risk regions.

Surveillance and Monitoring

Several initiatives have been taken lately to address LB in Europe and to promote monitoring and the sharing of information. The European Union Concerted Action on Lyme borreliosis (EUCALB) targets professional groups, ranging from scientific researchers to public health workers. Their website provides up-to-date information about scientific publications and other LB activities around Europe. The Network for Communicable Disease Control in Northern Europe (Epinorth) consists of infectious disease control institutes in Denmark, Norway, Iceland, Sweden, Finland, Russia, Estonia, Latvia and Lithuania. The aim of this network is to share information and register data on major infectious diseases, including LB in this region.

There is a need for the establishment and funding of several surveillance centres throughout Europe. These should be set up within a pan-European network to record changes in the reported number of human cases and in tick and pathogen prevalence in nature, including areas that have previously not been investigated. The following recommendations should be considered when data is collected:

▌ Standardized methods for tick collection should be employed. The methods should – if possible – not be changed over time to allow future comparisons. Sampling should be sufficiently frequent in time and space.

▌ Standardized methods for estimating spirochete prevalence in ticks in Europe should be employed.

▌ Other tick-borne pathogens, like the TBE virus, should be included in surveillance.

▌ Samples of ticks from a number of localities in Europe should be preserved at low temperature ($< -80\,^{\circ}$C) for future use (when new and better diagnostic methods are available) to allow future scientists to analyse potential changes of infection rates of ticks with "old" and "newly emerged" pathogens.

Future Research Needs

The impact of climatic factors (ambient temperatures, air and soil humidity etc.) on the life-cycle dynamics of the different vectors species of LB has been thoroughly documented in many laboratory studies. In addition, some parts of Europe have been rather well investigated with regard to tick distribution, habitat vegetation, host composition, infectivity of ticks/hosts/ humans and seasonality patterns. However, more of these latter types of local studies are needed for the whole European region. Also, better transdisciplinary-based mathematical scenario models need to be developed to address alterations in LB risk areas and disease burden in Europe from a future climate change.

6.2.8 Conclusions

It is likely that climate change has already led to changes in *I. ricinus* populations in Europe. Even if existing data are in general not reliable enough to allow comparisons over time and in space of changes in tick prevalence and disease incidence on a pan-European level, some studies from specific areas have been based on particularly reliable long-term data sets (Daniel et al., 2003). These studies have shown that recently observed increases in density and expansion in the distribution of *I. ricinus* into higher altitudes and latitudes are correlated to changes in local climate, just as observed variations in tick-borne disease incidence in places with long-term surveillance data have been shown to be linked to variations in local climatic conditions.

Based on the results of all the different studies that have been reviewed it can be concluded that future climate change will lead to an increase in the overall risk of LB in Europe. Changes in climate will facilitate a spread of LB into higher latitudes and altitudes, and contribute to an extended and more intense LB transmission season in some areas. In other areas, where future climate change will cause climate conditions too hot and dry for tick survival, LB will disappear.

References

Adler GH et al (1992) Vegetation structure influences the burden of immature *Ixodes dammini* on its main host. *Peromyscus leucopus*. Parasitology 105:105–110

Aeschlimann A et al (1986) *B. burgdorferi* in Switzerland. Zentralblatt für Bakteriologie, Mikrobiologie, und Hygiene, Series A 263:450–458

Alekseev AN et al (2001) Identification of *Ehrlichia* spp. and *Borrelia burgdorferi* in Ixodes ticks in the Baltic regions of Russia. Journal of Clinical Microbiology 39(6): 2237–2242

Åsbrink E, Olsson I, Hovmark A (1986) Erythema chronicum migrans Afzelius in Sweden. A study on 231 patients. Zentralblatt für Bakteriologie, Mikrobiologie, und Hygiene, Series A 263:229–236

Balashov YS (1972) Bloodsucking ticks (Ixodoidea) – vectors of diseases of man and animals. Miscellaneous Publications of the Entomological Society of America 8:163–176

Balmelli T, Piffaretti JC (1995) Association between different clinical manifestations of Lyme disease and different species of *Borrelia burgdorferi* sensu lato. Research in Microbiology 16:329–340

Barbour AG (1984) Isolation and cultivation of Lyme disease spirochetes. The Yale Journal of Biology and Medicine, 57(4):521–525

Begon M, Harper J, Townsend CR (1996) Ecology. Blackwell Science Ltd, Oxford

Beniston M, Tol RSJ (1998) Europe. In: The Regional impacts of climate change: An assessment of vulnerability. A Special Report of IPCC Working Group II. Cambridge University Press, New York, pp 149–187

Berglund J et al (1995) An epidemiologic study of Lyme disease in southern Sweden. New England Journal of Medicine 333(20):1319–1327

Berglund J, Eitrem R, Norrby SR (1996) Long-term study of Lyme borreliosis in a highly endemic area in Sweden. Scandinavian Journal of Infectious Diseases, 28(5):473–478

Berry MO (1981) Snow and Climate. In: Gray DM, Male DH (eds) Handbook of Snow. Pergamon Press Toronto, pp 32–59

Carlsson SA et al (1998) IgG seroprevalence of Lyme borreliosis in the population of the Åland Islands in Finland. Scandinavian Journal of Infectious Diseases, 30(5): 501–503

Cederlund G, Liberg O (1995) Rådjuret, Viltet, ekologin och jakten [The roe deer. Wildlife, ecology and hunting]. Almqvist and Wiksell, Uppsala

Christensen JH et al (2001) A synthesis of regional climate change simulations. A Scandinavian perspective. Geophysical Research Letters, 28:1003–1006

Ciceroni L, Ciarrocchi S (1998) Lyme disease in Italy, 1993–1996. The New Microbiologica 21(4):407–418

Ciceroni L et al (1996) Antibodies to *Borrelia burgdorferi* in sheep and goats. Alto Adige-South Tyrol, Italy. The New Microbiologica 19(2):171–174

Ciceroni L et al (2001) Isolation and characterization of *Borrelia burgdorferi* sensu lato strains in an area of Italy where Lyme borreliosis is endemic. Journal of Clinical Microbiology 39(6):2254–2260

Cizman M et al (2000) Seroprevalence of erlichiosis, Lyme borreliosis and tick-borne encephalitis infections in children and young adults in Slovenia. Wiener klinische Wochenschrift 112(19):842–845

Clark DD (1995) Lower temperature limits for activity of several Ixodid ticks (Acari: Ixodidae): effects of body size and rate of temperature change. Journal of Medical Entomology 32(4):449–452

Craine NG et al (1997) Role of grey squirrels and pheasants in the transmission of *Borrelia Burgdorferi* sensu lato, the Lyme disease spirochaete, in the UK. Folia Parasitologica 44:155–160

Cristofolini A, Bassetti D, Schallenberg G (1993) Zoonoses transmitted by ticks in forest workers (tick-borne encephalitis and Lyme borreliosis): preliminary results. La Medicina del Lavoro 84:394–402

Daniel M (1993) Influence of the microclimate on the vertical distribution of the tick *Ixodes ricinus* (L.) in central Europe. Acarologica XXXIV(2):105–113

Daniel M, Dusabek F (1994) Micrometeorlogical and microhabitat factors affecting maintenance and dissemination of tick-borne diseases in the environment. In: Sonenshine DE, Mather TN (eds) Ecological dynamics of tick-borne zoonoses. Oxford University Press, New York, pp 391–1138

Daniel M, Kolár J (1991) Using satellite data to forecast the occurrence of the common tick *Ixodes ricinus*. Modern Acarology 1:191–196

Daniel M et al (1977) Influence of microclimate on the life cycle of the common tick *Ixodes ricinus* (L.) in an open area in comparison with forest habitats. Folia Parasitologica 24:149–160

Daniel M et al (1999) Tick-borne encephalitis and Lyme borreliosis: comparison of habitat risk assessments using satellite data (an experience from the Central Bohemian region of the Czech Republic). Central European Journal of Public Health 7(1):35–39

Daniel M et al (2003) Shift of the tick *Ixodes ricinus* and tick-borne encephalitis to higher altitudes in central Europe. European Journal of Clinical Microbiology and Infectious Disease 22(5):327–328

Daniel M et al (2004) An attempt to elucidate the increased incidence of tick-borne encephalitis and its spread to higher altitudes in the Czech Republic. International Journal of Medical Microbiology 293(37):55–62

Daniels TJ, Fish D (1995) Effect of deer exclusion on the abundance of immature *Ixodes scapularis* (Acari: Ixodidae) parasitizing small and medium-sized mammals. Journal of Medical Entomology 32(1):5–11

Dautel H, Knülle W (1997) Cold hardiness, super cooling ability and causes of low-temperature mortality in the soft tick, *Argas reflexus*, and hard tick, *Ixodes ricinus* (Acari: Ixodoidea) from Central Europe. Journal of Insect Physiology 42(9):843–854

De Boer R et al (1993) The woodmouse (*Apodemus sylvaticus*) as a reservoir of tick-transmitted spirochetes (*Borrelia burgdorferi*) in The Netherlands. Zentralblatt für Bakteriologie: International Journal of Medical Microbiology 279(3):404–416

De Michelis S et al (2000) Genetic diversity of *Borrelia burgdorferi* sensu lato in ticks from mainland Portugal. Journal of Clinical Microbiology 38(6):2128–2133

De Mik EL et al (1997) The geographical distribution of tick bites and erythema migrans in general practice in The Netherlands. International Journal of Epidemiology 26(2):451–457

Dister SW et al (1997) Landscape characterization of peridomestic risk for Lyme disease using satellite imagery. American Journal of Tropical Medicine and Hygiene 57(6):687–692

Dobson A, Carper R (1993) Biodiversity. Lancet 342:1096–1099

Duffy DC, Campbell SR (1994) Ambient air temperature as a predictor of activity of adult *Ixodes scapularis* (Acari: Ixodidae). Journal of Medical Entomology 31:178–180

Easterling DR, Horton B, Jones FD (1997) Maximum and minimum temperature trends for the globe. Science 277:363–367

Epinorth (2003) A co-operation project for communicable disease control in Northern Europe. Epidata on Lyme borreliosis (http://www.epinorth.org, accessed 13 December 2003)

Epstein PR (1999) Climate and health. Science 285(5426):347–348

Estrada-Peña A (1997) Epidemiological surveillance of tick populations: A model to predict the colonization success of *Ixodes ricinus* (Acari: Ixodidae). European Journal of Epidemiology 13(5):573–580

Estrada-Peña A (1998) Geostatistics and remote sensing as predictive tools of tick distribution: a cokriging system to estimate *Ixodes scapularis* (Acari: Ixodidae) habitat suitability in the United States and Canada from advanced very high resolution radiometer satellite imagery. Journal of Medical Entomology 35(6):989–995

Estrada-Peña A (1999) Geostatistics as predictive tools to estimate *Ixodes ricinus* (Acari: Ixodidae) habitat suitability in the western palearctic from AVHRR satellite imager. Experimental and Applied Acarology 23(4):337–349

Estrada-Peña A (2002) Increasing habitat suitability in the United States for the tick that transmits Lyme disease: a remote sensing approach. Environmental Health Perspectives 110(7):635–640

EUCALB, European Union concerted action on Lyme borreliosis (http://www.dis.strath.ac.uk/vie/LymeEU/, accessed 13 December 2003)

Fingerle V et al (1999) Epidemiological aspects of human granulocytic Ehrlichiosis in southern Germany. Wiener Klinische Wochenschrift 111(22/23):1000–1004

Fraenkel CJ, Garpmo U, Berglund J (2002) Determination of novel borrelia genospecies in Swedish *Ixodes ricinus* ticks. Journal of Clinical Microbiology 40(9):3308–3312

Fujimoto K (1994) Comparison of the cold hardiness of *Ixodes nipponensis* and *I. persulcatus* (Acari: Ixodidae) in relation to the distribution patterns of both species in Chichibu Mountains. Japanese Journal of Sanitary Zoology 1:333–339

Gern L et al (1991) *Ixodes (Pholeoixodes) hexagonus*, an efficient vector of *Borrelia burgdorferi* in the laboratory. Medical and Veterinary Entomology 5:431–435

Gern L et al (1998) European reservoir hosts of *Borrelia burgdorferi* sensu lato. Zentralblatt fur Bakteriologie: International Journal of Medical Microbiology 287:196–204

Gigon F (1985) Biologie d'Ixodes ricinus L. sur le Plateau Suisse – une contribution à l'écologie de ce vecteur [Biology of Ixodes ricinus at the Plateau Suisse – a contribution to the vector's ecology] [Dissertation]. Faculty of Sciences, University of Neuchâtel

Glass GE et al (1995) Environmental risk factors for Lyme disease identified with geographic information systems. American Journal of Public Health 85(7):944–948

Goossens HA, van der Bogaard AE, Nohlmans MK (2001) Dogs as sentinels for human *Lyme borreliosis* in The Netherlands. Journal of Clinical Microbiology 39(3):844–848

Gorelova NB et al (1995) Small mammals as reservoir hosts for borrelia in Russia. Zentralblatt für Bakteriologie. International Journal of Medical Microbiology 282:315–322

Gray JS (1981) The fecundity of *Ixodes ricinus* (L.) (Acarina: Ixodidae) and the mortality of its development stages under field conditions. Bulletin of Entomological Research 71:533–542

Gray JS (1991) The development and seasonal activity of the tick, *Ixodes ricinus*: a vector of *Lyme borreliosis*. Review of Medical and Veterinary Entomology 79:323–333

Gray JS et al (1992) Studies on the ecology of Lyme disease in a deer forest in County Galway, Ireland. Journal of Medical Entomology 29(6):915–920

Gray JS et al (1994) Acquisition of *Borrelia burgdorferi* by *Ixodus ricinus* ticks fed on the European hedgehog, *Erinaceus europaeus* L. Experimental and Applied Acarology 18(8):485–491

Gray JS et al (1998a) Lyme borreliosis habitat assessment. Zentralblatt für Bakteriologie: International Journal of Medical Microbiology 287(3):211–228

Gray JS et al (1998b) Lyme borreliosis awareness. Zentralblatt für Bakteriologie: International Journal of Medical Microbiology 287(3):253–265

Gubler DJ et al (2001) Climate variability and change in the United States: potential impacts on vector- and rodent-borne diseases. Environmental Health Perspectives 109(Suppl 2):223–233

Gustafson R et al (1993) Clinical manifestations and antibody prevalence of Lyme borreliosis and tick-borne encephalitis in Sweden: a study in five endemic areas close to Stockholm. Scandinavian Journal of Infectious Diseases 25(5):595–603

Gylfe A et al (2000) Reactivation of borrelia infection in birds. Nature 403(6771):724–725

Heroldova M, Nemec M, Hubalek Z (1998) Growth parameters of Borrelia burgdorferi sensu stricto at various temperatures. Zentralblatt für Bakteriologie: International Journal of Medical Microbiology 288(4):451–455

Hill DE (1998) Entomopathogenic nematodes as control agents of developmental stages of the black-legged tick, Ixodes scapularis. Journal of Parasitology 84(6): 1124–1127

Hofmann H (2002) Early diagnosis of Lyme borreliosis. Do not look only for erythema migrans. MMW Fortschritte der Medizin 144(22):24–28

Hovmark A et al (1988) First isolation of Borrelia burgdorferi from rodents collected in northern Europe. Acta Pathologica et Microbiologica Scandinavica 96:917–920

Hristea A et al (2001) Seroprevalence of Borrelia burgdorferi in Romania. European Journal of Epidemiology 17(9):891–896

Hsia EC et al (2002) Cost-effectiveness analysis of the Lyme disease vaccine. Arthritis and Rheumatism 46(6):1651–1660

Hubalek Z, Halozka J, Heroldova M (1998) Growth temperature ranges of Borrelia burgdorferi sensu lato strains. Journal of Medical Microbiology 47(10):929–932

Humair PF, Gern L (2000) The wild hidden face of Lyme borreliosis in Europe. Microbes and Infection 2:915–922

Humair PF et al (1998) An avian reservoir (Turdus merula) of the Lyme disease spirochetes. Zentralblatt für Bakteriologie: International Journal of Medical Microbiology 287:521–538

Humair PF, Rais O, Gern L (1999) Transmission of Borrelia afzelii from Apodemus mice and Clethrionomys voles to Ixodes ricinus ticks: differential transmission pattern and overwintering maintenance. Parasitology 118:33–42

Huppertz HI et al (1999) Incidence of Lyme borreliosis in the Würzburg region of Germany. European Journal of Clinical Microbiology and Infectious Diseases 18(10):697–703

IPCC (2001) Climate Change 2001: The third assessment report of the Intergovernmental Panel on Climate Change. Cambridge University Press, Cambridge

Jaenson TGT (1991) The epidemiology of Lyme borreliosis. Parasitology Today 7:39–45

Jaenson TGT, Tälleklint L (1996) Lyme borreliosis spirochetes in Ixodes ricinus and the varying hare on isolated islands in the Baltic sea. Journal of Medical Entomology 33(3):339–343

Jaenson TGT, Tälleklint L (1999) The reservoir hosts of Borrelia burgdorferi sensu lato in Europe. In: Needham G, Mitchell R, Horn DJ, Welbourn WC (eds) Acarology IX. Vol 2, Symposia. Ohio Biological Survey, Columbus, pp 409–414

Jaenson TGT, Lindström A, Pålsson K (2003) Repellency of the mosquito repellent MyggA (N,N-diethyl-3-methyl-benzamide) to the common tick Ixodes ricinus (L.) (Acari: Ixodidae) in the laboratory and field. Entomologisk Tidskrift 124:245–251

Jaenson TGT et al (1991) Methods for control of tick vectors of Lyme borreliosis. Scandinavian Journal of Infectious Diseases 77(Suppl):151–157

Jaenson TGT et al (1994) Geographical distribution, host associations, and vector roles of ticks (Acari: Ixodidae, Argasidae) in Sweden. Journal of Medical Entomology 31(2):240–256

Jenkins A et al (2001) Borrelia burgdorferi sensu lato and *Ehrlichia* spp. in Ixodes ticks from southern Norway. Journal of Clinical Microbiology 39(10):3666–3671

Jensen PM, Frandsen F (2000) Temporal risk assessment for Lyme borreliosis in Denmark. Scandinavian Journal of Infectious Diseases 32(5):539–544

Jensen PM, Hansen H, Frandsen F (2000) Spatial risk assessment for Lyme borreliosis in Denmark. Scandinavian Journal of Infectious Diseases 32:545–550

Jones CG et al (1998) Chain reactions linking acorns to gypsy moth outbreaks and Lyme disease risk. Science 279(5353):1023–1026

Junttila J et al (1999) Prevalence of *Borrelia burgdorferi* in *Ixodes ricinus* ticks in urban recreational areas of Helsinki. Journal of Clinical Microbiology 37(5):1361–1365

Kahl O et al (1998) Risk of infection with *Borrelia burgdorferi* sensu lato for a host in relation to the duration of nymphal *Ixodes ricinus* feeding and the method of tick removal. Zentralblatt für Bakteriologie: International Journal of Medical Microbiology 287:41–52

Kahl O, Knülle W (1988) Water vapour uptake from subsaturated atmospheres by engorged immature ixodid ticks. Experimental and Applied Acarology 4(1):73–83

Kaiser R (1995) Tick-borne encephalitis in southern Germany. Lancet 345:463

Korenberg EI (1994) Problems of epizootiology, epidemiology and evolution associated with modern borrelia taxonomy. In: Yanagihara Y, Masuzawa T (eds) Present status of Lyme disease and biology of Lyme borrelia. Shizuoka, Japan, pp 18–47

Korenberg EI (2000) Seasonal population dynamics of *Ixodes* ticks and tick-borne encephalitis virus. Experimental and Applied Acarology 24:665–681

Krause PJ et al (1996) Concurrent Lyme disease and babesiosis. Evidence for increased severity and duration of illness. JAMA: the Journal of the American Medical Association 275(21):1657–1660

Krause PJ et al (2002) Deer-Associated Infection Study Group. Disease-specific diagnosis of coinfecting tickborne zoonoses: babesiosis, human granulocytic ehrlichiosis, and Lyme disease. Clinical Infectious Diseases 34(9):1184–1191

Kuiper H et al (1993) One year follow-up study to assess the prevalence and incidence of Lyme borreliosis among Dutch forestry workers. European Journal of Clinical Microbiology and Infectious Diseases 12(6):413–418

Kurtenbach K et al (1998a) Competence of pheasants as reservoirs for Lyme disease spirochetes. Journal of Medical Entomology 35(1):77–81

Kurtenbach K et al (1998b) Differential transmission of the genospecies of *Borrelia burgdorferi* sensu lato by game birds and small rodents in England. Applied and Environmental Microbiology 64(4):1169–1174

Kurtenbach K et al (2002) Host association of *Borrelia burgdorferi* sensu lato – the key role of host complement. Trends in Microbiology 10(2):74–79

Lindgren E (1998) Climate and tick-borne encephalitis. Conservation Ecology 2(1)5:1–14 (http://www.consecol.org/Journal/vol2/iss1/art5/, accessed 13 December 2003)

Lindgren E. Managing the human life-supporting environment for disease transmission. Forthcoming (in preparation)

Lindgren E, Gustafson R (2001) Tick-borne encephalitis in Sweden and climate change. Lancet 358:16–18

Lindgren E, Tälleklint L, Polfeldt T (2000) Impact of climatic change on the northern latitude limit and population density of the disease-transmitting European tick, *Ixodes ricinus*. Environmental Health Perspectives 108(2):119–123

Lindsay IR et al (1995) Survival and development of *Ixodes scapularis* (Acari: Ixodidae) under various climatic conditions in Ontario, Canada. Journal of Medical Entomology 32:143–152

Matuschka FR et al (1992) Capacity of European animals as reservoir hosts for the Lyme disease spirochete. The Journal of Infectious Diseases 165:479–483

Mejlon HA (1997) Diel activity of *Ixodes ricinus* Acari: Ixodidae at two locations near Stockholm, Sweden. Experimental and Applied Acarology 21:247–255

Mejlon HA, Jaenson TGT (1993) Seasonal prevalence of *Borrelia burgdorferi* in *Ixodes ricinus* (Acari: Ixodidae) in different vegetation types in Sweden. Scandinavian Journal of Infectious Diseases 25:449–456

Mejlon HA, Jaenson TGT (1997) Questing behaviour of *Ixodes ricinus* ticks (Acari: Ixodidae). Experimental and Applied Acarology 21:747–754

Memeteau S et al (1998) Assessment of the risk of infestation of pastures by *Ixodes ricinus* due to their phyto-ecological characteristics. Veterinary Research 29(5):487–496

Montiejunas L et al (1994) Lyme borreliosis in Lithuania. Scandinavian Journal of Infectious Diseases 26(2):149–155

Mulic R et al (2000) Lajmska borelioza u Hrvatskoj od 1987 do 1998 – epidemioloski aspekt [Lyme borreliosis in Croatia from 1987 to 1998 – epidemiological aspects]. Lijecnicki Vjesnik 122(9–10):214–217

Nilsson A (1988) Seasonal occurrence of *Ixodes ricinus* (Acari) in vegetation and on small mammals in southern Sweden. Holarctic Ecology 11:161–165

Nuti M et al (1993) Infection in an Alpine environment: antibodies to hantaviruses, leptospira, rickettsiae and *Borrelia burgdorferi* in defined Italian populations. The American Journal of Tropical Medicine and Hygiene 48(1):20–25

O'Connell S et al (1998) Epidemiology of European Lyme borreliosis. Zentralblatt für Bakteriologie: International Journal of Medical Microbiology 287:229–240

Oehme R et al (2002) Foci of tick-borne diseases in southwest Germany. International Journal of Medical Microbiology 291(33):22–29

Ogden NH, Nuttall PA, Randolph SE (1997) Natural Lyme disease cycles maintained via sheep by cofeeding ticks. Parasitology 115:591–599

Oksi J, Viljanen MK (1995) Tick bites, clinical symptoms of Lyme borreliosis, and borrelia antibody responses in Finnish army recruits training in an endemic region during summer. Military Medicine 160(9):453–456

Olsen B, Jaenson TGT, Bergström S (1995) Prevalence of *Borrelia burgdorferi* sensu lato-infected ticks on migrating birds. Applied and Environmental Microbiology 61:3082–3087

Olsen B et al (1993) A Lyme borreliosis cycle in seabirds and *Ixodes urinae* ticks. Nature 362:340–342

Ornstein K et al (2002) Three major Lyme Borrelia genospecies (*Borrelia burgdorferi* sensu stricto, *B. afzelii* and *B. garinii*) identified by PCR in cerebrospinal fluid from patients with neuroborreliosis in Sweden. Scandinavian Journal of Infectious Diseases 34(5):341–346

Pal E et al (1998) Neuroborreliosis in county Baranya, Hungary. Functional Neurology 13(1):104

Pancewicz SA et al (2001) Wybrane aspekty epidemiologiczne boreliozy z Lyme wsrod mieszkancow wojewodztwa podlaskiego [Epidemiologic aspect of lyme borreliosis among the inhabitants of Podlasie Province]. Przeglad epidemiologiczny 55(3):187–194

Piesman J, Gray JS (1994) Lyme disease/Lyme borreliosis. In: Sonenshine DE, Mather TN (eds) Ecological dynamics of tick-borne zoonoses. Oxford University Press, New York, pp 327–350

Randolph SE (2001) The shifting landscape of tick-borne zoonoses: tick-borne encephalitis and Lyme borreliosis in Europe. Philosophical Transactions of the Royal Society of London, Series B. Biological Sciences 356(1411):1045–1056

Randolph SE, Storey K (1999) Impact of microclimate on immature tick-rodent host interactions (Acari: Ixodidae): implications for parasite transmission. Journal of Medical Entomology 36(6):741–748

Randolph SE, Gern L, Nuttall PA (1996) Co-feeding ticks: epidemiological significance for tick-borne pathogen transmission. Parasitology Today 12:472–479

Randolph SE et al (2002) An empirical quantitative framework for the seasonal population dynamics of the tick *Ixodes ricinus*. International Journal for Parasitology 32(8):979–989

Reiffers-Mettelock J, Glaesener G, Schroell M (1986) Burgdorfer's borreliosis or Lyme disease. The first Luxembourg cases. Bulletin de la Société des sciences medicales du Grand-Duche de Luxembourg 123(2):103–110

Richter D, Allgower R, Matuschka FR (2002) Co-feeding transmission and its contribution to the perpetuation of the Lyme disease spirochete *Borrelia afzelii*. Emerging Infectious Diseases 8(12):1421–1425

Richter D et al (1999) Genospecies diversity of Lyme disease spirochetes in rodent reservoirs. Emerging Infectious Diseases 5:291–296

Rizzoli A et al (2002) Geographical information systems and bootstrap aggregation (bagging) of tree-based classifiers for Lyme disease risk prediction in Trentino, Italian Alps. Journal of Medical Entomology 39(3):485–492

Robertson JN et al (1998) Seroprevalence of *Borrelia burgdorferi* sensu lato infection in blood donors and park rangers in relation to local habitat. Zentralblatt für Bakteriologie, International Journal of Medical Microbiology 288(2):293–301

Robertson JN, Gray JS, Stewart P (2000) Tick bite and Lyme borreliosis risk at a recreational site in England. European Journal of Epidemiology 16(7):647–652

Ruzic-Sabljic E et al (2002) Characterization of *Borrelia burgdorferi* sensu lato strains isolated from human material in Slovenia. Wiener klinische Wochenschrift 114(13/14):544–550

Santino I et al (1997) Geographical incidence of infection with *Borrelia burgdorferi* in Europe. Panminerva Medica 39(3):208–214

Santino I et al (2002) Detection of four *Borrelia burgdorferi* genospecies and first report of human granulocytic ehrlichiosis agent in *Ixodes ricinus* ticks collected in central Italy. Epidemiology and Infection 129(1):93–97

Schaarschmidt D et al (2001) Detection and molecular typing of *Borrelia burgdorferi* sensu lato in *Ixodes ricinus* ticks and in different patient samples from southwest Germany. European Journal of Epidemiology 17(12):1067–1074

Schmidtmann ET (1994) Ecologically based strategies for controlling ticks. In: Sonenshine DE, Mather T (eds) Ecological dynamics of tick-borne zoonoses. Oxford University Press, New York

Schulze TL, Jordan RA (1995) Potential influence of leaf litter depth on effectiveness of granular carbaryl against subadult *Ixodes scapularis* (Acari: Ixodidae). Journal of Medical Entomology 32(2):205–208

Smith HV, Gray JS, Mckenzie G (1991) A Lyme borreliosis human serosurvey of asymptomatic adults in Ireland. Zentralblatt für Bakteriologie: International Journal of Medical Microbiology 275(3):382–389

Smith R, O'Connell S, Palmer S (2000) Lyme disease surveillance in England and Wales, 1986–1998. Emerging Infectious Diseases 6(4):404–407

Sonenshine DE (1991) Biology of ticks. Oxford University Press, Oxford

Sonenshine DE (1993) Biology of ticks, Vol 2. Oxford University Press, Oxford

Spielman A (1994) The emergence of Lyme disease and human babeosis in a changing environment. Annals of the New York Academy of Sciences 740:146–155

Stamouli M et al (2000) Very low seroprevalence in young Greek males. European Journal of Epidemiology 16(5):495–496

Stanek S et al (2002) History and characteristics of Lyme borreliosis. In: Gray JS et al (eds) Lyme borreliosis: Biology, Epidemiology and Control. CAB International, Wallingford, Oxon, UK, pp 1–28

Steere AC (2001) Lyme disease. The New England Journal of Medicine 345:115–125

Štefančiková A et al (2001) Epidemiological survey of human borreliosis diagnosed in Eastern Slovakia. Annals of Agricultural and Environmental Medicine 8:171–175

Tälleklint L, Jaenson TGT (1993) Maintenance by hares of European *Borrelia burgdorferi* in ecosystems without rodents. Journal of Medical Entomology 30:273–276

Tälleklint L, Jaenson TGT (1994) Transmission of *Borrelia burgdorferi* s.l. from mammal reservoirs to the primary vector of Lyme borreliosis, *Ixodes ricinus* (Acari: Ixodidae), in Sweden. Journal of Medical Entomology 31(6):880–886

Tälleklint L, Jaenson TGT (1995) Control of Lyme borreliosis in Sweden by reduction of tick vectors: An impossible task? International Journal of Angiology 4:34–37

Tälleklint L (1996) Transmission of Lyme borreliosis spirochetes at the tick vector – mammal reservoir interface [Dissertation]. Department of Zoology, Uppsala University, Sweden

Tälleklint L, Jaenson TGT (1997) Infestation of mammals by *Ixodes ricinus* ticks (Acari: Ixodidae) in south-central Sweden. Experimental and Applied Acarology, 21(12):755–771

Tälleklint L, Jaenson TGT (1998) Increasing geographical distribution and density of *Ixodes ricinus* (Acari: Ixodidae) in central and northern Sweden. Journal of Medical Entomology 35(4):521–526

Tazelaar DJ et al (1997) Detection of *Borrelia afzelii*, *Borrelia burgdorferi* sensu stricto, *Borrelia garinii* and group VS116 by PCR in skin biopsies of patients with erythema migrans and acrodermatitis chronica atrophicans. Clinical Microbiology and Infection 3(1):109–116

Thomas DR et al (1998) Low rates of ehrlichiosis and Lyme borreliosis in English farm workers. Epidemiology and Infection 121(3):609–614

Van Dam AP et al (1993) Different genospecies of *Borrelia burgdorferi* are associated with distinct clinical manifestations of Lyme Borreliosis. Clinical Infectious Diseases 17:708–717

WHO (1995) WHO Workshop on Lyme Borreliosis Diagnosis and Surveillance, Warsaw, Poland, 20–22 June 1995. World Health Organization, Geneva (document WHO/CDS/VPH/95.141)

Zakovska A (2000) Monitoring the presence of borreliae in *Ixodes ricinus* ticks in Brno Park Pisarky, Czech Republic. Biologia 55(6):661–666

Zeman P (1997) Objective assessment of risk maps of tick-borne encephalitis and Lyme borreliosis based on spatial patterns of located cases. International Journal of Epidemiology 26(5):1121–1129

Zeman P, Benes C (2004) A tick-borne encephalitis ceiling in Central Europe has moved upwards during the last 30 years: possible impact of global warming? International Journal of Medical Microbiology 293(37):48–54

Zeman P, Januska J (1999) Epizootic background of dissimilar distribution of human cases of Lyme borreliosis and tick-borne encephalitis in a joint endemic area. Comparative Immunology, Microbiology and Infectious Disease 22:247–260

Zhioua E et al (1997) Prevalence of antibodies to *Borrelia burgdorferi* in forestry workers of Ile de France, France. European Journal of Epidemiology 13(8):959–962

Zore A et al (1999) Infection of small mammals with *Borrelia burgdorferi* sensu lato in Slovenia as determined by polymerase chain reaction (PCR). Wiener Klinische Wochenschrift 111(22/23):997–999

6.3 Tick-borne Encephalitis

MILAN DANIEL, VLASTA DANIELOVÁ, BOHUMÍR KŘÍŽ,
ČESTMÍR BENEŠ

6.3.1 Introduction

By 1972, some 68 different *Flavivirus* viruses had been recorded from more than 80 tick species, some 20 of which were believed to cause disease in humans or domestic animals (Hoogstraal, 1973; Calisher, 1988; Calisher et al., 1989). Since the publication of Hoogstraal's review, many other viruses have been isolated from ticks, although their role as causative agents of human or animal disease is often unknown or uncertain. Many areas of Europe remain poorly surveyed, and more viruses will certainly be found in further studies.

Tick-borne encephalitis (TBE) is the most important and widespread of the arboviruses transmitted by ticks in Europe. It is a member of the family Flaviviridae. This virus was first isolated in Czechoslovakia in 1948 (Gallia et al., 1949; Krejčí, 1949a, b; Rampas and Gallia, 1949) and subsequently in other central European states. Tick-borne encephalitis should be considered a general term encompassing at least three diseases caused by similar flaviviruses, whose range spans an area from the British Isles (Louping ill), across Europe (central European tick-borne encephalitis), and to the Far East of Russia (Russian spring-summer encephalitis).

These three diseases differ in degree, with Louping ill the mildest and Russian spring-summer encephalitis the most severe. Humans are infected by the bite of infected ticks and, much more rarely, by the ingestion of unpasteurized milk from infected domestic animals (Dumpis et al., 1999). TBE is often the cause of a serious acute central nervous system (CNS) disease, which may result in death or long-term neurological sequellae for a considerable period after recovery from the initial infection. The disease may take the forms of meningitis, meningoencephalitis, meningoencephalomyelitis or meningoradiculoneuritis. About 40% of infected patients are left with a residual post-encephalitic syndrome. The course of the disease is more severe in the elderly than in young people. The mortality of the central European form of TBE is 0.7–2% (Ozdemir et al., 1999); this may be even higher in severe cases of infections. In regard to the Far Eastern form of the disease, the mortality rate may be as high as 25 to 30% (Gratz, 2004).

Within the cCASHh study, several assessments were carried out in the Czech Republic. This chapter summarizes some of the results (Daniel, Danielová et al., 2003; Materná, Danielova et al., 2003; Daniel, Danielová et al., 2004; Kříž, Beneš et al., 2004; Materna, Daniel et al., 2005; Danielová, Kříž et al., 2004).

6.3.2 Spatial and Temporal Distribution

TBE Epidemiology

Overall, 3000–4000 TBE cases are reported annually from the European countries including the Baltic States. An additional 6000–8000 cases are reported each year from the Russian Federation. The quality of reporting varies from country to country, depending on the availability of laboratory diagnostic facilities and the surveillance system in place. Nevertheless, there is an increasing trend of TBE incidences in some European countries, as described in Gratz (2004, pages 45–47) and shown in Table 1.

Morbidity rates of TBE vary between 2 and 8 per 100 000 (Fig. 1) in the Czech Republic. There is a wide heterogeneity of TBE in the country (Fig. 2). Furthermore, over recent decades, increasing TBE trends were associated with higher numbers of cases in areas well known for TBE occurrence in humans; with the re-emergence in areas where TBE cases had not been observed any more, or only sporadically; and with the emergence of TBE in sites unknown previously (including high elevated areas) (see 6.3.4).

A high incidence of this disease was recorded from 1951–1953, peaking in the latter year (19.7/100 000 population). The highest numbers of meningococcal encephalitis (14.8/100 000) and poliomyelitis (20.6/100 000) were likewise recorded in 1953. Analysis of this epidemiological situation has become the subject of research. High incidences were also reported during the same period in Denmark. One of the theories is that TBE was associated with the increase in mean annual temperatures at the end of the 1940s and early 1950s. Comparison of the monthly number of cases with average monthly air temperatures in the Czech Republic is shown in Figure 3. However in-

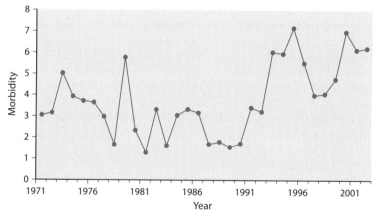

Fig. 1. Tick-borne encephalitis, Czech Republic, 1971–2002. Morbidity per 100 000 population

Table 1. Incidence of tick-borne encephalitis (cases per 100 000 population) (WHO, 2005)

	1993	1994	1995	1996	1997	1998	1999	2000	2001	2002	2003
Albania	0	0	0.03	0.06	0	0	0.03	0	0	0	0
Austria	0.19	0.48	0.64	0.94	0.65	0.76	0.07	0.15	0.28	0.45	0.55
Belarus							0.25	0.23	0.6	0.18	0.53
Bosnia and Herzegovina										0.02	0
Bulgaria	0.18	0.53	0.4	0.43	0.17	0.3	0.27	0		0	
Croatia					0.54	0.52	0.56	0.39	0.58	0.64	0.77
Czech Republic	6.09	5.99	7.19	5.53	4.02	4.1	4.76	7	6.17	6.31	5.92
Estonia						27.11	13.12	19.52	15.62	6.61	17.62
Finland			0.1	0.16	0.37	0.31	0.23	0.79	0.64	0.73	0.31
Germany							0		0.31	0.29	0.34
Hungary	3.2	2.71	2.35	2.49	1.06	0.83	0.56	0.46	0.55	0.61	0.74
Kazakhstan											0.35
Kyrgyzstan								0.2	0.72	0.35	0.1
Latvia	29.79	53.58	53.31	29.57	35.42	41.99	14.37	22.47	12.59	6.4	15.34
Lithuania	5.31	7.63	11.49	8.36	17.4	14.79	4.62	11.34	8.08	4.56	20.77
Norway	0	0	0	0	0	0	0.02	0.02	0.02	0.04	0.02
Poland	0.65	0.47	0.69	0.67	0.52	0.54	0.26	0.44	0.54	0.33	0.88
Russian Federation									4.51		
Serbia and Montenegro								0.01			
Slovakia	0.96	1.12	1.66	1.97	1.41	1	1.17	1.7	1.39	1.15	1.37
Slovenia	10.03	26.87	14.22	20.35	13.73	7.67	7.59	9.86	13.1	13.21	14.23
Sweden							0.6	2.24	1.32		1.21
Switzerland					1.72	0.95	1.56	1.27	1.51	0.74	1.62
Ukraine						0.05	0.09	0.09		0.06	0.06

Fig. 2. Map of tick-borne encephalitis human cases incidence in the Czech Republic from 1971–2002 (EPIDAT, NIPH, Prague)

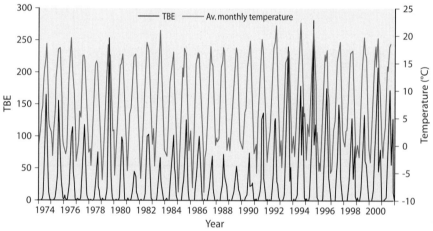

Fig. 3. Average monthly temperature and number of TBE cases in the Bohemia region from 1974 to 2001

terpretation is difficult. In the 1960s and 1970s, there was a decline in the incidence of TBE, although the values remained relatively high (2–7/ 100 000). Lower and higher incidence alternated in 3–4 year intervals. In the 1980s, there was another decline in morbidity; this period has had the lowest number of recorded cases (1.3/100 000 in 1981).

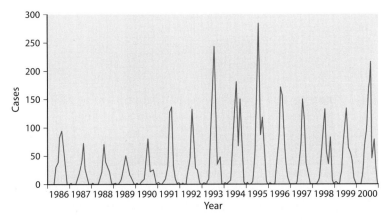

Fig. 4. The seasonality of tick-borne encephalitis in the Czech Republic from 1986 to 2000

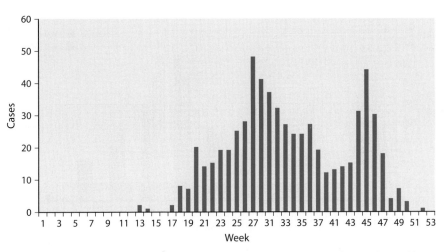

Fig. 5. Number of cases of tick-borne encephalitis by week of onset in the Czech Republic in 2002

Radical changes in the incidence of TBE occurred in the 1990s. These changes manifested themselves in three ways. First, there was an increased incidence of TBE throughout this whole period (the highest morbidity was 7.2/100 000 in 1995). Second, during most of these years the incidence was twin-peaked. The first peak of morbidity occurred in June and July, and a second, smaller peak occurred in October (Figs. 4 and 5). Third, during the last two decades there has been observed extension of TBE season towards spring (April, May) and autumn (October, November) months (Figs. 6 and 7).

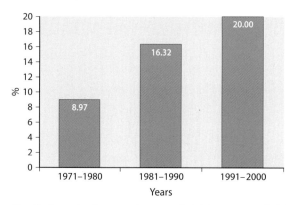

Fig. 6. Percent of reported cases of tick-borne encephalitis in the Czech Republic for the months 4, 5, 10, and 11

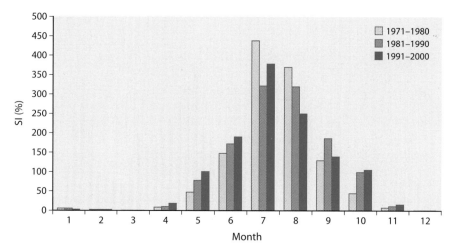

Fig. 7. Index of seasonality of tick-borne encephalitis in the Czech Republic from 1971 to 2000

Persons at Risk

Infection by TBE may be caused by an infected tick or by ingestion of un-pasteurized milk or milk products (cheese) made from unpasteurized milk. Infected ticks may attach themselves to persons visiting endemic locations, or during removal from, for instance, a dog that has entered a natural focus of infection. Persons at greatest risk are those frequently visiting or working in such areas, mainly adults, with the incidence being higher in men than in women (men to women ratio 1.5:1) in the last few decades. However, the age-specific incidence over recent years has increased steadily in children and adolescent age groups, while in the ten-year age groups from 25 to 65, it has remained practically at the same level (6–8/100 000).

In older persons it has fallen to 2–3/100000. Tick bites infect patients mainly during their recreational activities. A very small proportion (less than 1%) acquires the infection through the alimentary route. We have tested the hypothesis whether the increase in TBE incidence in the 1990s was due to economical or social changes after the velvet revolution of 1989 in the Czech Republic. Between 1991 and 1995 unemployment largely remained at the same level (between 2% and 3%). Over the next few years the percentage of unemployed persons increased rapidly to 9.3% in 1999 (7.8% in 2001). This trend differs significantly from the trend of TBE incidence that peaked in 1995. No correlation between the district incidence of TBE and the district percentage of unemployment in the years 1997–2001 was found (r = –0.20). The percentage of unemployed persons among the TBE cases was 1–3% in contrast to the Czech Republic figures which were 5–9% for the same period. The gross domestic product in USD per capita increased from US $ 2600 in 1991 to US $ 5000 in 1995. Since then it has varied between US $ 4800 and US $ 5600. This trend, therefore, also differs from the trend of TBE incidence. Among the TBE cases, the percentage of foresters and other persons working in the forests in the years 1997–2001 was 0.5–1%. The behavioural and socioeconomic aspects of TBE cases has remained stable despite the political changes which have taken place in the Czech Republic since the beginning of the 1990s. They are not, therefore, responsible for the increased TBE incidence (Kříž, Beneś et al., 2004).

6.3.3 TBE Vector Ecology

Characteristics of *Ixodes ricinus*

The main vector of TBE (CEE virus) in Europe is the tick *Ixodes ricinus*. However, it has also been isolated in other tick genera and species: *Ixodes hexagonus, I. gibbosus, Haemaphysalis inermis, H. concinna, H. punctata* and *Dermacentor marginatus*. In the Mediterranean, *I. gibbosus* is described as a vicarious vector of the virus. Nevertheless, the ecology of the *I. ricinus* tick fundamentally affects the TBE distribution and epidemiology in all aspects. The *I. ricinus* tick is an external (exophilic) species, meaning that their development occurs in open nature away from the burrows and nests of its hosts in contrast to nidicolous (endophilic) species. These ticks are, therefore, exposed to direct abiotic and biotic conditions in the environment. Chapter 6.2 gives a detailed description of *I. ricinus*. Its ontogenetic development has alternating phases of parasitic and free modes of life. It is a three-host tick.

The three-host developmental cycle is as follows:

Egg – larva (L) – nymph (N) – adult (imago I – female F or male M).

This covers three active stages (L, N, I), of which each stage feeds only once, each time on a new host, generally of a different species and from a different ecological group. The larvae mainly feed on small terrestrial mammals (murine, soricine), or birds foraging on the ground (very often Turdidae) in the Czech Republic. Nymphs primarily attack medium-sized mammals (squirrels, hedgehogs, rabbits, hares, and occasionally birds). The main source of blood for adult ticks are large mammals – hoofed game, foxes, domestically grazed animals, etc. In an urban environment, the hosts are often cats and dogs.

The duration of the development of *I. ricinus* depends on the geographic location and the local climate. In central Europe, the tick's development cycle is two to three years and is considerably affected by the microclimate of the habitat and long-term meteorological changes.

Seasonality of *I. ricinus* Tick

I. ricinus activity has a distinct annual cycle in central Europe: starting usually in March/April, tick activity (all stages simultaneously) rapidly increases to culminate at the end of the spring/beginning of the summer. Then activity gradually decreases during the main summer months. A new short increase occurs at the end of the summer. All developmental stages are capable of overwintering in either the engorged or the starving condition. Full normal activity of *I. ricinus* ranges between 18 °C and 25 °C at nearly 100% relative humidity; although limited activity begins at temperatures as low as 5 °C.

Habitats of *I. ricinus*

A strong connection with certain types of plant association is characteristic of *I. ricinus* ticks. This particular plant association secures suitable microclimatic conditions for their existence and development, as well as for animal hosts that provide a source of blood and, thus, ensure the viral cycle in nature. In central Europe, the most common habitat of *I. ricinus* includes broad-leaved and mixed woodlands, such as oak and hornbeam groves (*Querceto-Carpinetum; Acereto-Carpinetum*) etc. The vegetation is a suitable indicator for the type of ecosystem, of which *I. ricinus* ticks and TBE virus are integral parts. Thus, plant communities are important bioindicators of the TBE natural foci.

Current Distribution of Disease/Vector

The spread of *I. ricinus* is illustrated in Figure 1 in Chapter 6.2. The territory of the CEE virus distribution does not exceed that of the main vector *I. ricinus*; in fact, it does not even cover all of the territory occupied by this tick in Europe. This fact has not as yet been univocally explained, but it has been suggested that the tick itself, and not the TBE virus, plays an important role.

Infection rates in ticks can be very high in TBE-endemic high-risk areas; Danielová et al. (2002) examined the rate of TBE infections in ticks in two

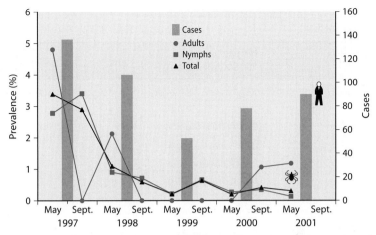

Fig. 8. TBEV prevalence in ticks (PCR) in Baden-Württemberg 1997–2001 and the number of TBE cases (Süss, 2003)

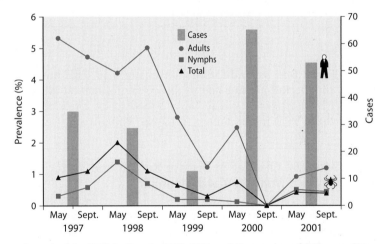

Fig. 9. TBE prevalence in ticks (PCR) in Bavaria 1997–2001 and the number of TBE cases (Süss, 2003)

high risk districts of the South-Bohemian region of the Czech Republic. TBE virus was found in 17 of 187 pooled samples, consisting of a total of 2968 *I. ricinus* ticks. The mean minimum infection rate was 0.6% for all tick stages combined. Infection rates in nymphs collected in different locations varied between 0.2% and 1.3% and between 5.9% and 11.1% in adult ticks.

Direct impacts on transmission are best detected by comparisons of TBE prevalence in ticks. Recent studies by Süss and colleagues found no evidence for any increase in virus prevalence in ticks between 1997 and

Table 2. Tick infection rates by using a standardized methodology, nRT-PCR (Süss, personal communication)

Country	Incidence/ 100 000 inhabitants	Prevalence (%) nRT-PCR
Latvia	53.6	2.82 *I. ricinus* 5.05 *I. persulcatus*
Finland	0.48 (whole country)	0.34
Germany	0.27 (whole country)	0–3.4
	0.5–5 (endemic areas)	0–3.4
	Baden-Württemberg	0–2.3
	Bavaria	0–2.0
	Low endemic areas	
	– Area Odenwald	0.5
	– District Birkenfeld (single cases)	0.0
	– District Marburg-Biedenkopf (single cases)	0.0
	– District Kitzingen (single cases)	0.1
	– District Saale-Holzland (single cases)	0.0

2001 (Figs. 8 and 9), indicating that changes in TBE human case reports during this period are not due to direct environmental impacts on the zoonotic TBE transmission cycle. Table 2 shows tick infection rates by using a standardized PCR methodology in Latvia, Finland and Germany.

6.3.4 Potential Changes in Disease Burden and Vector Distribution. Findings from the Czech Republic

In recent decades, changes in the geographic and temporal distribution of the disease and vector have been observed (see also Chapter 6.2) and changes in TBE transmission seasonality and intensity.

Altitudinal Distribution Changes in the Czech Republic

Early signs of effects from changes in climate are more easily recognized in areas located close to the geographical distribution limits (altitudinal or latitudinal) of an organism (see Chapter 6.2). Mountain studies on *I. ricinus* populations have been performed in the same locations in the Czech Republic in 1957, 1979–1980 and 2001–2002. A shift in the upper altitude boundary of permanent tick population from 700 m to 1100 m a.s.l. has been observed (Daniel et al., 2003). Specifically, tick surveys (on permanently resident dogs and by flagging) were carried out between 2001 and 2002 at altitudes between 700 m and 1200 m a.s.l. in the Šumava mountains. Ticks were found on all dogs up to 1100 m a.s.l. and similarly up to this altitude by flagging. These findings contrasted with analogous surveys carried out in the same region in 1957 at altitudes between 780 m and 1200 m a.s.l., when no *I. ricinus* ticks were detected above 800 m a.s.l., and few between 780 m and 800 m a.s.l. Similarly no ticks were detected in the same region during surveys at 760 m a.s.l. between 1979 and 1980. These tick distribution changes have been shown to be linked to changes in climate (Daniel et al., 2004). Ticks have been also found in abundance up to 1300 m a.s.l. in the Italian Alps (Rizzoli et al., 2002) and along the Baltic Sea coastline up to latitude 65 °N in Sweden (Jaenson et al., 1994; Tälleklint and Jaenson, 1998).

Recent studies suggest a linkage between observed changes in climate and changes in *I. ricinus* distribution and TBE incidence (Daniel et al., 2003). *I. ricinus* tick activity, its daily pattern, and day-to-day differences in this activity are significantly influenced by the weather. The relationship between weather and ticks are solid enough to be used for the prediction of tick activity, and for the prediction of risk of TBE infection as a part of the anticipated warning system in the future.

Changes in TBE Transmission Intensity and Seasonality

In the Czech Republic increasing TBE trends were associated with higher number of cases in areas well known for TBE occurrence in humans; with re-emergence in areas where TBE cases were not observed, or only sporadically, for a long time; and with emergence of TBE in sites unknown previously (including high elevated areas). A significant shift of TBE incidence towards spring and autumn months was recorded during the last decade.

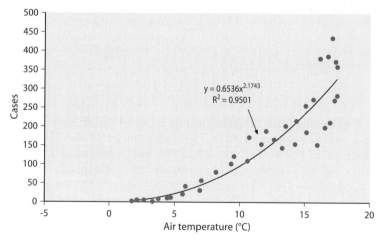

Fig. 10. Relationship between air temperature and occurrence of TBE from 1993 to 2002 in the Czech Republic, n = 5873 cases

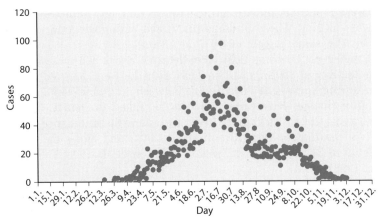

Fig. 11. TBE number of cases by onset of illness in the Czech Republic (1993–2002), n = 5873 cases

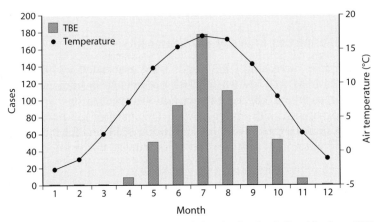

Fig. 12. Seasonality of tick-borne encephalitis in the Czech Republic from 1993 to 2000, and mean air temperature (1961–1990)

Figure 10 shows the relationship between mean weekly air temperature and weekly number of 5873 TBE cases (onset of the disease) which were recorded in the years 1993–2002. The best correlation $r^2 = 0.95$ was found for a lag of 1 week which approximately corresponds to the incubation period. First cases were recorded at average weekly temperatures of 3.3 °C to 4.4 °C (8 cases in 10 years). At the average weekly temperatures of 5.6 °C = 12 cases, 6.9 °C = 22 cases, 8.1 °C = 32 cases, 9.3 °C = 80 cases, 10.5 °C = 103 cases. Highest mean weekly temperature when TBE cases were recorded was 17.4 °C with 732 cases. Further ecological studies are needed to elucidate this situation.

Figure 11 shows the TBE cases recorded in the Czech Republic during the years 1993–2002 by onset of the disease. The first case occurred on March 15, the last case on December 2. The maximum number of cases (92) was recorded on July 20.

Seasonal incidence of TBE in the long-term correlates highly (rs –0.98) with the average temperatures during the year. In the case of the second peak, a more important factor is the rainfall during August (Danielová, Beneš, 1997). Temperature in the autumn months is also involved, but only as a second-rate factor (Fig. 12).

6.3.5 Adaptation and Preventive Measures

Current Available Prevention

No specific treatment is available. An effective vaccine is available for prevention of infection, and is recommended for persons who are at particular risk, such as foresters and persons involved in outdoor recreational activities. Until the early 1980s, TBE was a frequent cause of central nervous system infectious disease in Austria. From 1981, vaccination was encouraged by intensive media campaigns, but it was voluntary rather than compulsory. As a result, the number of hospitalized cases due to TBE declined significantly from 1981 to 1990, with important savings in health care costs (Schwarz, 1993). A study of the rate of vaccination among school children in Austria revealed that the prevalence of at least one TBE vaccination was 91.4% for 7-year-olds, 97.3% for 10-year-olds, and 97.1% for 13-year-olds. The prevalence of basic TBE immunization was 84.0%, 91.7% and 92.3% respectively. The lowest vaccination rates were found in families with four or more children and in children with mothers of the lowest educational level (Stronegger et al., 1998).

Protection against Tick Bites

Protection against tick bites is also very important for TBE prevention. Two kinds of protection are proposed: collective and individual. To decrease the risk of tick bites during tick activity periods, it is necessary to wear suitable clothing in TBE-risk areas. Self-inspection and prompt removal of ticks should always be performed after outdoor activities in risk sites during the tick activity seasons. Education and information must be part of the prevention programme. Effective tools for this could be television, radio, press, film, lectures, informational pamphlets and posters.

6.3.6 Future Research Needs

There is a need to study the effect of temperature on virus amplification in the tick, the differences in susceptibility of tick populations for the virus and virus properties (virulence).

For better surveillance in the future, several surveillance centres should be set up on a pan-European level to record changes in tick and pathogen prevalence and in disease distribution. These centres should emphasize the following points:

- Standardized methods for tick collection should be used and the methods should not be changed over time. Sampling should be sufficiently frequent in time and space.
- Standardized methods for measuring virus prevalence and virus quantity in ticks in Europe should be used: nRT-PCR and standardized real time PCR technologies for quantitative measurement of TBEV in ticks.
- TBE case definitions and diagnostic tools used for humans should be similar throughout Europe and cases reported and registered on a pan-European level.
- For early detection of changes in TBE distribution, serological TBE surveillance should also include people in currently non-TBE-endemic locations if they are at risk of exposure for *I. ricinus* bites in the area.
- Attention should be paid to the occurrence of *I. ricinus* ticks and TBE virus in urban environments.
- Spatial analyses of TBE distributional data based on remote sensing data with a resolution of at least 30 m should be initiated.
- Promote studies that cover the potential impacts of socioeconomic and demographic factors, and environmental and climatic changes as predictors of TBE risk, and use the outcomes of such studies for scenario simulations of future trends in disease risk and changes in risk areas.

6.3.7 Future Perspectives of TBE Incidence Prediction

Prediction of situations exposing humans to the risk of *I. ricinus* tick attacks and, thus, the TBE virus infection is a basic condition for the successful prevention of TBE disease. In principle, such prediction can be:

- **spatial** – specifying the high-risk parts of the landscape and determining the level of risk thereof (limiting habitats that provide conditions for the existence of *I. ricinus* populations and circulation of TBE virus in the animal sphere)
- **temporal** – predicting the period of increased risk:
  - *Long-term prediction* – predicts seasonal patterns or multi-year changes
  - *Short-term prediction* – changes in the behaviour of ticks, in particular their aggressiveness, on the short-term horizon.

Both categories of predictions are dependent on climatic changes. The microclimatic component, which determines the conditions of the ticks' immediate environment (superficial layer of soil, forest litter, fallen leaf layer, ground layer of vegetation) is particularly of interest. The microclimate, which is a derivative of the macroclimate, is conditioned by the local type of habitat and at the same time has an influence on that very same habitat.

Spatial prediction examines the distribution of suitable ecosystems along with risk evaluations. In this way, it maps the natural foci of TBE and their individual spatial components (elementary foci, their nuclei, coats etc.). Geographical information system (GIS) and GIS examination of changes in land-cover and land-use – caused either by natural succession or by human interventions – represent a suitable methodological approach.

Long-term time prediction estimates long-term trends based on the analysis of long-term series of climatological data and their comparisons with epidemiological and environmental data. Knowledge of dependence of phenological manifestations of additional components in the forest ecosystems on long-term climate changes can serve as an important factor providing more details.

Short-term time prediction informs about the changes in the level of risk in natural foci of TBE according to the anticipated changes in the behaviour of *I. ricinus*, depending on the weather, in particular with respect to host-seeking and feeding activities. Understanding of the relations between the micro- and macroclimates is of key importance in this category of prediction. The accuracy of the prediction is determined by the level of the general weather forecasts that are produced 5 to 10 days ahead by the Meteorological Service. This type of prediction is important, for example, when decisions need to be made upon which type of outdoor activities that should be carried out in high-risk areas at a particular time.

6.3.8 Future Risks

Recent studies suggest a linkage between observed changes in climate and changes in the vector *I. ricinus* distribution and TBE incidence described above (Daniel et al., 2003). *I. ricinus* tick activity, its daily pattern, and day-to-day differences in this activity are significantly influenced by the weather. The relationship between weather and ticks is solid enough to be used for the prediction of tick activity and for the prediction of TBE infection risk as a part of the anticipated warning system in the future.

References

Calisher ChH et al (1989) Antigenic relationships between flaviviruses as determined by cross-neutralization tests with polyclonal antisera. J Gen Virol 70:37–43

Calisher ChH (1988) Antigenic classification and taxonomy of flaviviruses (family Flaviviridae) emphasizing a universal system for the taxonomy of viruses causing tick-borne encephalitis. Acta virol 32:469–478

Daniel M, Danielová V, et al. (2003) Shift of the tick Ixodes ricinus and tick-borne encephalitis to higher altitudes in central Europe. Eur J Clin Microbiol Infect Dis 22(5):327–328

Daniel M, Danielová V et al (2004) An attempt to elucidate the increased incidence of tick-borne encephalitis and its spread to higher altitudes in the Czech Republic. Int J Med Microbiol 293 (Suppl 37):55–62

Danielová V, Beneš Č (1997) Possible role of rainfall in the epidemiology of tick-borne encephalitis. Centr Eur J Publ Hlth 5:151–154

Danielová V, Holubová J, Daniel M (2002) Tick-borne encephalitis virus prevalence in Ixodes ricinus ticks collected in high risk habitats of the South-Bohemian region of the Czech Republic. Experimental and Applied Acarology 26:145–151

Danielová V, Kříž B, et al. (2004) Climate changes influencing the tick-borne encephalitis incidence on the Czech Republic in last two decades. Epid Microb Immunol, 53:174–181 (in Czech)

Dumpis U, Crook D, Oksi J (1999) Tick-borne encephalitis. Clinical Infectious Diseases 28(4):882–890

Gallia F, Rampas J, Hollender L (1949) Laboratory infection with encephalitis virus. Čas Lék čes 88:224–229 (in Czech)

Gratz N (2004) Vector borne diseases in Europe. World Health Organization, Regional office for Europe, Copenhagen

Hoogstraal H (1973) Viruses and ticks. In: Gibbs AJ (ed) Viruses and invertebrates, Chapter 18. North-Holland Publ Co, Amsterdam, 349–417

Jaenson TGT, Tälleklint L, et al. (1994) Geographical distribution, host associations, and vector roles of ticks (Acari: Hodidae, Argasidae) in Sweden. J Med Entomol 31:240–256

Krejčí J (1949a) Isolement d'un virus nouveau en course d'une epidemie de meningoencephalitis dans la region de Vyškov (Moravie). Presse méd (Paris) 74:1084

Krejčí J (1949b) Epidemics of viral meningoencephalitis in Výskov area (Moravia, Czech Republic) Lék. Listy, 4:73–75, 112–116, 132–134 (in Czech)

Kříž B, Beneš Č et al (2004) Socio-economic conditions and other anthropogenic factors influencing tick-borne encephalitis incidence in the Czech Republic. Int J Med Microbiol 293 (Suppl 37):63–68

Materna J, Daniel M, Danielová V (2005) Altitudinal distribution limit of the tick Ixodes ricinus shifted considerably towards higher altitudes in Central Europe: results of the three years monitioring in the Krkonose Mountains (Czech Republic). Cent Eur J Public Health 13:24–28

Materna J, Danielova V, Daniel M (2003) Results of monitoring of Ixodes ricinus tick distribution on the territory of Krkonose National Park. Journal Krkonose 1(19) (in Czech)

Ozdemir FA, Rosenow F, Slenczka W, Kleine TO, Oertel WH (1999) [Early summer meningoencephalitis. Extension of the endemic area to mid-Hessia]. Nervenarzt 70(2):119–122

Rampas J, Gallia F (1949) Isolation of an encephalitis virus from Ixodes ricinus ticks. Čas lék českých 88:1179–1180 (in Czech)

Rizzoli A, Merler S, Furlanello C, Genchi C (2002) Geographic information systems and bootstrap aggregation (bagging) of three-based classifiers for Lyme Diseases Risk Prediction in Trentino, Italian Alps. J Med Entomol 39:485–492
Süss J (2003) Epidemiology and ecology of TBE relevant to the production of effective vaccines. Vaccine 21:19–35
Tälleklint L, Jaenson TGT (1998) Increasing geographical distribution and density of Ixodes ricinus (Acari: Ixodidae) in central and northern Sweden? J Med Entomol 35:521–526

6.4　Malaria

Katrin Kuhn

Malaria is recognized as the most important parasitic infection in the world, causing an estimated 300 million acute cases and at least 1 million deaths annually (Roll Back Malaria, 2002). Approximately 40% of the world's population currently lives in areas at risk of malaria (Mendis et al., 2001). Disease in humans, caused by one of four species of the *Plasmodium* parasite, is transmitted by female *Anopheles* mosquitoes. Worldwide approximately 400 species of anophelines are natural vectors (i.e. mosquitoes which transmit the parasite in nature without manipulation).

The transmission of malaria is intricately linked to climatic factors such as temperature and precipitation as well as a range of agricultural and socioeconomic issues (e.g. land use, health system infrastructure, use of antimalarial medicine). Temperature directly influences the duration of the extrinsic incubation period (parasite development time in the mosquito), rate of vector development and the frequency of blood feeding (McDonald, 1957; Jetten and Takken, 1994). The amount of rainfall as well as the locations of lakes, rivers and other water bodies play a significant role in determining the distribution and abundance of vectors since all anophelines breed in water (Smith et al., 1995). Thus, in areas where temperatures and precipitation are below or above certain thresholds, malaria transmission does not occur (Hay et al., 1998) as parasites can not develop or vector populations not be maintained. Recently, much research activity has focussed on predicting whether global climate changes may alter the transmission risk in these areas by creating a climate more suitable for transmission. In these cases, Europe is often considered potentially vulnerable due to the previous existence of malaria throughout the continent and the continued presence of competent malaria vectors in all countries (Kuhn et al., 2002).

6.4.1　Historical Malaria in Europe

The historical distribution of malaria throughout the world was much more widespread than at present, including endemic areas throughout Europe, North America and Australia (Fig. 1). In Europe, evidence suggests that malaria transmission was most likely established during the Neolithic period (8000 to 5000 BC). Then followed a geographical spread of the disease and the most important vectors until the 13th century where malaria was present throughout most of the continent, including Scandinavia and large parts of northern Russia (Hackett, 1937). At this time, *Plasmodium vivax* was the most predominant parasite transmitted throughout all endemic

☐ Areas with previously endemic malaria which has been eradicated or disappeared spontaneously
■ Areas with current malaria transmission

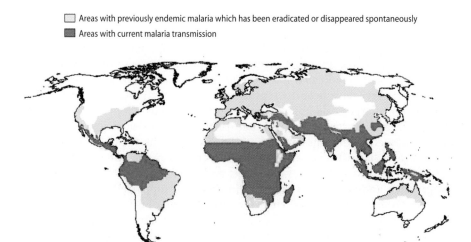

Fig. 1. Worldwide historical and present distribution of malaria (based on Bruce-Chwatt and de Zulueta, 1980; Molineaux, 1988; WHO, 2001)

areas, while a European strain (now extinct) of *P. falciparum* was present only in countries along the Mediterranean Sea (Italy, Greece and Albania; de Zulueta, 1994). Transmission of *P. malariae* was reported primarily from Mediterranean countries (Balfour, 1935); however, the relative importance of each parasite in areas where all or only some species were transmitted is difficult to document due to lack of data. Thus, for the next 5–6 centuries the disease was a health problem in many areas, particularly in the Mediterranean and inland parts of eastern Europe caused by a combination of virulent *P. falciparum* parasites, highly efficient vectors (primarily *An. sacharovi* and *An. atroparvus*), suitable breeding sites and favourable climatic factors.

In the early twentieth century, malaria spontaneously started disappearing from northern Europe (Bruce-Chwatt and de Zulueta, 1980). This decline has mainly been attributed to changes in agricultural practices as well as improved housing, health care and overall socioeconomical factors. A recent study has confirmed that the decline of malaria in England was significantly associated with decreases in marsh wetlands as well as increased cattle densities (Kuhn et al., 2003).

In the Mediterranean, prevalences remained high and no apparent decline was observed until the initiation of vector control (disease eradication) programmes using DDT residual spraying. The establishment of the WHO eradication programme in 1955 was followed by a string of country-wide eradications ranging from Czechoslovakia in 1963, to the Netherlands and Italy in 1970 until the last focus of malaria in the Greek republic of Macedonia was officially eradicated and Europe declared malaria free in 1975.

6.4.2 Malaria in Present Day Europe

Malaria is today distributed in large areas throughout South America, Africa and Asia (Fig. 1). Despite vast investments and strenuous efforts, the disease was never totally eradicated from the former Soviet Union and Turkey, and prevalences here have been increasing since the early 1990s. Endemic transmission with large-scale epidemics have been reported from Turkey, Armenia, Azerbaijan, Kyrgyzstan, Tajikistan, Turkmenistan, Uzbekistan. Bordering countries such as the Russian Federation, Georgia, the Republic of Moldova and Bulgaria have also shown significant increases in the number of cases of malaria in the last few years. In the most afflicted areas, the Roll Back Malaria Programme and the WHO Office for Europe have supported a wide range of interventions (WHO, 1999) such as the introduction of new antimalarials and local vector control in close collaboration with non-governmental health institutions. The effect of the Roll Back Malaria programme has been apparent in the decreasing reported prevalences (from 90712 in 1995 to 23832 in 2001); however, it is thought that the magnitude of the malaria problem is much greater than these reports indicate.

At present, there is no endemic malaria transmission within the EU region (i.e. not including the former Soviet Union and the Asian part of Turkey); however, three important issues relating to this region have recently been identified: sporadic autochthonous cases, airport malaria and imported cases.

Autochthonous transmission of malaria by local vectors in western Europe has been reported from Italy, Germany and Greece. The cases in Germany and Italy (Baldari et al., 1998; Mantel et al., 1995; Kruger et al., 2001; Anselmo et al., 1996) are thought to represent true autochthonous transmission where a local malaria vector had picked up the infection from a human infected with a tropical strain and transmitted this parasite on to a new susceptible person. An interesting fact is that both Italian (Baldari et al., 1998) and German (Kruger et al., 2001) cases occurred during the 1997 heat-wave and in locations which had previously been highly malarious. This could indicate that higher than normal temperatures (and global warming) may indeed play a significant role in determining the risk of malaria re-emergence in these areas.

Few isolated cases reported in the United Kingdom, Luxembourg, France and Belgium (Peleman et al., 2000; Van den Ende, 1998; Bouvier, 1990) were confirmed to be airport malaria – i.e. disease in people who work in or live near airports and become infected by tropical vectors transported in luggage. During the European heat-wave of 1994, six new cases of airport malaria were described in and around Charles de Gaulle airport, Paris, and consequently aircraft spraying using residual pyrethroids is now regularly carried out at this airport (Giacomini et al., 1995).

With the advance of commercial travel and increasing levels of immigration, Europe has experienced a steady rise in the number of imported cases of malaria (i.e. infections acquired in the tropics). In the period from 1992

Table 1. Imported cases in four European countries in 1992 and 2002

Country	Total cases in 1992 (% *P. falciparum*)	Total cases in 2000 (% *P. falciparum*)
France	3430 (79)	6846 (96)
Germany	773 (57)	732 (78)
Italy	497 (74)	986 (81)
United Kingdom	1629 (57)	2069 (76)

Source: WHO Europe Centralized Information System for Infectious Diseases, 2004

to 2000, there was a 70% increase in the number of imported cases reported from France, the United Kingdom, Germany and Italy alone (Table 1). These countries account for the majority of malaria cases imported into the European Union region. Most of the imported cases are due to overseas visits by immigrants settled in Europe, from European tourists travelling abroad and foreign visitors who fall ill while in one of the European countries (Bradley, 1989).

In spite of the increase in imported malaria, the risk of these infections being transmitted by local vectors is probably very low in most central and western countries – primarily because of rapid detection and treatment of cases which keeps the human reservoir to a minimum. This is evident in the fact that so few autochthonous cases are reported, despite the presence of competent vectors.

Underreporting of cases, however, is a problem and this may particularly be serious in countries bordering the currently endemic regions. For instance, the steady influx of cases from the Former Russian Federation into socially vulnerable areas of Bulgaria and Romania is causing concern about re-emergence of local transmission (Vutchev, 2001; Nicolaiciuc et al., 1999).

6.4.3 Preventive and Adaptive Measures and Strategies

Malaria is a notifiable disease in all countries of the WHO European region where health departments are asked to report the annual number of malaria cases to the WHO who maintains a database on all reported cases. This continuous surveillance of the disease throughout the region facilitates the monitoring of malaria re-emergence and enables us to assess which areas may be at future risk of outbreaks.

Non-endemic Regions

In these areas, imported malaria may become a particular problem with respect to the re-emergence of indigenous transmission (i.e. patients with imported infections carry parasites which can be picked up by local mosquitoes). The length of time during which infected people can act as reservoirs is a crucial point but, in most European countries, the high personal medical awareness and health system quality ensure that this period is usually short. On average in the United Kingdom, an infected patient will be diagnosed, treated and parasites cleared from the blood within 7–12 days (D. Warhurst, personal communication).

However, misdiagnoses do occur and information aimed at both health personnel and the general public about signs, symptoms and treatment of malaria is essential.

An important factor in preventing the re-emergence of indigenous malaria in non-endemic parts of Europe is the status of the vector population in vulnerable areas (i.e. near wetlands and other known breeding grounds). The initial elimination of European malaria (in the South and East) was mainly due to large scale DDT spraying, but the current susceptibility of European *Anopheles* to DDT is unknown. There is some indication of resistance in Spain and Italy as well as the currently endemic regions of Russia, Central Asia and Turkey (e.g. Hemingway et al., 1992; Prokhorova et al., 1992; Encinas Grandes and Astudillo Sagrado, 1988). Ideally, the sensitivity of the major European vectors to the most commonly used insecticides should be re-assessed for developing guidelines for future *Anopheles* control.

At present in many areas of western and southern Europe, larval stages of nuisance mosquitoes (not anophelines) are controlled using biological insecticides, but unfortunately any potential effect on *Anopheles* species has not been demonstrated.

Currently Endemic Regions

The Roll Back Malaria Programme has been implemented in all countries reporting endemic transmission of malaria where focus is placed on prompt diagnosis and treatment of cases, education of health personnel and provision of personal protection such as insecticide-treated bed nets. Vector control by residual house spraying is also undertaken with the aim of reducing transmission (Aliev and Saparova, 2001; Bismil'din et al., 2001; Gockchinar and Kalipsi, 2001). However, these efforts have primarily been initiated by various non-governmental organizations undertaking research programmes in the areas, but these usually do not cover all endemic areas and are only in operation for relatively short periods. Here, there should be future emphasis on developing country-wide and centrally organized control efforts.

6.4.4 The Relationship between Environmental and Climatic Factors and the Transmission of Malaria in Europe

Malaria has not been present in central and western Europe for almost a century and at present there is a lack of knowledge about how climatic and environmental factors were related to disease transmission at the peak of its distribution. However, from limited observational studies performed on European malaria vectors and parasites, a number of climate links have been inferred:

▌ The geographical distribution may have been strongly related to temperature. Malaria was only present within the 15 °C July isotherm (i.e. in places where the average temperature during the month of July reached 15 °C or above).

▌ Parasite development thresholds lie between 14.5–16 °C and 18 °C for tropical *P. vivax* and *P. falciparum*, respectively (MacDonald, 1957). The threshold for the now extinct European strains is likely to have been less considering the past distribution of the disease.

▌ There is no evidence for a link between European malaria parasites and humidity (or rainfall), but because the distribution of mosquitoes is related to rainfall patterns (see below), this variable should also be taken into consideration.

▌ The length of the mosquito season is determined by latitude – probably a proxy for temperature (Hackett and Missiroli, 1935).

In addition, a study recently found significant relationships between temperature and precipitation and the presence of five major malaria vectors in Europe (Kuhn et al., 2002) and identified the optimum temperatures and diurnal temperature ranges for each species. The presence of four main malaria vectors in Europe was also found to be positively correlated with rainfall.

6.4.5 Possible Effects of Future Climate Change in Europe

There has been much speculation about the role of climate change in the potential re-emergence of malaria within the EU region. Mathematical and biological models of climate-disease relationships (e.g. Martens et al., 1995; Martens et al., 1999; Rogers and Randolph, 2000) have produced contradictory results and no conclusions have yet been reached due to the many unresolved questions regarding the epidemiology and biology of European malaria and its mosquito vectors.

From a scientific perspective, several factors could favour the re-establishment of malaria in the EC region with climate changes. First, the recent autochthonous cases where *P. vivax* and *P. falciparum* were transmitted by local vectors demonstrate that malaria transmission is still feasible. The predicted increases in temperatures will speed up *Anopheles* development rates and increase the seasonal abundance of mosquitoes, creating a larger vector reservoir. Accordingly Kuhn (2002) showed that the projected climate changes for Europe can significantly increase the abundance of some of the major former malaria vectors. In addition, the parasite development cycle will be shorter, increasing the probability that parasites can develop successfully within the mosquito during its life span (Bradley, 1993). In combination with the increase in the number of imported parasites, this is a scenario which could potentially cause the re-emergence of malaria transmission.

On the other hand, there are considerable factors acting against this picture. First, the chance of a mosquito encountering an infected blood meal will be exceedingly low due to rapid detection and treatment of cases in most western European countries. In addition, we know that socioeconomical and agricultural factors were strongly linked to the distribution of historical European malaria. The changes in these elements during the 19th century played important roles in the natural disappearance of malaria from most of the European continent. Already, the re-establishment of transmission in the currently endemic areas has been linked to decaying social structures and population movements (Bismil'din et al., 2000) rather than changes in climate. In a recent paper, Kuhn et al. (2003) demonstrated the strong link between non-climatic factors and malaria in the United Kingdom and showed that climate changes alone are unlikely to have a significant effect on malaria in this country. It is likely that the situation in eastern Europe will be different, however, as the increase in mosquito abundances will create an increased possibility that imported parasites are transmitted locally (because detection and treatment of cases is comparatively slower in these areas).

While there has not been much work to identify links between European malaria and climate factors, many studies have focussed on first identifying relationships between the current distribution of tropical malaria and climate, and second predicting any effects of climate changes on disease patterns (e.g. Gagnon et al., 2002; Kleinschmidt et al., 2000; Barrera et al., 1999; Craig et al., 1999; MARA/ARMA, 1998). Other work has examined the correlation between long-term fluctuations in malaria trends and climate variations (Bouma et al., 1997; Bouma and Van der Kaay, 1996; Hay et al., 2002; Patz et al., 2002) and evidence indicates that, in some areas, rising malaria trends might be related to increases in temperatures (Patz et al., 2002).

6.4.6 Future Research Needs

An important aspect of predicting the potential effect of climate changes is first to describe (using statistical or biological models) the current relationship between climatic factors and the distribution of disease or vectors. In Europe, this approach can only be undertaken for the currently endemic areas where disease data are available. Within the non-endemic EU region, we need to focus on historical patterns of disease and the current distribution of mosquitoes and their respective relationship with climate variables of interest. The main issues which should be addressed are, therefore,

- Collate available long-term data on historical malaria prevalence, at varying resolutions (country and region level) in non-endemic countries.
- Collate available long-term data on malaria prevalence, at varying resolutions in currently endemic countries.
- Collate historical data on the distribution of important vectors throughout the WHO European region.
- Collect new field data on the current distribution of important anophelines throughout the WHO European region.
- Use above data in combination with climate, environmental and socioeconomic variables to (a) identify the relationship between the geographical distribution of malaria and mosquitoes and climate/environment, (b) identify the relationship between the temporal distribution (i.e. inter-annual variation) of malaria and climate/environment and (c) assess the importance of climate in relation to non-climatic factors in determining the geographical and inter-annual variation in disease.
- Collect new field data on mosquito biological parameters for all relevant species (including activity patterns, seasonality, survival and infection rates) in relation to defined climatic and environmental variables in transect regions over the full seasons.
- Develop climate change scenarios at a regional scale to account for variations in vector species and malaria parasites.

In conclusion, the current level of our knowledge about any potential impact of climate changes on European malaria is still low. There is no convincing evidence that recent observed climate changes have impacted on the re-emergence of malaria in the WHO European region, but published studies indicate a possible link between temperature increases and rising malaria incidences or shifting geographical distributions in the tropics.

Malaria in Europe is a unique example due to the many factors involved in the disappearance of the disease from this region (e.g. agricultural, socioeconomic, control measures). The issue of climate changes in relation to European malaria in the future is best resolved by thoroughly studying disease and climate relationships in the currently endemic regions and, using long-term historical data, examining which factors influenced past transmission in areas where the disease is now extinct.

Acknowledgements. This chapter is based on the WHO document "Malaria in Europe: the potential effects of climate change, ecology, epidemiology and adaptation measures" by Katrin Kuhn et al., forthcoming. The author is grateful to David Bradley, Diarmid Campbell-Lendrum, Clive Davies, Sari Kovats, Pim Martens, David Warhurst, Pernille Joergensen, Milan Daniel, Michael van Lieshout, Elisabet Lindgren, Bettina Menne and Kris Ebi for invaluable comments and suggestions.

References

Aliev S, Saparova N (2001) Current malaria situation and its control in Tajikistan. Med Parazitol Mosk 1:35–37

Anselmo M, De Leo P, Rosone A, Minetti F, Cutillo A, Vaira C, Menardo G (1996) Port malaria caused by Plasmodium falciparum a case report. Infez Med 4:45–47

Baldari MA, Tamburro A, Sabatinelli G, Romi R, Severini C, Cuccagna C, Fiorilli G, Allegri MP, Buriani C, Toti M (1998) Malaria in Maremma, Italy. Lancet 351:1246–1247

Balfour MC (1935) Malaria studies in Greece. Am J Trop Med Hyg 3:301–303

Barrera R, Grillet ME, Rangel Y, Berti J, Ache A (1999) Temporal and spatial patterns of malaria infection in north-eastern Venezuela. Am J Trop Med Hyg 61:784–790

Bismil'din FB, Shapieva ZZ, Anpilova EN (2001) Current malaria situation in the Republic of Kazakhstan. Med Parazitol Mosk 1:24–33

Bouma M, Rowland M (1995) Failure of passive zooprophylaxis: cattle ownership in Pakistan is associated with a higher prevalence of malaria. Trans R Soc Trop Med Hyg 89:351–353

Bouma MJ, van der Kaay (1996) The El Nino Southern Oscillation and the historic malaria epidemics on the Indian subcontinent and Sri Lanka: an early warning system for future epidemics? Trop Med Int Health 1:86–96

Bouma MJ, Poveda G, Rojas W, Chavasse D, Quinones M, Cox J, Patz J (1997) Predicting high risk years for malaria in Colombia using parameters of El Nino Southern Oscillation. Trop Med Int Health 2:1122–1127

Bouvier M, Pittet D, Loutan L, Starobinski M (1990) Airport malaria: mini-epidemic in Switzerland. Schweiz Med Wochenschr 120:1217–1222

Bradley DJ (1989) Current trends in malaria in Britain. J R Soc Med 82:S8–S13

Bruce-Chwatt LJ, de Zulueta J (1980) The rise and fall of malaria in Europe. A historico-epidemiological study. Oxford University Press, Oxford

Craig MH, Snow RW, le Sueur D (1999) A climate-based distribution model of malaria transmission in sub-Saharan Africa. Parasitology Today 15:105–111

de Zulueta J (1994) Malaria and ecosystems: from prehistory to posteradication. Parassitologia 36:7–15

Gagnon AS, Smoyer-Tomic KE, Bush AB (2002) The El Nino southern oscillation and malaria epidemics in South America. Int J Biometeorol 46:81–89

Giacomini T, Axler O, Mouchet J, Lebrin P, Carlioz R, Paugam B, Brassier D, Donetti L, Giacomini F, Vachon F (1997) Pitfalls in the diagnosis of airport malaria. Seven cases observed in the Paris area in 1994. Scand J Infect Dis 29:433–435

Gockchinar T, Kalipsi S (2001) Current malaria situation in Turkey. Med Parazitol Mosk 1:44–45

Encinas Grandes A, Astudillo Sagrado E (1999) The susceptibilty of mosquitoes to insecticides in Salamanca province, Spain. J Am Mosq Control Assoc 4:167–171

Hackett LW (1937) Malaria in Europe: an ecological study. Oxford University Press, Oxford

Hackett LW, Missiroli A (1935) The varieties of An. maculipennis and their relation to the distribution of malaria in Europe. Riv Mal 14:45–109

Hay SI, Cox J, Rogers DJ, Randolph SE, Stern DI, Shanks GD, Myers MF, Snow RW (2002) Climate change and the resurgence of malaria in the East African highlands. Nature 415:905–909

Hay SI, Snow RW, Rogers DJ (1998) Predicting malaria seasons in Kenya using multi-temporal meteorological satellite data. Trans R Soc Trop Med Hyg 92:12–20

Hemingway J, Small GJ, Monro A, Sawyer BV, Kasap H (1992) Insecticide resistance gene frequencies in Anopheles sacharovi populations of the Cukurova plain, Adana province, Turkey. Med Vet Entomol 6:342–348

Jetten TH, Takken W (1994) Anophelism without malaria in Europe: A review of the ecology and distribution of the genus Anopheles in Europe. Wageningen Agricultural university, The Netherlands

Kleinschmidt I, Bagayoko M, Clarke GPY, Craig M, le Sueur D (2000) A spatial statistical approach to malaria mapping. Int J Epid 29:355–361

Kruger A, Rech A, Su XZ, Tannich E (2001) Two cases of autochthonous Plasmodium falciparum malaria in Germany with evidence for local transmission by indigenous Anopheles plumbeus. Trop Med Int Health 6:983–985

Kuhn KG, Campbell-Lendrum DH, Armstrong B, Davies CR (2003) Malaria in Britain: past, present and future. PNAS 9997–10001

Kuhn KG, Campbell-Lendrum DH, Davies CR (2002) A continental risk map for malaria mosquito vectors in Europe. J Med Ent 39:621–630

Kuhn KG (2002) Environmental determinants of malaria risk in Europe: past, present and future. PhD thesis, University of London

Mantel CF, Klose C, Scheurer S, Vogel R, Wesirow AL, Bienzle U (1995) Plasmodium falciparum malaria acquired in Berlin, Germany. Lancet 346(8970):320–321

MARA/ARMA (1998) Towards an atlas of malaria risk in Africa: first technical report of the MARA/ARMA collaboration. MARA/ARMA, Durban, South Africa

Martens P, Niessen LW, Rotmans J, Jetten TH, McMichael AJ (1995) Potential impact of global climate change on malaria risk. Env Health Perspect 103:458–464

Martens P, Kovats RS, Nijhof S, de Vries P, Livermore MTJ, Bradley DJ, Cox J, McMichael AJ (1999) Climate change and future populations at risk of malaria. Global Env Change 9:S89–S107

McDonald G (1957) The epidemiology and control of malaria. Oxford University Press, Oxford

Mendis K, Sina BJ, Marchesini P, Carter R (2001) The neglected burden of Plasmodium vivax malaria. Am J Trop Med Hyg 64:97–106

Molineaux L (1988) The epidemiology of human malaria as an explanation of its distribution, including some implications for its control. In: Wernsdorfer WH, McGregor I (eds) Malaria; Principles and practice of malariology. Churchill Livingstone, Edinburgh

Nicolaiciuc D, Popa MI, Popa L (1999) Malaria in the whole world and Romania. Roum Arch Microbiol Immunol 58:289–296

Patz JA, Hulme M, Rosenzweig C, Mitchell TD, Goldberg RA, Githeko AK, Lele S, McMichael AJ, Le Sueur D (2002) Regional warming and resurgence. Nature 420:627–628

Peleman R, Benoit D, Goossens L, Bouttens F, Puydt HD, Vogelaers D, Colardyn F, Van de Woude K (2000) Indigenous malaria in a suburb of Ghent, Belgium. J Travel Med 7:48–49

Prokhorova IN, Kuz'menko Iiu, Artem've MM, Levitin AI, Kulikova LN (1992) The sensitivity of malarial mosquitoes to DDT and malathion in the southern European part of Russia. Med Parazitol Mosk 3:60–61

Ramsdale CD, Wilkes TJ (1985) Some aspects of overwintering in southern England of the mosquitoes Anopheles atroparvus and Culiseta annulata (Diptera: Culicidae). Ecol Ent 10:449–454

Rogers DJ, Randolph SE (2000) The global spread of malaria in a future, warmer world. Science 289:1763–1766

Roll Back Malaria (2002) (http://www.rbm.who.int/newdesign2/)

Smith T, Charlwood JD, Takken W, Tanner M, Spiegelhalter DJ (1995) Mapping the densities of malaria vectors within a single village. Acta Trop 59:1–18

Van den Ende J, Lynen L, Elsen P, Colebunders R, Demey H, Depraetere K, De Schrijver K, Peetermans WE, Pereira de Almeida P, Vogelaers D (1998) A cluster of airport malaria in Belgium in 1995. Acta Clin Belg 53:259–263

Vutchev D (2001) Tertian malaria outbreak three decades after its eradication. Jpn J Infect Dis 54:79–80

WHO Europe Centralized Information System for Infectious Diseases (http://data.euro.who.int/cisid/)

WHO (2001) (http://www.who.int/ith/chapter07_01.html)

World Health Organisation (1999) Strategy to Roll Back Malaria in the WHO European Region. Regional Office for Europe, Copenhagen

6.5 West Nile Virus: Ecology, Epidemiology and Prevention

Zdenek Hubálek, Bohumír Kříž, Bettina Menne

6.5.1 Introduction

Within a European-Commission funded project (EVK2-2000-00070), the National Institute of Public Health of the Czech Republic, the Stockholm University and the WHO assessed the influences of climate change on vector-borne diseases. This document is the result of the literature review of the presence of West Nile virus in Europe, the ecology, the potential influences of climate variability and change, and prevention measures.

The aim of this review is to contribute to a better understanding of temporal and spatial distribution of this vector-borne disease in Europe, of the influences of environmental factors, in particular the climate and weather, and to discuss potential risks. West Nile fever is re-emerging in Europe; therefore, this review also addresses measures to reduce disease burden and vector control measures.

West Nile virus (WNV) is the etiologic agent of West Nile fever (WNF: Goldblum et al., 1954; the code ICD-10-A92.3). The most serious manifestation of WNV infection is fatal encephalitis (inflammation of the brain) and encephalomyelitis in humans and horses, respectively, as well as mortality in certain domestic and wild birds. The term West Nile encephalitis is sometimes used for the disease instead of the more general term WNF, e.g. in North America.

WNV is a member of the Japanese encephalitis antigenic group of the genus *Flavivirus*, family *Flaviviridae*. The other members of this group are Japanese encephalitis, Koutango, Murray Valley encephalitis (subtype Alfuy), St. Louis encephalitis, Usutu and Yaounde viruses (van Regenmortel et al., 2000). All viruses of this group are transmissible by mosquitoes and most of them can cause febrile, sometimes severe and even fatal illnesses in humans.

WNV can be divided by the nucleotide sequence of a part of the envelope glycoprotein *env* gene into two major lineages: Lineage–1 strains, present in Africa, Middle East, Europe, India, Australia and the United States, and Lineage–2 strains (less virulent, they have not been involved in significant human or horse outbreaks) present in Subsaharan Africa. Lineage–1 virus includes Kunjin and Indian subtypes.

6.5.2 Current distribution and risks

West Nile flavivirus (WNV) has emerged or re-emerged in recent years in temperate regions of Europe, North Africa and North America, presenting a threat to public, equine, and animal health. It was first isolated from the blood of a febrile woman in the West Nile district of Uganda in 1937. A number of other WNV isolates were subsequently recovered from patients (including the topotype strain EV-101, relabelled later as Eg-101), birds and mosquitoes in Egypt. The virus was soon recognized as one of the most widespread species of *Flaviviridae*, with geographic distribution involving Africa, Eurasia, Australia (Kunjin) and, since 1999, also North America. Geographical distribution of WNV in Europe is shown in Figure 1 (Hubálek and Halouzka, 1999; Hubálek 2000).

WNV is transmitted infrequently to humans through the bite of an infected mosquito, usually of the genus *Culex*, but sometimes also by non-*Culex* mosquito bridge vectors. The virus can be transmitted also from person to person through blood transfusion or organ transplantation

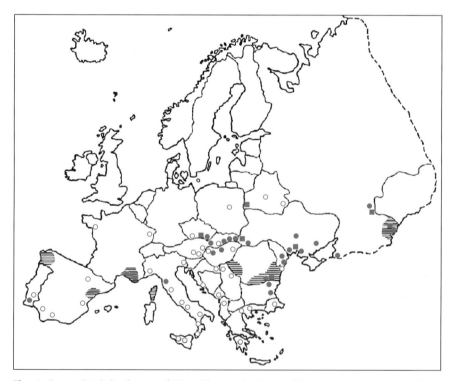

Fig. 1. Geographical distribution of West Nile virus in Europe: full circles, the virus isolated from mosquitoes or vertebrates; full squares, laboratory/confirmed human or equine cases of West Nile fever; circles and hatched areas, presence of specific antibodies (Hubálek and Halouzka, 1999)

(CDC, 2002). Infections by aerosol, e.g. in the laboratory have also been reported (Nir, 1959; Umrigar and Pavri, 1977; B.K. Johnson, personal communication). Moreover, WNV was isolated from a newborn breastfeeding baby (the mother had been infected by blood transfusion) in the United States (CDC, 2002).

Human West Nile Fever in Europe

In Europe, WNV presence was first indicated in 1958, when 1.8% of 112 healthy Albanian citizens revealed antibodies neutralizing WNV; the two sera did not cross-react with tick-borne Central European encephalitis flavivirus (Bárdoš et al., 1959). The incidence of human WNV infections in Europe remains largely unknown, and only some of the epidemics with tens or hundreds of WNF cases have probably been recorded (Hubálek and Halouzka, 1999). Larger outbreaks were described in the 1960s in southern France (the Rhone delta, mainly in 1962: Hannoun et al., 1964, 1969), Spain (Filipe and De Andrade, 1990), southern Russia (the Volga delta: Chumakov et al., 1964) and southwestern Romania (Topciu et al., 1971); in the 1970s in southern Ukraine and Belarus (Voinov et al., 1981); 1985 in western Ukraine (Buletsa et al., 1989); 1996–97 in southeastern Romania (Le Guenno et al., 1996); and 1999 in southern Russia again (Platonov et al., 2001). The epidemiological situation concerning WNV circulation in particular European countries is described below. Results of some serologic surveys are also mentioned. It is well known that WNV cross-reacts with antibodies to other flaviviruses in haemagglutination-inhibition test (HIT), whereas neutralization test (NT) including plaque-reduction neutralization test (PRNT) give specific results. Of the neutralization tests, $PRNT_{90}$ (i.e. that using a 90% reduction in plaque-forming units of the virus as the titre) is more specific than $PRNT_{50}$. Table 1 contains examples of antibody examinations.

It is important to add that in 1998 a large WNF epidemic of 624 cases in western Georgia preceded the 1999 Southern-Russian outbreak (Tsereteli et al., 1998). There has also been a significant WNV activity in Israel since 1998. A high morbidity and mortality was reported on goose farms, with the lethality 40%. Three WNV isolates were obtained from dead birds (a domestic goose and a stork *Ciconia ciconia*) in 1998, and a high prevalence rate of WNV antibodies was found among 38 goose raisers and avian veterinarians (84%), 83 persons living in localities frequently visited by storks (41%), while a lower rate in the general population (7% of 25 persons). In addition, 34% of 263 hospitalized children with unexplained meningoencephalitis or rash were seropositive. Three cases (two fatal) of WNV encephalitis were diagnosed in elderly people between August and December 1999 (Bin et al., 2000; Malkinson et al., 2002).

Table 1. Distribution of West Nile virus and West Nile fever in Europe

Country	Virus detection/ isolation (no. isol.)	Antibodies detected	WN fever
▌ Albania	–	Humans, 1958 (2%, NT)	–
▌ Austria	–	Humans, 1962–75 (1–6%, HIT) Wetland passerines, 1971 (1–3%, HIT) Wild mammals, 1970's (1–5%, HIT) Domestic mammals, 1964–77 (2–33%, HIT)	–
▌ Belarus	–	Humans, 1970–71 (1%, HIT) Wild birds, 1972–73 (3%, HIT)	Humans, 1970s (a few cases)
▌ Bulgaria	Mosquitoes, 1978 (3)	Humans (2–3%, HIT & NT) Wetland birds (2% HIT, 10% NT) Domestic animals (1%, HIT)	–
▌ Cyprus	Migrating birds (1)	–	–
▌ Czech Republic (southern)	Mosquitoes, 1997–01 (2)	Humans, 1988–89 (1%, NT) Humans, 1997 (2%, NT) Domestic mammals, 1978 (2%, HIT) Game animals, 1978–1980s (4–8%, HIT) Wetland birds, 1985–90 (4–10%, HIT)	Humans, 1997 (5 cases)
▌ France (southern)	Mosquitoes, 1964 (2)	Humans, 1963–65 (20%, HIT)	Humans, 1962–65 (> 12) Humans, 1963 (2)
		Horses, 1963–66 (30% HIT, 59% NT)	Horses, 1962–65 (c.50)
	Horse, 1965 (1)	Wetland birds, 1963–66 (6%, HIT)	Horses, 2000 (58)

Table 1 (continued)

Country	Virus detection/ isolation (no. isol.)	Antibodies detected	WN fever
		Dom.mammals, 1963–66 (17%, HIT) Game animals, 1963–66 (9%, HIT) Wild rodents, 1963–66 (4%, HIT) Humans, 1975–80 (5%, HIT) Horses, 1975–80 (2%, HIT)	(Horses, 1942: epizootic?)
Greece	–	Humans, 1988 (1%, NT) Domestic animals, 1970–78 (1–19%, HIT) Birds, 1970–78 (20%, HIT)	–
Hungary	Wild rodents, 1972 (2)	Humans, 1970s (4–6%, HIT) Cattle, 1970's (4–9%, HIT) Horses, 2000 (4 reactors, NT)	–
Italy	Horse, 1998 (1)	Horses, 1998 (40%, ELISA) Humans, 1965–70 (2–11%, HIT) Dom.mammals, 1965–70s (2–13%, HIT) Wild rodents, 1967–81 (1–8%, HIT) Migr.birds, 1967–69 (10%, HIT) Chickens, 1965–70s (0–20%, HIT)	Horses, 1998 (14)
Moldavia	Ixodid ticks, 1973 (1)	Humans, 1970s (3%, HIT)	–
Poland	–	Wild birds, 1996 (3–12%, HIT)	–
Portugal (southern)	Mosquitoes, 1971 (1)	Humans, 1967–70 (3%, HIT) Dom.ruminants, 1967–70 (15%, HIT) Horses, 1970 (29%, NT) Wild birds, 1967–70 (5%, HIT)	Horses, 1960's (epiz.)

Table 1 (continued)

Country	Virus detection/ isolation (no. isol.)	Antibodies detected	WN fever
Romania (southern)	Mosquitoes, 1960 (1)	Humans, 1980–95 (2–12%, HIT & NT)	Humans, 1966–70 (tens)
	Mosquitoes, 1996 (1)	Dom.mammals, 1980s (2–16%, HIT)	Humans, 1996–98 (408)
	Human, 1996 (1)	Wild mammals, 1980s (2–23%, HIT) Dogs, 1980s (19–45%, HIT) Wild birds, 1980s (up to 22%, HIT) Chickens, 1996–97 (41%, NT)	
Russia (southern)	Humans, 1963–65 (3)	Humans, 1963–65 (12%, HIT)	Humans, 1963–68 (>10)
	Humans, 1999 (2) Ixodid ticks, 1963–65 (4) Water birds, 1963–67 (2) Mosquitoes, 1967 (3)	Humans, 1999 Wetland birds, 1963–68 (4–59%, HIT) Sentinel birds, 1968 (12%, HIT)	Humans, 1999 (c.900)
Slovakia	Mosquitoes, 1972 (1)	Humans, 1962–64 (2%, HIT)	–
	Migratory birds, 1970–73 (4)	Migratory birds, 1970–73 (1–13%, NT) Game animals, 1970–80s (2%, NT) Domestic mammals, 1970–80s (1–8%, HIT) Pigeons, 1975 (5%, HIT)	
Spain	–	Humans, 1973–76 (8–30%, HIT) Wild rodents, 1979 (3%, HIT)	Humans, 1960–70s (epid.)
Ukraine (southern)	Humans, 1970s (4)	–	Humans, 1970–85 (>40)
	Wild birds, 1970s (7) Mosquitoes, 1970s (5)		
Yugoslavia (former)	–	Humans, 1970–80s (1–8%, HIT)	–

HIT haemagglutination-inhibition test; *NT* neutralization test

Imported Human Cases of West Nile Fever

There have been described few cases of an introduced WNF in Europe. A man contracted WNV meningitis during his stay in Syria in September 1984. The disease was diagnosed only after his return to Russia, and WNV was isolated from the blood on the 22nd day of the clinical disease, which included triphasic fever and necessitated a 31-day hospitalization (Ivanov et al., 1986).

An unusual WNF outbreak occurred in the crew on a Romanian ship (14 cases; one fatality); the index case appeared 9 days after the ship had left Constanca on 10 August 1975 (Draganescu et al., 1977). The disease had possibly been contracted in Romania or transmitted by mosquitoes breeding on the ship, and not during passing the Suez Canal as the authors believed.

In the Czech Republic, one case of WNF in a Czech citizen returning from the United States in 2002 was diagnosed and confirmed serologically in our laboratory (unpublished). He had acquired the infection in the United States.

Equine West Nile Virus Disease in Europe

Equine (horse, mule, donkey) disease, caused by WNV and called 'lourdige' in France or 'Near Eastern equine encephalitis' in Egypt, was described and experimentally reproduced as fever and diffuse encephalomyelitis with a moderate to heavy fatality rate in a few European countries (Box 1): southern France (Panthier et al., 1966; Guillon et al., 1968; Hannoun et al., 1969), southern Portugal (Filipe et al., 1973) and northern Italy (Cantile et al., 2000). Viremia is low in adult equines but moderate in young animals (Hannoun et al., 1969, Hannoun, 1971); foals cannot be therefore excluded as occasional donors of WNV for competent mosquito vectors.

Box 1: Equine West Nile virus identification cases

Portugal: An epizootic of encephalomyelitis in horses, most probably due to WNV, was recorded at Beja in southern Portugal before 1970: 29% of surviving horses had neutralizing antibodies against WNV (Filipe et al., 1973; Filipe and De Andrade, 1990). WNV was also isolated from local mosquitoes in 1971 (Table 1).

France: An epizootic occurred in the Rhone Delta (Camargue), with about 50 encephalomyelitic cases among domestic and semiferal equines in 1962 (Panthier et al., 1966; Guillon et al., 1968; Hannoun et al., 1969; Hannoun, 1971). Febrile form of the disease was not easily discovered among horses living in semiwild conditions here, but the biphasic, encephalomyelitic form was very apparent. However, epizootics of equine encephalomyelitis that had been probably caused by WNV were observed in Camargue much earlier, in 1942 (Hannoun, 1971). The virus was also isolated from mosquitoes

(1964: Table 1) and horses (1965) in Camargue. A few equine cases were observed from 1963 through 1965, but no cases were recorded in Camargue later. Concurrently, a much lower frequency of seropositive animals (2% of 99 horses) was revealed in the area later, between 1975 and 1980 (Rollin et al., 1982); this is a very low seroprevalence rate when compared to the mean 30% seropositive horses in this region between 1963 and 1965 (Hannoun et al., 1969). Another outbreak of WNF occurred among horses in the same area between September and November 2000, with at least 76 equine patients having neurologic disorders (28% died: Murgue et al., 2001; Durand et al., 2002).

█ **Italy:** An epizootic of WNV infection occurred in Tuscany near Pistoia ('Padule di Fucecchio' wetland) from August through October 1998 (Cantile et al., 2000). Fourteen horses from nine stables were diagnosed as having WNF with neurologic disorders (ataxia, loss of balance, depression): two horses died and four were euthanatized, whereas eight animals recovered. Histopathology revealed aseptic polioencephalomyelitis, with lesions mainly situated in the ventral horns of the thoracic and lumbar segments of the spinal cord. In the affected stables, the morbidity rate varied between 0.4 and 20%, and 39.6% of 159 local horses examined during the epizootic had antibodies to WNV. WNV was isolated from the CNS of a horse; the isolate shares 99.2% protein E gene nucleotide identity with a Senegal WNV isolate from *Culex neavei* in 1993 (V. Deubel, personal communication).

6.5.3 West Nile Virus Ecology in Europe

Arthropod Vectors of West Nile Virus in Europe

WNV has been isolated from eight species of mosquitoes and two species of hard ticks in Europe (Table 2). Principal vectors of WNV in Europe are largely ornithophilic species *Culex pipiens* and *C. modestus*. Exceptionally, in the warm and dry climatic zone of southern Russia, ixodid ticks (*Hyalomma marginatum*) were found to carry the virus (Butenko et al., 1967, 1968, Chumakov et al., 1968, 1974). The Russian virologists have also isolated WNV from soft ticks.

Of the species given in Table 2, a successful experimental transmission of WNV was described in mosquitoes *Culex pipiens* (Work et al., 1955; Hurlbut, 1956; Kostyukov et al., 1986), *Aedes caspius* (Hurlbut et al., 1956), and a long-term infection of ticks with WNV (but usually without subsequent regular transmission by bites) in *Dermacentor marginatus* (Chumakov et al., 1974). Some of the competent mosquito vector species of WNV occur throughout Europe (e.g. *Culex pipiens*).

Table 2. Isolation of West Nile virus from arthropods in Europe

Species	No. of isolations	Country
Mosquitoes		
▌ *Culex modestus*	3	France, Southern Russia
▌ *C. pipiens*	3	Romania, Czech Republic
▌ *Mansonia richiardii*	1	Southern Russia
▌ *Aedes cantans*	2	Slovakia, Western Ukraine
▌ *A. caspius*	1	Southern Ukraine
▌ *A. excrucians*	1	Southern Ukraine
▌ *A. vexans*	1	Southern Russia
▌ *Anopheles maculipennis group*	3	Portugal, Ukraine
Hard ticks		
▌ *Hyalomma marginatum*	4	Southern Russia
▌ *Dermacentor marginatus*	1	Moldavia

Sources: Hannoun et al., 1964, 1969; Butenko et al., 1967, 1968; Chumakov et al., 1968, 1974; Berezin et al., 1971, 1972; Filipe, 1972, Labuda et al., 1974; Lvov and Ilyichev, 1979; Vinograd et al., 1982; Savage et al., 1999; Hubálek et al., 1998, 2000

Vertebrate Hosts of West Nile Virus in Europe

Free-living birds are principal vertebrate hosts of WNV, with viraemia sufficient to infect mosquitoes. The virus was isolated from 11 wild aquatic
and terrestrial avian species in Europe (Table 3). All of them are migrants
except for the crow *(Corvus corone)* which is usually vagrant (East-Europe-
an populations are, however, migratory); five species (*Ardeola ralloides,
Plegadis falcinellus, Anas querquedula, Tringa ochropus, Streptopelia turtur*)
are members of the Palearctic-African bird migration system, having their
winter ranges in sub-Saharan Africa; these birds could occasionally transport WNV between Africa and Europe. The other species (*Fulica atra, Vanellus vanellus, Larus ridibundus, Sturnus vulgaris*) overwinter in Mediterranean countries. In addition, WNV was recovered from other European
bird species on other continents: *Ixobrychus minutus* in Central Asia (Chunikhin, 1973), *Botaurus stellaris* in Azerbaijan (Lvov and Ilyichev, 1979),
migrating *Ciconia ciconia* in Israel (Malkinson et al., 2002), *Anas platyrhynchos* in Tajikistan (Gordeeva, 1980), *Larus argentatus/cachinnans* in
Azerbaijan (Lvov and Ilyichev, 1979), *Sterna albifrons* and *Pica pica* in Tajikistan (Gordeeva, 1980), *Columba livia* and *Corvus corone* in Egypt (Work
et al., 1953), *Turdus merula* and *Sitta europaea* in Azerbaijan (Gaidamovich
and Sokhey, 1973), migrating *Sylvia nisoria* in Cyprus (Watson et al., 1972)
and *Motacilla alba* in Israel (Nir et al., 1972).

Viraemia sufficient to infect competent mosquito vectors has been observed in experimentally infected birds of a number of European avian spe-

Table 3. Isolations of West Nile virus from wild birds in Europe

Species	No. isolations	Country
Ardeola ralloides	1	Southern Ukraine
Plegadis falcinellus	1	Southern Russia
Anas querquedula	1	Southern Ukraine
Fulica atra	1	Southern Ukraine
Tringa ochropus	1	Slovakia
Vanellus vanellus	2	Slovakia, Southern Ukraine
Larus ridibundus	1	Slovakia
Streptopelia turtur	1	Slovakia
Corvus corone	1	Southern Russia
Corvus frugilegus	1	Southern Ukraine
Sturnus vulgaris	1	Southern Ukraine

Source: Butenko et al., 1968; Berezin et al., 1971; Sidenko et al., 1973; Grešíková et al., 1975; Ernek et al., 1977; Lvov and Ilyichev, 1979; Vinograd et al., 1982

cies (Hubálek and Halouzka, 1996), e.g. *Ardea cinerea, Anas acuta, Aythya ferina, Falco tinnunculus, Phasianus colchicus, Columba livia, Riparia riparia, Corvus corone, C. frugilegus* and *Passer domesticus*. Moreover, WNV antibodies have often been detected in waterbirds breeding in colonies (*Nycticorax nycticorax, Ardea cinerea, A. purpurea, Egretta garzetta, Ardeola ralloides, Phalacrocorax carbo, P. pygmaeus, Platalea leucorodia, Plegadis falcinellus, Sterna hirundo, Chlidonias nigra, Larus minutus, L. cachinnans*) and other European wetland birds (*Podiceps cristatus, Gavia arctica, Ciconia ciconia, Anas penelope, A. platyrhynchos, A. querquedula, Fulica atra, Gallinula chloropus, Vanellus vanellus, Philomachus pugnax, Recurvirostra avosetta, Himantopus himantopus, Gallinago gallinago, Calidris ferruginea, Sterna caspia, Chlidonias nigra, Locustella luscinioides, Acrocephalus* spp.).

Some terrestrial European avian species have also frequently revealed WNV antibodies (e.g. *Falco tinnunculus, F. vespertinus, Milvus migrans, Columba livia, Streptopelia senegalensis, S. decaocto, S. turtur, Erithacus rubecula, Corvus corone, C. frugilegus, Pica pica, Passer domesticus, P. montanus, Sturnus vulgaris*).

Migratory birds are obviously involved in the WNV dissemination and variability in the "Old World"; they are regarded instrumental in periodical re-introduction of WNV to temperate areas of Eurasia during spring migrations (Nir et al., 1968; Hannoun et al., 1972; Chunikhin, 1973; Ernek et al., 1977; Lvov and Ilyichev, 1979; Morvan et al., 1990; Hubálek, 1994; Hubálek and Halouzka, 1996; Savage et al., 1999; Rappole et al., 2000; Zeller and Murgue, 2001).

WNV isolations from mammals have only infrequently been reported in Europe (Hubálek and Halouzka, 1996): *Apodemus flavicollis* and *Clethrion-*

omys glareolus (Hungary), *Lepus europaeus* (S. Russia), horse (S. France, Italy), humans (France, Russia, Ukraine, Romania). Mammals, including man, are much less important than birds or unimportant as amplifying hosts of WNV in transmission cycles.

Only a few species of mammals are susceptible to WNV infection, first of all equines (see below). Out of the other mammals, morbidity and mortality has been documented in certain species in North America, but always at a very low incidence compared to birds, humans and horses. Experimental infection of sheep resulted in fever, abortion and infrequent encephalitis (Oudar et al., 1972; Barnard and Voges, 1986). Laboratory mice and Syrian hamsters are susceptible to WNV infection and often develop fatal encephalitis, even when inoculated peripherally (Karabatsos, 1985). Rodents stressed and/or immunosuppressed by cold, isolation, cyclophosphamide, corticosterone or bacterial lipopolysccaride (LPS) endotoxin injections develop fatal encephalitis even when attenuated WNV is given (Cole and Nathanson, 1968; Ben-Nathan et al., 1989, 1994; Lustig et al., 1992). Experimental infection of monkeys can also cause fever with occasional encephalitis, fatality or long-term virus persistence (Smithburn et al., 1940; Goverdhan et al., 1992). Only subclinical infection was observed in pigs (Oudar et al., 1972; Ilkal et al., 1994) and dogs (Peiris and Amerasinghe, 1994).

Interestingly, some amphibia also can harbour WNV: three isolates were reported from *Rana ridibunda* frog in Asian Tajikistan (Kostyukov et al., 1985), and the frog viraemia and donor ability were confirmed experimentally (Kostyukov et al., 1985). WNV reproduces very easily with formation of cytopathic effect in a clawed toad cell line at 28 °C (Leake et al., 1977). WNV was also detected in alligators and crocodiles in the United States and Israel. HI antibodies to WNV were reported in a wetland reptile *Natrix natrix* in Austria: 24% of 25 individuals had HI antibody, whereas terrestrial species of reptiles did not react with WNV antigen (Sixl et al., 1973). However, the role of amphibians and reptiles in the circulation of WNV in wetland ecosystems remains to be elucidated.

6.5.4 Transmission Cycle of West Nile Virus

The West Nile cycle involves the virus, a vertebrate host (many species of birds and some mammals) and a vector (a mosquito). The various vectors have been explained in the paragraphs above, each with their own particular biology. Humans become involved in this cycle by accident, they are regarded as incidental hosts. Most ecological requirements of WNV are similar to those of other arbovirus transmission. For successful transmission there is a need of a competent mosquito population, a susceptible amplifying vertebrate host population, the pathogen and appropriate environmental conditions. Temperature and humidity (Reiter, 1988), food and space resources, predators and parasites influence significantly the transmission (Fig. 2).

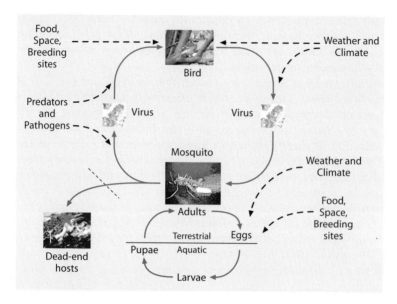

Fig. 2. Principal cycle of West Nile virus in Europe and North America (Source: CDC)

WNV circulation is confined to two basic cycles and ecosystems in Europe:

▌ Exoanthropic (rural, sylvatic) cycle involving wild, largely wetland birds and bird-feeding mosquitoes (*Culex pipiens pipiens, C. modestus, Mansonia richiardii*); Palearctic natural foci of WNV infections are mainly situated in wetland ecosystems (river deltas, floodplain ecosystems) and characterized by the 'bird-mosquito cycle';

▌ Synanthropic (urban) cycle involving synanthropic (and domestic) birds and mosquitoes feeding on both birds and humans (so-called bridge vectors, in Europe e.g. *C. pipiens molestus*).

In southern Russia only, an alternative and infrequent amblyommine tick-bird cycle has been described, where ticks may serve as substitute vectors in certain dry and warm habitats lacking mosquitoes and constitute the 'bird-tick cycle' (Lvov et al., 1989).

The principal cycle is exoanthropic (e.g. France 1962–65, Italy 1998). The synanthropic cycle predominated, e.g. in Bucharest during the 1996–97 epidemic (Savage et al., 1999; Tsai et al., 1998) or in Volgograd in 1999 (Lvov et al., 2000). Natural (exoanthropic, sylvatic) foci of WNV characterized by the wild bird-ornithophilic mosquito cycle probably occur in many wetlands of climatically warm and some temperate parts of Europe; these foci remain silent but could activate under circumstances supporting an enhanced virus circulation due to appropriate abiotic (weather) and biotic (increased populations of vector mosquitoes and susceptible avian hosts) factors.

6.5.5 Influence of Environmental Factors on Disease Risk

Many environmental factors including human activities might enhance vector population density (for example irrigation, heavy rains followed by floods and subsequent dry and warm weather, higher than usual temperature (Reeves et al., 1994; Reiter, 2001) suitable breeding habitats for mosquitoes, e.g. flooded basements in blockhouses (Bucharest 1996–98), the rainwater sewage system with sediment traps (New York 1999–2000)). Possibly also warm air currents can carry infected mosquitoes northwards (Sellers and Maarouf, 1990).

Additional factors might contribute such as the introduction of new potential vectors, virus adaptation to new vectors and/or hosts, virus genetic alterations and mutations, or social changes and human behaviour.

6.5.6 Possible Impact of Future Climate Risks

There is some indication that epidemics of WNF and ecologically similar St. Louis encephalitis in North America occur often after long periods of dry summers and one wet summer (Epstein 2001). The data are, however, insufficient at present.

The main season for WNF among humans and horses in Europe (as well as in North America) is from July to October, with a peak in August and September, due to the enhanced seasonal activity of vector mosquitoes.

Principal vectors of WNF in Europe are mosquitoes of the genus *Culex*: *Cx. pipiens* complex, and *Cx. modestus*. Whereas *Cx. pipiens* is distributed widely over all of Europe, *Cx. modestus* is confined to wetland and fish-pond areas of southern and central Europe and only lives in reedbelt (*Phragmites communis*, *Typha* spp.) habitats. Global climate changes can hardly affect the geographic range of *Cx. pipiens* in Europe considerably but, if the scenario of global warming is correct, the anticipated increased temperature could enhance population density of vector mosquitoes and, importantly, cause an advanced reproduction rate of WNV in the vector. However, it should be noted that large mosquito populations alone are not enough to cause epidemic transmission; the virulent strain must also be present.

6.5.7 Symptoms, Diagnosis and Treatment

Clinical Symptoms

West Nile fever in man is a febrile, influenza-like illness simulating sandfly fever or dengue. WNF is characterized by an abrupt onset (after an incubation period of 3–6 d) of moderate to high fever (3–5 d, infrequently biphasic), headache (often frontal and severe), sore throat, myalgia, backache, arthralgia, fatigue, conjunctivitis, ocular (retrobulbar) pain, photophobia, anorexia, nausea, abdominal pain, diarrhoea, distress, maculopapular rash (spreading from the trunk to the head and extremities; often a flushed face), slightly enlarged lymph nodes, respiratory involvement and, occasionally (up to 15% of cases), vomiting, meningism, encephalitis or anterior myelitis, confusion, insomnia or somnolence, mild hepatosplenomegaly, hepatitis, pancreatitis and myocarditis (Goldblum et al., 1954; Hannoun et al., 1969; Ivanov et al., 1986; Buletsa et al., 1989; Peiris and Amerasinghe, 1994; Hubálek, 2001). Common laboratory findings include slightly increased RBC sedimentation rate and mild leukocytosis; cerebrospinal fluid in patients with the CNS involvement is clear, showing moderate pleocytosis and elevated protein. Mortality is usually rare, occurring mainly in elderly patients. Recovery is complete (quite rapid in children but slow in adults, often accompanied by long-term myalgia and weakness), permanent sequelae are unknown. Many of the WNF symptoms were observed in 78 volunteers with underlying neoplastic disease who were experimentally treated by infection with Eg-101 strain of WNV to achieve pyrexia and oncolysis (Southam and Moore, 1954): e.g. 12% of them revealed diffuse encephalitis, and four patients died (WNV was recovered from their brains).

Virus Isolation and Virus RNA Detection

Demonstration of the presence of West Nile virus or virus-derived RNA in serum, cerebrospinal fluid (CSF), or other tissue confirms the diagnosis of a current infection. WNV can be recovered from the blood of a febrile human patient between the 1st and 4th day of the disease (fever) but occasionally as late as 18 days after the onset; prolonged viraemia is often observed in immunocompromised patients (Southam and Moore, 1954; Ivanov et al., 1986; Peiris and Amerasinghe, 1994). Isolation of the virus by growth in cell culture is insensitive, especially if specimens have not been fresh-frozen. Isolation of WNV can take three to seven days, depending on the amount of virus present.

RNA of West Nile virus can be identified by qualitative reverse-transcriptase polymerase chain reaction (RT-PCR) and by quantitative real-time PCR. However, the use of RNA-detection assays alone is not an appropriate diagnostic approach, because of their lower sensitivity. WNV antigen can be identified rapidly in postmortem brain specimens with the use of immunohistochemical staining.

Serological Tests

Because of the limitations of virus-detection assays, WNV infections in humans are most frequently diagnosed by assessment of the antibody response in an IgM and/or IgG antibody-capture enzyme-linked immunosorbent assay (ELISA) or haemagglutination-inhibition test (HIT, using goose erythrocytes), which can be performed in state public health laboratories in 24 to 36 hours. Approximately 75% of patients with flavivirus encephalitis have detectable IgM in serum or cerebrospinal fluid during the first four days of illness, and nearly all test positive by seven to eight days after the onset of illness. IgM antibodies may persist for a year or more after infection. Although identification of WNV–specific IgM in CSF confirms the presence of a current WNV infection, identification of virus-specific IgM in serum indicates only a probable infection and necessitates further testing, including the use of serum specimens from the acute and convalescent phases of illness to identify a change by a factor of four or more in the antibody titer. Because of serologic cross-reactions with other closely related flaviviruses (e.g. St. Louis encephalitis and dengue viruses) in both ELISA and HIT, virus-neutralization tests are used to confirm a diagnosis of WNV infection when only serum specimens are available. The most specific (and, therefore, the best confirmatory) serologic test is plaque-reduction neutralization test (PRNT$_{90}$) on VERO cells at present.

Humoral antibodies against WNV appear in the blood of patients since about the 5th day post infection – IgM earlier than IgG, and the latter persisting for extended periods (several years) contrary to the former (these persist for a few weeks).

Case Definition

The WNF case definition is based on the CDC Guidelines (2001) with some minor modifications.

Confirmed case: A confirmed case of WNF is defined as a febrile illness usually associated with neurologic manifestations ranging from headache to aseptic meningitis or encephalitis, combined with at least one of the following:
- Isolation of WNV from, or detection of WNV antigen or RNA in, tissue, blood, CSF or other body fluids
- Demonstration of IgM antibody to WNV in CSF
- At least fourfold increase in plaque-reduction neutralizing (PRNT) antibody titre to WNV in paired blood serum or CSF samples, taken at 2–4-week interval.

Probable case: A probable case of WNF is defined as a febrile illness usually associated with neurologic manifestations ranging from headache to aseptic meningitis or encephalitis, combined with at least one of the following:
- demonstration of IgM antibody to WNV in the blood serum
- demonstration of an elevated titre of WNV-specific antibody in convalescent-phase serum confirmed by PRNT.

Treatment

There is only symptomatic treatment for WNF at present.

6.5.8 Adaptation and Preventive Measures

Epidemiological Surveillance and Monitoring

Enhanced surveillance might become a priority for those countries that are affected or that are at higher risk for being affected by WNV because of bird migration patterns and virus spread. Depending on the geographic location of the country state, it would be necessary to decide if and when active surveillance should be implemented.

The surveillance should be preferentially local, concentrated on specific potential endemic foci of WNF. All significant outbreaks of WNF in humans or equines in Europe have occurred or originated in wetland areas, especially in river deltas harbouring dense populations of both water birds and ornithophilic mosquitoes, e.g. Rhone delta in southern France, Volga delta in southern Russia, Danube delta in Romania. There are many other similar wetland areas in Europe (Ebro and Donana deltas in Spain, Dnestr delta, but also continental wetlands like South-Moravian and South-Bohemian fishpond reserves in the Czech Republic) and these areas should be monitored preferentially.

Main components of the surveillance system for WNF in Europe should include (similarly as in North America) monitoring of:

▌ Vector mosquito populations for the infection rate with WNV and vector density
▌ Wild vertebrates for antibodies (serological surveys)
▌ the extent of its transmission outside the bird-mosquito cycle, such as passive surveillance (passive surveillance enhanced by general alerts to veterinarians) for neurologic disease in horses and other animals should be implemented
▌ Active ill and dead wild birds, as well as sentinel birds (ducks or chickens) monitoring for seroconversion
▌ Human population for disease (summer influenza-like symptoms, measles-like exanthema symptoms, aseptic meningoencephalitis).

Although the European outbreaks of WNF have not been associated with a mass dying of wild birds (e.g. corvids) as it is the case in North America, this absence of recorded dying birds might be biased because of the lack of surveillance in place. For instance, the American crow (*Corvus brachyrhynchos*) is a common, synanthropic species and detection of their carrions in towns and other human settlements is much more straightforward than it could be in the case of a relatively shy, exoanthropic European hooded crow (*Corvus corone*) whose corpses distributed in the field might easily escape the vigilance. This can be demonstrated on an outbreak of mass dying of synanthropic blackbirds (*Turdus merula*) and barn swallows (*Hirundo rustica*) due to Usutu flavivirus in Austria during 2001, when the ill and dead birds in the towns and villages were detected easily (Weissenböck et al., 2002).

Public Health Infrastructure

Effective surveillance, prevention and control of vector-borne diseases, including disease caused by WN virus, may require a re-evaluation of resource priorities.

Specific laboratory infrastructure, biosafety level 3 is required in reference laboratories. The arbovirus facilities should be supported to combat the new or re-emerging mosquito-borne diseases including WNF, which might spread further.

Unequivocal diagnosis of WN virus or other arbovirus infection requires specialized laboratory diagnostic tests. Success of surveillance activities is, therefore, dependent on the availability of laboratories that can provide diagnostic support especially in areas where WNF was already detected or in areas with favourable conditions for its occurrence.

Prevention and Control

Currently, the most effective way to prevent transmission of WN virus and other arboviruses to humans and other animals, or to control an epidemic once transmission has begun, is to reduce human exposure via mosquito control.

Mosquito Abatement
The most effective and economical way to suppress mosquitoes is by larval control. Experience suggests that this is best done through locally funded abatement programs that monitor mosquito populations and initiate control before disease transmission to humans and domestic animals occurs. These programs can also be used as the first-line emergency response for mosquito control if and when virus activity is detected in an area or human disease is reported. Preferentially, larval mosquito vectors should be controlled by *Bacillus thuringiensis* var. *israelensis* or *B. sphaericus* toxins. Permethrin preparations (Aqua Reslin Super etc.) and other insecticides to kill adult mosquitoes should only be used in emergency situations.

Public Outreach
A critical component of any prevention and control program for vector-borne diseases is public education about these diseases, how they are transmitted and how to prevent or reduce risk of exposure. Public education should utilize behavioural science and social marketing methods to effectively communicate information to target populations.

Immunization

Immunization is generally the best prevention of a virus disease, but there is no WNF vaccine at present. In the case of persisting high risk of WNF for humans, equids and precious bird species in certain enzootic areas, immunization with a cross-protective vaccine against Japanese encephalitis should be considered (Hubálek, 2000; Tesh et al., 2002b). It was previously found that animals (mice, monkeys, pigs) immunized with Japanese encephalitis virus were fully protected against WNV challenge (Goverdhan et al., 1992; Ilkal et al., 1994). There are two commercially available Japanese encephalitis vaccines which are used in Asia for immunization of children: (i) mouse-brain derived and formalin-inactivated Nakayama vaccine, used in Japan and South Korea; (ii) a live attenuated SA14-14-2 vaccine, used in China since 1988 (Sohn, 2000).

In addition, formalin-killed WNV vaccine (Fort Dodge Animal Health, Iowa) can be used for immunization of animals (horses, precious birds etc.). A live attenuated chimeric virus vaccine candidate (ChimerVax – WNV, Acambis Inc., Cambridge, MA) is in the state of preclinical testing (Tesh et al., 2002a).

Data Sharing

WNF is a zoonosis that affects a number of animal species, including humans. Effective monitoring and response require close coordination and data exchange between many agencies, including national and local public health agencies, veterinarian and biological institutions for vector control, agriculture and wildlife departments. Information and data exchange can be facilitated through a system of secure electronic communication, e.g. list servers and web sites that can be accessed by authorized users.

6.5.9 Future Research Needs

It has been demonstrated that WNV causes human cases and outbreaks of WNF even in temperate Europe. It is obvious that small clusters of WNF cases can only be detected by a surveillance effort and normally escapes the reporting (e.g. Czech Republic 1997). Sylvatic foci of WNV circulation are usually silent and remain undetected but they could activate and 'expand' under circumstances supporting the enhanced virus circulation due to appropriate abiotic (weather) and biotic (higher than normal populations of vector mosquitoes and susceptible avian hosts) factors. Sylvatic foci can only be detected by a systematic virological field investigation. It is still unsure how much outbreaks can be attributed to climatic factors or climate change. This problem presents a challenge for arbovirologists, infectionists and epidemiologists.

The mechanism of persistence of WNV in endemic foci of temperate Europe presents a significant challenge for further research. General hypotheses of how an arbovirus could overwinter under adverse climatic conditions have already been postulated (Reeves et al. 1994). According to them, WNV could persist in:

▌ hibernating female *Culex* spp. mosquitoes

▌ transovarially infected progeny of *Culex* spp.
▌ chronically infected vertebrates (birds, frogs?).

These issues have to be addressed; present data partially support any of the particular mechanisms or their combinations as well. For instance, the 'hibernating vector' idea has been supported by a few field and experimental data on female *Culex* (Taylor et al., 1956; Cornel et al., 1993; Nasci et al., 2001), and also documented in North America.

Another long-discussed but plausible scheme is the occasional re-introduction of the virus by infected migratory birds from (sub)tropical foci at irregular intervals (Hannoun et al., 1972; Chunikhin, 1973; Ernek et al., 1977; Lvov and Ilyichev, 1979). On the other hand, the birds might also spread the virus the other way, southwards, during their fall migration (Malkinson et al. 2002). The role of birds in the distribution of WNV is probably greater than that of the weather, in that principal vector species of mosquitoes occur over Europe. Up to date, no WNV strain was isolated nor antibody to WNV detected in Northern Europe, e.g. in Scandinavia (but serosurveys in these countries have been limited). However, Sindbis alphavirus (*Togaviridae*), with an ecological transmission cycle very similar to WNV (principal hosts are wild birds and the vectors ornithophilic mosquitoes predominantly of the *Culex* genus: Hubálek and Halouzka, 1996), causing hundreds of human cases of a febrile illness with headache, myalgia, arthralgia and rash, has appeared quite surprisingly in Scandinavia and Russian Karelia since 1981, although it normally occurs in tropical and subtropical countries of Africa, Asia and, less so, southern Europe. Migratory birds have played an important role in the geographical distribution of Sindbis virus and have caused its introduction in Northern Europe (Uryvaev et al., 1992; Hubálek, 1994). Experimental viraemia in some avian species lasts three days or more, and even a chronic course of infection was demonstrated experimentally in birds (Semenov et al., 1973).

6.5.10 Conclusions

The re-emergence of WNF in Europe has probably been caused by a whole complex of ecological (abiotic and biotic) factors, including simultaneous presence in a specific geographic area of:
▌ Favourable weather conditions (temperature, humidity and their appropriate annual distribution)
▌ Abundant competent mosquito vectors (*Culex pipiens pipiens, Cx. modestus, Mansonia richiardii*), which requires the presence of favourable habitats
▌ Infected (viraemic) migrating birds
▌ Abundant competent (amplifying, viraemic) local avian hosts
▌ Abundant competent bridge vectors (*Culex pipiens molestus, Aedes* spp. etc.) able to feed on both birds and mammals

▌ Susceptible population of equines and/or humans (as 'dead-end' or occasional vertebrate hosts).

Obviously, there is a need to start a surveillance programme for WNF targeted at least for selected countries of the European Region, and to address many research questions that have been mentioned in the last paragraph.

References

Antipa C et al (1984) Serological investigations concerning the presence of antibodies to arboviruses in wild birds. Revue Roumaine de Medecine, Virologie 35:5–9

Antoniadis A et al (1990) Seroepidemiological survey for antibodies to arboviruses in Greece. Archives of Virology, 1(Suppl):277–285

Aspöck H et al (1973) Virologische und serologische Untersuchungen über die Rolle von Vögeln als Wirte von Arboviren in Österreich. Zentralblatt für Bakteriologie, I Originale, A224:156–167

Balducci M et al (1967) Survey for antibodies against arthropod-borne viruses in man and animals in Italy, II. American Journal of Tropical Medicine and Hygiene, 16:211–215

Bárdoš V et al (1959) Neutralizing antibodies against some neurotropic viruses determined in human sera in Albania. Journal of Hygiene, Epidemiology, Microbiology and Immunology 3:277–282

Berezin VV et al (1971) Investigation on the ecology of mosquito-borne viruses using sentinel animals in the Volga Delta (in Russian). Voprosy Virusologii, p 739–745

Berezin VV et al (1972) Importance of birds in natural cycle of mosquito-borne viruses in the Volga delta (in Russian). In: Cherepanov AI (ed) Transcontinental Connections of Migratory Birds and their Role in the Distribution of Arboviruses. Novosibirsk, Nauka, pp 310–313

Bin H et al (2000) Resurgence of virulent West Nile virus infections in humans in Israel. Abstracts of the Third International Conference on Emerging Infectious Diseases. Atlanta, GA, p 252

Buletsa BA et al (1989) Neurological manifestations of West Nile fever in the Transcarpathian region, Ukrainian SSR (in Russian). Zhurnal Nevropatologii i Psichiatrii, 89(2):29–30

Butenko AM, Chumakov MP, Stolbov DN (1967) Serological and virological examinations in a natural focus of West Nile fever in the Astrakhan region (in Russian). Voprosy Medicinskoi Virusologii 1:208–211

Butenko AM et al (1968) New investigations of West Nile virus infections in the USSR – Astrakhan region (in Russian). Materialy XV Nauchnoi Sessii Instituta Poliomielita i Virusnykh Encefalitov (Moskva), 3:175–176

Cantile C et al (2000) Clinical and neuropathological features of West Nile virus equine encephalomyelitis in Italy. Equine Veterinary Journal, 32:31–35

Centers for Disease Control and Prevention. Epidemic/epizootic West Nile virus in the United States: revised guidelines for surveillance, prevention, and control, 2001 (http://www.cdc.gov/ncidod/dvbid/westnile/publications.htm)

Cernescu C et al (2000) Continued transmission of West Nile virus to humans in southeastern Romania, 1997–1998. Journal of Infectious Diseases, 181:710–712

Chumakov MP et al (1974) Isolation and identification of Crimean hemorrhagic fever and West Nile fever viruses from ticks collected in Moldavia (in Russian). Medicinskaya Virusologia, 22(no 2):45–49

Chumakov MP, Belyaeva AP, Butenko AM (1964) Isolation and study of an original virus from Hyalomma plumbeum plumbeum ticks and from the blood of a febrile patient in the Astrakhan region (in Russian). Materialy XI Nauchnoi Sessii Instituta Poliomielita i Virusnykh Encefalitov (Moskva), pp 5–7

Chumakov MP et al (1968) Isolation of West Nile virus from Hyalomma plumbeum plumbeum Panz. ticks (in Russian). Trudy Instituta Poliomielita i Virusnykh Encefalitov (Moskva) 13:365

Chunikhin SP (1973) Introduction to arbovirus ecology (in Russian). Medicinskaya Virusologia (Moskva), 21(no 1):7–88

Draganescu N et al (1977) Epidemic outbreak caused by West Nile virus in the crew of a Romanian cargo ship passing the Suez canal and the Red Sea on route to Yokohama. Revue Roumaine de Médecine, Virologie, 28:259–262

Draganescu N, Gheorghiu V (1968) On the presence of group B arbovirus infections in Romania. Investigations on the incidence of West Nile antibodies in humans and certain domestic animals. Revue Roumaine de Inframicrobiologie 5:255–258

Draganescu N, Girjabu E, Iftimovici R (1993) Investigations sérologiques chez l'homme et chez diférentes espéces d'animaux domestiques de la région de sylvosteppe de Roumanie, pour le dépistage de quelques Togaviridae et Bunyaviridae. Revue Roumaine de Virologie 44:207–210

Draganescu N et al (1975) Investigations on the presence of antibodies to several flaviviruses in humans and some domestic animals in a biotope with a high frequency of migratory birds. Revue Roumaine de Médecine, Virologie 26:103–108

Duca M et al (1989) A seroepidemiologic and virological study on the presence of arboviruses in Moldavia in 1961–1982 (in Romanian). Revue de Médecine Chirurgie de la Societe Medical Nationale Iasi 93:719–733

Durand B et al (2002) West Nile virus outbreak in horses, southern France, 2000: results of a serosurvey. Emerging Infectious Diseases 8:777–782

Epstein PR (2001) West Nile virus and the climate. Journal of Urban Health 78:367–371

Ernek E et al (1977) Arboviruses in birds captured in Slovakia. Journal of Hygiene, Epidemiology, Microbiology and Immunology 21:353–359

Filipe AR (1972) Isolation in Portugal of West Nile virus from Anopheles maculipennis mosquitoes. Acta Virologica 16:361

Filipe AR, De Andrade HR (1990) Arboviruses in the Iberian Peninsula. Acta Virologica 34:582–591

Filipe AR, Pinto MR (1969) Survey for antibodies to arboviruses in serum of animals from southern Portugal. American Journal of Tropical Medicine and Hygiene 18:423–426

Filipe AR, Sobral M, Campanico FC (1973) Encefalomielite equina por arbovirus. A propósito de una epizootia presumtiva causada pelo virus West Nile. Revista Portuguesa de Ciencias Veterinarias 68:90–101

Gaidamovich SY, Sokhey J (1973) Studies on antigenic peculiarities of West Nile virus strains isolated in the USSR by three serological tests. Acta Virologica 17:343–350

Garea González MT, Filipe AR (1977) Antibodies to arboviruses in northwestern Spain. American Journal of Tropical Medicine and Hygiene 26:792–797

Goldblum N, Sterk VV, Padersky B (1954) West Nile fever – the clinical features of the disease and the isolation of West Nile virus from the blood of nine human cases. American Journal of Hygiene 59:89–103

Gordeeva ZE (1980) Ecological connections between arboviruses and wild birds in Tajikistan (in Russian). In: Lvov DK (ed) Ecology of Viruses. Moskva, Institut Virusologii 126–130

Goverdhan MK et al (1992) Two-way cross-protection between West Nile and Japanese
encephalitis viruses in bonnet macaques. Acta Virologica 36:277–283

Grešíková M, Sekeyová M (1967) Haemagglutination-inhibiting antibodies against arboviruses in the population of Slovakia. Journal of Hygiene, Epidemiology, Microbiology and Immunology 11:278–285

Grešíková M, Sekeyová M, Prazniaková E (1975) Isolation and identification of group
B arboviruses from the blood of birds captured in Czechoslovakia. Acta Virologica,
19:162–164

Grešíková M et al (1979) Haemagglutination-inhibiting antibodies to some arboviruses
in human and animal sera from České Budějovice. In: Sixl W (ed) Naturherde von
Infektionskrankheiten in Zentraleuropa. Hygiene Institut der Universität Graz-Seg-
gau, S 25–29

Grešíková M et al (1973) Haemagglutination-inhibiting antibodies against arboviruses
in human sera from different regions in Steiermark (Austria). Zentralblatt für Bakteriologie. I Originale 224:298–302

Gubler DJ (2002) The global emergence/resurgence of arboviral diseases as public
health problems. Archives of Medical Research, 33:330–342

Gubler DJ et al (2000) West Nile virus in the United States: guidelines for detection,
prevention, and control. Viral Immunology 13:469–475

Guillon JC et al (1968) Lésions histologiques du systéme nerveux dans l'infection,
virus West Nile chez le cheval. Annales de l'Institut Pasteur 114:539–550

Han LL et al (1999) Risk factors for West Nile virus infection and meningoencephalitis, Romania, 1996. Journal of Infectious Diseases 179:230–233

Hannoun C (1971) Progrés récents dans l'étude des arbovirus. Bulletin de l'Institut
Pasteur 69:241–278

Hannoun C, Corniou B, Mouchet J (1972) Role of migrating birds in arbovirus transfer between Africa and Europe. In: Cherepanov AI (ed) Transcontinental connections of migratory birds and their role in the distribution of arboviruses. Novosibirsk, Nauka, pp 167–172

Hannoun C, Panthier R, Corniou B (1969) Epidemiology of West Nile infections in the
South of France. In: Bárdős V (ed) Arboviruses of the California complex and the
Bunyamwera group. Bratislava, Publishing House SAS, pp 379–387

Hannoun C et al (1964) Isolement en France du virus West Nile a partir de malades et
du vecteur Culex molestus Ficalbi. Compte Rendus de l'Académie Des Sciences,
Serie D, 259:4170–4172

Hubálek Z (1994) Pathogenic microorganisms associated with free-living birds. Acta
Scientiarum Naturalium Brno, 28(5):1–74

Hubálek Z (2000) European experience with the West Nile virus ecology and epidemiology: could it be relevant for the New World? Viral Immunology 13:415–426

Hubálek Z (2001) Comparative symptomatology of West Nile fever. Lancet 358:254–
255

Hubálek Z, Halouzka J (1996) Arthropod-borne viruses of vertebrates in Europe. Acta
Scientiarum Naturalium Brno 30(4–5):1–95

Hubálek Z, Halouzka J (1999) West Nile fever – a reemerging mosquito-borne viral
disease in Europe. Emerging Infectious Diseases 5:643–650

Hubálek Z, Halouzka J, Juřicová Z (1999) West Nile fever in Czech Republic. Emerging
Infectious Diseases 5:594–595

Hubálek Z et al (1998) First isolation of mosquito-borne West Nile virus in the Czech
Republic. Acta Virologica 42:119–120

Hubálek Z et al (1989) Arboviruses associated with birds in southern Moravia, Czechoslovakia. Acta Scientiarum Naturalium Brno 23(7):1–50

Hurlbut HS (1956) West Nile virus infection in arthropods. American Journal of Tropical Medicine and Hygiene 5:76–85

Ilkal MA et al (1994) Experimental studies on the susceptibility of domestic pigs to West Nile virus followed by Japanese encephalitis virus infection and vice versa. Acta Virologica 38:157–161

Ivanov KS, Lobzin YV, Nikolaev VP (1986) West Nile fever (West Nile encephalitis) (in Russian). Zhurnal Mikrobiologii Epidemiologii Immunobiology 7:110–113

Juřicová Z (1992) Antibodies to arboviruses in game animals in Moravia, Czech Republic (in Czech). Veterinární Medicina (Brno) 37:633–636

Juřicová Z, Halouzka J (1993) Serological examination of domestic ducks in southern Moravia for antibodies against arboviruses of the groups A, B, California and Bunyamwera (in Czech). Biológia (Bratislava) 48:481–484

Juřicová Z et al (1993) Virological examination of cormorants for arboviruses (in Czech). Veterinární Medicína (Praha) 38:375–379

Katsarov G et al (1980) Serological studies on the distribution of some arboviruses in Bulgaria. Problems of Infectious and Parasitic Diseases 8:32–35

Khodko LP et al (1979) Serological investigations in arbovirus natural foci in Belarus (in Russian). Proceedings of the Conference "Prirodnoochagovye Infekcii i Invazii", Vilnyus, pp 85–87

Koptopoulos G, Papadopoulos O (1980) A serological survey for tick-borne encephalitis and West Nile viruses in Greece. Zentralblatt für Bakteriologie (Suppl 9):185–188

Kostyukov MA et al (1986) Experimental infection of Culex pipiens mosquitoes with West Nile by feeding on infected Rana ridibunda frogs and its subsequent transmission (in Russian). Medicinskaya Parazitologia 64(6):76–78

Kostyukov MA et al (1985) The frog Rana ridibunda, a host of hematophagous mosquitoes in Tajikistan, as the West Nile virus reservoir (in Russian). Medicinskaya Parazitologia 63(3):49–50

Kožuch O et al (1976) Surveillance on mosquito-borne natural focus in Záhorská lowland. In: Sixl W, Troger H (eds) Naturherde von Infektionskrankheiten in Zentraleuropa. Hygiene Institut der Universität Graz, S 115–118

Kunz C (1963) Nachweis hämagglutinations-hemmender Antikörper gegen Arbo-Viren in der Bevölkerung Österreichs. Zentralblatt für Bakteriologie, I Originale A, 190: 174–182

Labuda M, Kožuch O, Grešíková M (1974) Isolation of West Nile virus from Aedes cantans mosquitoes in west Slovakia. Acta Virologica 18:429–433

Leake CJ, Varma MGR, Pudney M (1977) Cytopathic effect and plaque formation by arboviruses in a continuous cell line (XTC-2) from the toad Xenopus laevis. Journal of General Virology 35:335–339

Le Guenno B et al (1996) West Nile: a deadly virus? Lancet 348:1315

Lopes MC et al (1970) Survey for antibodies against arthropod-borne viruses in man and animals in Italy, III. Annali del Istituto Superiore di Sanita 6:121–124

Lvov DK et al (2000) Isolation of two strains of West Nile virus during an outbreak in southern Russia, 1999. Emerging Infectious Diseases 6:373–376

Lvov DK, Ilyichev VD (1979) Bird Migration and Transport of Pathogenic Agents (in Russian). Moskva, Nauka

Lvov DK, Klimenko SM, Gaidamovich SY (1989) Arboviruses and Arbovirus Infections (in Russian). Moskva, Medicina

Malkinson M et al (2002) Introduction of West Nile virus in the Middle East by migrating white storks. Emerging Infectious Diseases 8:392–397

Molnár E et al (1973) Arboviruses in Hungary. Journal of Hygiene, Epidemiology, Microbiology and Immunology 17:1–10

Molnár E et al (1976) Studies on the occurrence of tick-borne encephalitis in Hungary. Acta Veterinaria Academiae Scientiarum Hungaricae 26:419–438

Murgue B et al (2000) West Nile in the Mediterranean basin: 1950–2000. Annals of the New York Academy of Sciences 951:117–126

Murgue B et al (2001) West Nile outbreak in horses in southern France, 2000: the return after 35 years. Emerging Infectious Diseases 7:692–696

Murgue B et al (2002) The ecology and epidemiology of West Nile virus in Africa, Europe and Asia. Current Topics in Microbiology and Immunology 267:195–221

Murphy FA (1998) Emerging zoonoses. Emerging Infectious Diseases 4:429–435

Nasci RS et al (2001) West Nile virus in overwintering Culex mosquitoes, New York City, 2000. Emerging Infectious Diseases 7:742–744

Nir Y (1959) Airborne West Nile virus infection. American Journal of Tropical Medicine and Hygiene 8:537–539

Nir Y et al (1972) Arbovirus activity in Israel. Israel Journal of Medical Sciences 8:1695–1701

Panthier R et al (1968) Épidemiologie du virus West Nile: étude d'un foyer en Camargue. Les maladies humaines. Annales de l'Institut Pasteur 115:435–445

Panthier R et al (1966) Isolement du virus West Nile chez un cheval de Camargue atteint d'encéphalomyélite. Compte Rendus de l'Académie des Sciences, Serie D 262:1308–1310

Peiris JSM, Amerasinghe FP (1994) West Nile fever. In: Beran GW, Steele JH (eds) Handbook of Zoonoses: Viral, 2nd. Boca Raton, FL, CRC Press, pp 139–148

Petrescu A et al (1993) The evidence of circulation of some viruses in a great town of Romania during the first half of 1992. Revue Roumaine de Virologie 44:243–251

Platonov AE et al (2001) Outbreak of West Nile virus infection, Volgograd region, Russia, 1999. Emerging Infectious Diseases 7:128–132

Rappole JH, Derrickson SR, Hubálek Z (2000) Migratory birds and spread of West Nile virus in the Western Hemisphere. Emerging Infectious Diseases 6:319–328

Rappole JH, Hubálek Z (2003) Migratory birds and West Nile virus. Journal of Applied Microbiology 94(Suppl):47–58

Reeves WC et al (1994) Potential effect of global warming on mosquito-borne arboviruses. Journal of Medical Entomology 310:323–332

Reiter P (1988) Weather, vector biology and arboviral recrudescence. In: Monath TP (ed) Arboviruses: Epidemiology and Ecology, vol. 1. Boca Raton, FL, CRC Press, pp 215–255

Reiter P (2001) Climate change and mosquito-borne disease. Environmental Health Perspectives 109(Suppl 1):141–161

Rollin PE et al (1982) Résultats d'enquetes séroépidemiologiques récentes sur les arboviroses en Camargue: populations humaines, équines, bovines et aviaires. Médecine et Maladies Infectieuses 12:77–80

Rusakiev M (1969) Studies on the distribution of arboviruses transmitted by mosquitoes in Bulgaria. In: Bárdoš V (ed) Arboviruses of the California Complex and the Bunyamwera Group. Bratislava, Publishing House SAS, pp 389–392

Savage HM et al (1999) Entomologic and avian investigations of an epidemic of West Nile fever in Romania in 1996, with serologic and molecular characterization of a virus isolate from mosquitoes. American Journal of Tropical Medicine and Hygiene 61:600–611

Sekeyová M, Grešíková M (1967) Haemagglutination-inhibiting antibodies against arboviruses in cattle sera. Journal of Hygiene, Epidemiology, Microbiology and Immunology 11:417–421

Sekeyová M, Grešíková M, Kouch O (1976) Haemagglutination-inhibition antibodies to some arboviruses in sera of pigeons trapped in Bratislava. In: Sixl W, Troger H (eds) Naturherde von Infektionskrankheiten in Zentraleuropa. Hygiene Institut der Universität Graz, S 187–189

Sellers RF, Maarouf AR (1990) Trajectory analysis of winds and Eastern equine encephalitis in USA 1980–85. Epidemiology and Infection 104:329–343

Semenov BF et al (1973) Studies on chronic arbovirus infections in birds (in Russian). Vestnik Akademii Medicinskikh Nauk SSSR 2:79–83

Sidenko VP et al (1973) Birds of southern Ukraine – a reservoir of tick-borne encephalitis (in Russian). Ekologia Virusov 1:164–169

Sidenko VP et al (1976) Prediction of West Nile fever morbidity (in Russian). Virusy i Virusnye Zabolevania 4:95–96

Sixl W et al (1973) Haemagglutination-inhibiting antibodies against arboviruses in animal sera, collected in some regions of Austria. Zentralblatt für Bakteriologie I. Originale 224:303–308

Sixl W, Stünzner D, Withalm H (1976) Untersuchungen auf Anthropozoonosen bei Landarbeitern in Österreich. Zentralblatt für Bakteriologie I Originale A 234:265–270

Sohn YM (2000) Japanese encephalitis immunization in South Korea: past, present, and future. Emerging Infectious Diseases 6:17–24

Southam CM, Moore AE (1954) Induced virus infections in man by the Egypt isolates of West Nile virus. American Journal of Tropical Medicine and Hygiene 3:19–50

Stünzner D et al (1973) Hunde und Katzen als Indikatoren von Arboviren. Wissenschaftliche Arbeiten aus Burgenland, Sonderheft 1:53–56

Tesh RB et al (2002 a) Efficacy of killed virus vaccine, live attenuated chimeric virus vaccine, and passive immunization for prevention of West Nile virus encephalitis in hamster model. Emerging Infectious Diseases 8:1392–1397

Tesh RB et al (2002 b) Immunization with heterologous flaviviruses protective against fatal West Nile encephalitis. Emerging Infectious Diseases 8:245–251

Topciu V, Rosiu N, Arcan P (1971) Contribution to the study of arboviruses in Banat. Revue Roumaine de Inframicrobiologie 8:101–106

Topciu V et al (1971) Existence des arbovirus de groupe B (Casals) décelée par sondages sérologiques chey quelques espces animales de la province du Banat (Roumanie). Archives Roumaines de Pathologie Expérimentale et de Microbiologie 30:231–236

Tsai TF et al (1998) West Nile encephalitis epidemic in southeastern Romania. Lancet 352:767–771

Tsereteli DG, Tsiklauri RA, Ivanidze EA (1998) West Nile fever in West Georgia. Abstracts of the Eight International Congress of Infectious Diseases, Boston, MA, pp 206

Umrigar MD, Pavri KM (1977) Comparative biological studies on Indian strains of West Nile virus isolated from different sources. Indian Journal of Medical Research 65:596–602

Ungureanu A, Popovici V, Nicolescu G, Tutoveanu A, Ionita I, Safta M et al (1989) Arbovirus infection prevalence in Romania: preliminary data (in Romanian). Revista de Igiena Bacteriologie Virusologie Parazitologie Epidemiologie Pneumoftiziologie, Seria Bacteriologia, Virusologia Parazitologia Epidemiologia 33:341–346

Uryvaev LV et al (1992) Isolation and identification of Sindbis virus from migratory birds in Estonia (in Russian). Voprosy Virusologii 37:67–70

van Regenmortel MHV et al (2000) Virus Taxonomy. Academic Press, San Diego

Verani P et al (1967) Survey for antibodies against arthropod-borne viruses in man and animals in Italy. I. American Journal of Tropical Medicine and Hygiene 16:203–210

Verani P et al (1971) Arbovirus investigations in southern Italy (Calabria). Journal of Hygiene Epidemiology Microbiology and Immunology 15:405–416

Vesenjak-Hirjan J, Punda-Polič V, Dobec M (1991) Geographical distribution of arboviruses in Yugoslavia. Journal of Hygiene Epidemiology Microbiology and Immunology 35:129–140

Vinograd IA et al (1982) Isolation of West Nile virus in southern Ukrainian SSR (in Russian). Voprosy Virusologii 27:567–569

Vinograd IA, Obukhova VR (1975) Isolation of arboviruses from birds in western Ukraine (in Russian). Sbornik Trudov Instituta Virusologii (Moskva) 3:84–87

Voinov IN, Rytik PG, Grigoriev AI (1981) Arbovirus infections in Belarus (in Russian). In: Drozdov SG et al (eds) Virusy i Virusnyje Infekcii. Moskva, Institut Poliomielita i Virusnykh Encefalitov, pp 86–87

Watson GE, Shope RE, Kaiser MN (1972) An ectoparasite and virus survey of migratory birds in the eastern Mediterranean. In: Cherepanov AI (ed) Transcontinental Connections of Migratory Birds and their Role in the Distribution of Arboviruses. Novosibirsk, Nauka, pp 176–180

Weissenböck H et al (2002) Emergence of Usutu virus, an African mosquito-borne Flavivirus of the Japanese encephalitis virus group, Central Europe. Emerging Infectious Diseases 8:652–656

Work TH, Hurlbut HS, Taylor RM (1953) Isolation of West Nile virus from hooded crow and rock pigeon in the Nile Delta. Proceedings of the Society for Experimental Biology and Medicine 84:719–722

Work TH, Hurlbut HS, Taylor RM (1955) Indigenous wild birds of the Nile Delta as potential West Nile virus circulating reservoirs. American Journal of Tropical Medicine and Hygiene 4:872–878

Zeller HG, Murgue B (2001) Role des oiseaux migrateurs dans l'épidémiologie du virus West Nile. Médecine et des Maladies Infectieuses, 31(Suppl 2):168–174

6.6 Ecology, Epidemiology and Prevention of Hantaviruses in Europe

Milan Pejcoch, Bohumír Kříž

6.6.1 Introduction

Although hantaviruses were not classified as an independent taxonomic category until 1978, knowledge of these viruses has increased rapidly. The most serious diseases caused by hantaviruses occur in Asia and the Americas. However, thousands of newly infected people are also reported every year in Europe, particularly in the northern parts of the continent and in the Balkans. In Europe, hantavirus-caused diseases have been known since the early 20th century, although the causative agents were not identified at that time. Currently, a total of four hantavirus genotypes are known in Europe: Puumala, Dobrava, Saaremaa, and Tula. In addition, the Topografov virus seems to be circulating among lemmings in northern Russia. Of these, at least the Puumala, Dobrava and Saaremaa are human pathogens.

Hantaviruses circulate among live, wild rodents in nature, and the spread of these viruses is directly and indirectly influenced by the climate. Climatic changes alter the ecology and phenology of hantaviruses in their natural reservoirs. There are many scientific works describing the influence of the nature of biocenoses on hantavirus foci, and the influence of climate on the distribution and evolution not only of modern hantaviruses, but also those of the geological past.

Hantaviruses circulate in natural foci with no vectors, and the infection is transmitted most frequently by aerosol, the contaminated excreta of infected rodents or by accidental injury involving an infected animal. The virus is excreted in the urine, stools and saliva, and the respiratory tract serves as an entry point for the infection of a new host (Tsai, 1987). Asymptomatic and persistent infection occurs in rodents. Humans can be infected upon coincidental entrance into a natural focus, usually during leisure activities, agricultural or forest work, or during military exercises. A risk is present everywhere humans are exposed to the excreta of infected rodents or with any other subjects contaminated by rodents.

Hantaviruses are classified in the order *Mononegavirales,* family *Bunyaviridae, genus Hantavirus.* Hantavirus infections have been known for centuries under various names. In 1976 Dr Ho Wang Lee and his coworkers from Korea discovered the causative agent of Korean haemorrhagic fever in a striped field mouse (*Apodemus agrarius*) (Lee et al., 1978). The mouse was captured near the Hantaan River, which crosses the 38th parallel. Nobody knew then that the name of an unknown Korean river would be quoted in almost 2000 scientific publications on hantaviruses. Ho Wang Lee named the virus after the river, thus, establishing a genus *Hantavirus,* which now includes over 20 distinct genotypes. The name of the diseases caused by the hantavirus in the European region was unified as "haemorrhagic fever with renal syndrome", abbreviated as HFRS. The hantaviruses in Europe have an affinity towards renal tissue, while those in the Americas have an affinity towards pulmonary tissues, causing a severe and violent disease called hantavirus pulmonary syndrome (HPS) or hantavirus cardiopulmonary syndrome (HCPS). With respect to human infections, hantaviruses range from nonpathogenic genotypes to genotypes that cause severe diseases with high mortality rates. The highest mortality rate reported for hantavirus was discovered in the United States in 1993, when the mortality rate reached 40%.

Hantaviruses are classified among the so-called rodent-borne viruses, which circulate in the natural foci of infections among their natural hosts from the order of rodents (*Rodentia*). The only exception is the Thottapalayam virus, which has been confirmed in only one insectivore species (*Insectivora*), the musk shrew *Suncus murinus* from India (Carey et al., 1971). According to Sironen, Vaheri and Plyusnin (2001) the evolution of hantaviruses follows several routes:

- genetic drift
- genetic shift
- recombination
- reassortment.

During the historical long-term evolution of hantaviruses, the respective hantavirus genotypes separated into distinct host species. However, hantaviruses can also infect secondary hosts such as bats, elks and humans. The individual evolutionary branches of hantaviruses separated hundreds of thousands of years ago. During the coevolution of hantaviruses with their specific hosts, three groups of hantaviruses were formed with links to three different groups of rodents. Within the groups, both the rodents and the hantaviruses are related to each other (Table 1).

6.6.2 Description of the Disease

In the human body, hantaviruses attack the endothelial cells, causing acute thrombocytopenia. Both pathogenic and non-pathogenic hantaviruses replicate in the pulmonary endothelial cells and in the macrophages of the lungs in particular. The virus antigen, however, can be detected in many organs, especially in the spleen, kidneys and lungs (Mackow and Gavrilovskaya, 2000; Nolte et al., 1995; Yanagihara et al., 1990).

Hantaviral diseases include at least two major syndromes: haemorrhagic fever with renal syndrome in the Old World, with the prevailing affinity of the etiologic agent towards the renal tissue, and hantavirus pulmonary syndrome in the New World, with the prevailing affinity of the etiologic agent towards the pulmonary tissue. Both diseases have immunopathological backgrounds, and the clinical symptoms are caused by inflammatory mediators.

Table 1. Summary of species from genus *Hantavirus,* family *Bunyaviridae* (Lee et al., 1981)

Species	Abbreviation	Hosts
Andes	AND	*Oligoryzomys longicaudatus, O. chacoensis, flavescens, Bolomys obscurus, Akadon azarae*
Bayou	BAY	*Oligoryzomys palustris*
Black Creek Canal	BCC	*Sigmodon hispidus*
Cano Delgadito	CAD	*Sigmodon alstoni*
Dobrava-Belgrade	DOB	*Apodemus flavicollis*
El Moro Canyon	ELMC	*Reithrodontomys megalotis*
Hantaan	HTN	*Apodemus agrarius*
Isla Vista	ISLA	*Microtus californicus*
Khabarovsk	KHA	*Microtus fortis*
Laguna Negra	LAN	*Calomys laucha*
Muleshoe	MUL	*Sigmodon hispidus*
New York	NY	*Peromyscus leucopus*
Prospect Hill	PH	*Microtus pennsylvanicus*
Bloodland Lake	BLL	*Microtus ochrogaster*
Puumala	PUU	*Clethrionomys glareolus*
Rio Mamore	RIOM	*Oligoyomys microtis*
Rio Segundo	RIOS	*Reithrodontomys mexicanus*
Saaremaa	SAA	*Apodemus agrarius*
Seoul	SEO	*Rattus norvegicus, R. rattus*
Sin Nombre	SN	*Peromyscus maniculatus, P. leucopus*
Thailand	THAI	*Bandicota indica*
Thottapalayam	TPM	*Suncus murinus*
Topografov	TOP	*Lemmus sibiricus*
Tula	TUL	*Microtus arvalis, M. rossiaemeridionalis*

Haemorrhagic fever with renal syndrome has characteristic systemic manifestations. The kidneys show signs of tubular necrosis. In hantavirus pulmonary syndrome, T-lymphocytes affect the pulmonary endothelium, while interferon-γ and tumor necrosis factor are the main agents causing a reversible increase in vascular permeability, leading to non-cardiogenic pulmonary edema. Both syndromes are accompanied by myocardial depression and hypotension or shock (Peters et al., 1999). Death occurs as a result of uremia, shock, and pulmonary edema (Hjelle et al., 1995).

Haemorrhagic fever with renal syndrome occurs in Eurasia (with the exception of diseases caused by the Seoul virus, which have worldwide occurrence, although they occur predominantly in the Far East). The disease is caused by the following genotypes: Hantaan, Seoul, Puumala, Dobrava and probably Saaremaa (Linderholm and Elgh, 2000). The Hantaan virus is spread in Asia, especially in the Far East, and causes a severe form of the disease with a mortality rate of 5% to 10%, especially as a result of haemorrhagic shock and renal failure. Related to Hantaan virus in terms of antigen properties, Dobrava virus causes the most severe form of disease throughout Europe. Dobrava occurs primarily in the Balkan Peninsula, and extends into Central Europe. The Saaremaa virus has been separated from the Dobrava genotype as a distinct genotype in Estonia. First described in Finland, Puumala causes a mild form of the disease, often called epidemic nephropathy. Its mortality rate is less than 1%.

Haemorrhagic fever with renal syndrome is a febrile disease with a variable level of damage to hemostasis and renal functions. It includes a wide range of symptoms from asymptomatic infections up to fulminant haemorrhagic shock and death. The disease is characterized by fever, thrombocytopenia, systolic hypotension, or internal bleeding, and renal dysfunction (Linderholm and Elgh, 2000). The average incubation period is 14 days, ranging from 5 to 42 days (Hjelle et al, 1995).

Hantavirus Ecology

Hantaviruses are known to have high host specificity. Each genotype has its distinct host from the order of rodents (*Rodentia*) with which it has evolved in the so-called coevolution process that lasted for over 20 million years (Marshall, 1993; Nichol, 2000).

The following three subfamilies of rodents have been hosts for hantaviruses: *Murinae, Arvicolinae* and *Sigmodontinae*. There is a related group of hantaviruses, corresponding in both genomic and antigenic terms, to every rodent subfamily. The subfamily *Murinae* hosts the hantaviruses of the Old World: Hantaan, Dobrava and Saaremaa, transmitted by the genus *Apodemus*. The genus *Rattus* (brown rat, black rat) is a host of the Seoul virus, which has a worldwide occurrence because of the distribution of brown and black rats. In both the western and eastern hemispheres, voles and lemmings from the *Arvicolinae* subfamily host distinct groups of hantaviruses, such as Puumala and Tula in Eurasia, and Prospect Hill in North America. Members of the *Sigmodontinae* subfamily occur only in the western hemisphere, in North and South America, and host the Sin Nombre and Andes viruses and related genotypes causing HPS. The Sin Nombre virus is both genetically and antigenically closer to Puumala virus than to Hantaan (Linderholm and Elgh, 2000). An as yet unknown relationship seems to exist between the Thottapalayam virus and its host from the order of insectivores, *Suncus murinus*. It is unclear whether this is an isolated finding and the primary host remains to be identified among rodents, or whether the Thottapalayam virus is an exception among the hantaviruses.

In addition to these reservoir hosts, hantaviruses can infect other animal groups such as carnivores, insectivores, bats and humans.

The coevolution of hantaviruses with their hosts has important implications.

- The distinct properties of the various hantaviruses have developed as an adaptation to the distinct genetic environment of their hosts.
- The current occurrence of distinct hantaviruses reflects the complex history of cospeciation and the migration of rodents (the last major migration of rodents was caused because of glaciation and deglaciation in the northern hemisphere). This has created the basis for the circulation of hantaviruses on different continents, their coexistence in several geographic areas, and for the geographic grouping of genetic variants of hantaviruses.
- Humans are usually end hosts for hantaviruses in terms of evolution, and outbreaks of infection in humans provides no contribution to the evolutionary process of hantaviruses.

Over 100 000 human infections caused by hantaviruses are reported worldwide every year (Elgh et al., 1997). Today, the epidemiologic evaluation of old data regarding the seroprevalence in people or animals is often complicated because it is not always clear which of the current genotypes corresponds to past findings. The so-called western serotype corresponds most probably to the current Puumala genotype, while the so-called eastern serotype can be identified as the Dobrava genotype, cross-reacting with the Hantaan virus.

The seroprevalence of Puumala infections ranges from 5% to 9% in Northern Sweden (Ahlm, 1994), and from 5.1% in Estonia (Golovljova et al., 2002) up to 0.5–2% in Central and Western Europe. In Southern Europe, hantavirus antibody seroprevalences have been reported to be 5% in Bosnia, 4% in Greece, 1.6% in Croatia, and 1.7% in Slovenia (Avsic-Zupanc, 1998). The incidence of haemorrhagic fever with renal syndrome in the European part of Russia is on average 5/10 000, caused mainly by the Puumala virus, with the highest morbidity in Bashkirostan (58/100 000) (Tkachenko et al., 1999). The Dobrava virus is the main cause of haemorrhagic fever with renal syndrome in the Balkans (Lundkvist et al., 1997). The seroprevalence of antibodies to the Saaremaa virus in Estonia is 3.4% in a healthy population (Golovljova et al., 2002). The infection of humans with the Tula virus is exceptional (Vapalahti et al., 1996).

The Situation in the Czech Republic

The history of hantavirus research in the Czech Republic goes back to the former Czechoslovakia because many research projects were performed together with Slovak scientists. The migration of people between the Czech Republic and Slovakia continues to be rather high. The first reports of a disease corresponding to the hantavirus infection in the former Czechoslovakia were reported by Plank et al. in 1955 and by Poljak in 1956. The reports describe patients from eastern Slovakia who died from the disease. During the years 1956 and 1957, a working group studied the natural focus of haemorrhagic nephroso-nephritis (probably caused by hantaviruses) in

Ruská Poruba, eastern Slovakia (Plank, Režucha, and Rojkovič, 1961; Rosický et al., 1961; Rosický and Bárdoš, 1961). Seven cases of the disease were reported in Slovakia by the year 1966, of which only the last reported patient recovered (Bilčíková et al., 1990; Grešíková et al., 1988b).

After being discovered, hantaviruses were studied by Grešíková et al. from the Institute of Virology at the Slovak Academy of Sciences in Bratislava (Grešíková et al., 1984, 1986, 1988b). Their work resulted in the publication of a monograph (Grešíková et al., 1988a). In 1984, serological analysis confirmed the occurrence of antibodies against hantaviruses among the population of eastern Slovakia (Bilčíková et al., 1990; Grešíková et al., 1988b, 1994). This finding resulted in the logical assumption that hantaviruses can also cause human disease in this area. This hypothesis was confirmed in 1987 and 1988, when the disease was reported in four people. Of these four, antibodies against the eastern type (current HTN-like) were confirmed in two patients, and antibodies against the western type (current PUU) were confirmed in the other two. All patients suffered from the following symptoms: fever, headache, pain in the lumbosacral region, nausea, conjunctivitis, proteinuria, hepatomegaly and reddening of mucous membranes. All patients recovered.

Daneš began research into hantaviruses in the Czech Republic in 1984 (Daneš, 1985, 1987; Daneš et al., 1986, 1991, 1992a, b). The first two cases of mild disease were reported in the Břeclav region. Fever, headache, proteinuria, hematuria and back-pain were the prevailing clinical signs (Kobzík et al., 1992). Human sera and sera taken from both domestic and wild animals, including small rodents, were examined for the presence of specific antibodies. In addition, the lungs of rodents were examined for the presence of the hantavirus antigen. A total of 5228 human sera samples were examined and positive titers were identified in approximately 1% of these. Daneš et al. (1992a) reported an occurrence of 9.9% and 29.4%, respectively, in two groups of elderly farmers from southern Moravia. Antibodies against the eastern type were found in some people, while antibodies against the western type or both types were detected in others. Thus, it is obvious that the infection of humans occurs in the natural foci of hantavirus in the Czech Republic, although there are only sporadic reports of disease. The infection usually goes unnoticed, or is mild and often not recognized.

A case of imported HFRS was reported in a soldier in the United Nations Protection Force (UNPROFOR) serving in the Balkans (Pejčoch et al., 1996; Petrů et al., 1997). Vacková et al. (2000, 2002) published two other studies relating to the army and Dušek et al. (2001) described two cases of interstitial nephritis in paediatric patients.

The epidemiologic situation can be illustrated by the results from the examination of animals. Hantavirus antigens, in particular of the western type (perhaps Puumala), were seen in five rodent species in the Czech Republic: the yellow-necked field mouse, common field mouse, common vole, European pine vole and bank vole (Daneš et al., 1991; Pejčoch et al., 1992). Antibodies against these viruses in the Czech Republic were also found in hares, roe deer, fallow deer and even in domestic cattle. Positive findings and results were most frequently reported in southern Moravia.

During the years 1991 and 2001, a total of 1494 small mammals were trapped in the Czech Republic and examined for the presence of hantavirus antigen in the lungs. The hantavirus antigen was found in 101 animals (Pejč och et al., 2003). Most of the positive findings were reported for *Microtus arvalis* (97 samples). In addition, positive results were found twice in *Microtus subterraneus* and once in *Apodemus sylvaticus* and in *Clethrionomys glareolus*. Positive findings in *M. arvalis* were reported in 12 districts of the Czech Republic. When assessing the influence of sex on the positivity of results in the common vole, there is a predominance of positive females, but this is, however, non-significant ($\chi^2 = 2.33$; $p = 0.127$). The mean positivity during the entire period of monitoring in common voles was only 14.12%. However, this amount fluctuated over the course of a year, with the highest values during the spring (23.91% in March). The second highest yearly maximum was found during the autumn (18.27% in October). An increasing tendency towards positive results was found with the increase of weight in the individual animals examined.

Hantaviruses are known to cause acute kidney disorders in humans. However, data in the literature also describes the chronic effects of hantavirus infections. At the Hemodialysis Centre in west Bohemia, 301 people who being treated, ranging in age from 20 to 87 years old, were examined for anitbodies. Antibodies against hantavirus were confirmed in five people (1.7%) (Pejčoch et al., unpublished data, 2002).

In this work supported by the Grant Agency Ministry of Health of the Czech Republic grant number NI 5896-3 and the European Commission cCASHh EVK-CT-2000-00070 grant, a total of 710 blood sera samples, taken from randomly selected people who were over 20 years old and originating from the Czech Republic, were examined for the presence of antibodies against the Puumala and Hantaan antigens. Hantaan was used for its antigenic similarity with the Dobrava virus, because the latter is not supplied in commercial enzyme-linked immunosorbent assay (ELISA) kits.

It was found that five people generated IgG antibodies and two of the five also generated IgM antibodies against the Hantaan antigen cross-reacting with the Dobrava virus. In addition, two people were found to be only positive in the IgM class. Therefore, in total, seven people responded to the Hantaan antigen (1.0%). Eight people responded by generating IgG but not IgM to the Puumala antigen. Two people had a positive response to the Puumala antigen only in the IgM class. In total, 10 people (1.4%) responded to the Puumala antigen. Three people responded to both antigens.

The results suggest that Puumala and Hantaan-like genotypes circulate in the Czech population. The Tula virus was detected in approximately one tenth of common voles trapped in fields. However, the Tula virus plays no role in human pathology, even though a single human infection was identified with this genotype in southern Moravia: a healthy blood donor who was engaged in fishing (Vapalahti et al., 1996). Also, the occurrence of a clinical disease following infection with the Tula genotype was recorded in 2002, manifested as fever and exanthema (Schultze et al., 2002).

Based on the results of this research, we can suggest that the following three hantavirus genotypes circulate in the Czech Republic: Tula, Dobrava and Puumala. Found in virtually all populations of common voles examined in the Czech Republic, Tula is the most frequently occurring genotype. Although this virus can infect humans, the infections caused by it are likely to be asymptomatic. The most severe disease is caused by the Dobrava genotype, with genus *Apodemus flavicollis* being the host. Scientists in south Bohemia confirmed the circulation of this genotype in rodents (Křivanec, personal communication, 2002). Seroprevalence in the adult human population of the Czech Republic is approximately 1% (determined using the Hantaan antigen). Using molecular genetics techniques, the Puumala virus was detected in the common vole in Košíky in the Uherské Hradiště district. We identified antibodies in approximately 1% of the adult human population. However, the clinical course of the infection has not yet been very well examined in the Czech Republic, even though several cases of interstitial nephritis have been reported for this genotype.

The Situation in Europe

Hantavirus is an important human health issue in northern Europe and in Fennoscandia. Swedish authors described a novel disease from northern Scandinavia in 1934, giving an account of a mild form of haemorrhagic fever with renal syndrome, named nephropathia epidemica (Myhrman, 1951). The first account of the European hantavirus, Puumala, discovered in the bank vole (*C. glareolus*), also comes from this region (Brummer-Korvenkontio et al., 1980). Brummer-Korvenkontio et al. describe the discovery of the hantavirus antigen in bank voles and the reaction of this antigen with the sera of people suffering from nephropathia epidemica.

Research on hantaviruses in central Europe began in Czechoslovakia in 1984 (Daneš et al., 1986; Grešíková et al., 1984) and in Germany in 1985 (Pilaski, Zöller and Blenk, 1986). Other findings concerning hantaviruses were made on the other side of Europe, in the Balkan Peninsula. The Dobrava genotype (synonym Belgrade) was found in the former Yugoslavia (Avsic-Zupanc et al., 1992). In 1994, the third European hantavirus was described and named after the central Russian town of Tula (Plyusnin et al., 1994). The last distinct genotype from Europe described to date originates from and bears the name of Saaremaa Island in Estonia (Sjölander et al., 2001). In addition, the Topografov genotype was reported in western Siberia, with lemmings being the hosts (Escutenaire and Pastoret, 2000).

Geographic Distribution of Hantaviruses in Europe. There are four genospecies of hantaviruses in Europe:

Puumala Virus: The occurrence of the Puumala virus was first reported in Finland (Brummer-Korvenkontio et al., 1980). The occurrence of the Puumala virus is reported virtually throughout Europe, excluding the Mediterranean coastal regions, most of the Iberian Peninsula, and Greece. The Puumala virus is highly variable and constitutes several lines in Europe.

▌ *Dobrava Virus:* The Dobrava virus has been confirmed in Europe in the Balkans, Hungary, the Czech Republic, Germany, the Netherlands, Belgium, Fennoscandia, and Estonia. The earlier findings should be designated as Dobrava-like, because the said authors separated the Estonian strains into the distinct Saaremaa serotype.

▌ *Saaremaa Virus:* The Saaremaa virus (SAA) was described in 2001 by Sjölander et al. (Sjölander et al., 2001) from Estonia, who separated Dobrava as a distinct genotype with the striped field mouse (*Apodemus agrarius*) as a host. As mentioned above, the earlier findings of the Dobrava virus in *A. agrarius* should be re-evaluated as, in fact, the Saaremaa virus. The distribution area of the Saaremaa hantavirus is probably within the distribution area of the striped field mouse.

▌ *Tula Virus:* The Tula hantavirus has been detected in various European countries. The first confirmation of its genome was performed in Russia (Plyusnin et al., 1994). The first isolation was carried out successfully from common voles caught in Tvrdonice, South Moravia (Plyusnin et al., 1995). Other findings have been reported in Slovakia, Belgium, Switzerland, Austria, and Poland. In Serbia, the Tula virus was identified in *Microtus subterraneus* (Song, Gligic and Yanagihara, 2002).

▌ *Topografov Virus:* The Topografov virus was discovered in lemmings from the locality bearing the same name in western Siberia (*Lemmus sibiricus*) (Plyusnin et al., 1996b). However, given the distribution area of its host, the occurrence of Topografov can also be expected in the northernmost parts of Russia.

6.6.3 Climate–Disease Relationships

Climatic changes have considerable implications for infections bound to the natural environment. It is, however, difficult to disentangle the climate effects from the many other factors influencing disease distribution, intensity and frequency. With regard to hantaviruses, some research was carried out after the Sin Nombre virus outbreak in 1993. Nichol et al. (1996) examined the influence of landscape changes on the emergence of RNA viruses. The outbreak of HPS in the southwestern United States in 1993, caused by the Sin Nombre hantavirus, is often referred to in this connection. The participation of a climatic phenomenon, El Niño, that causes increased precipitation and excessive proliferation of deer mice (*Peromyscus maniculatus*, a host of Sin Nombre), is also considered. One of the theories is that the increased density of rodent populations resulted in the easier propagation of the virus and increased the risk of human infections.

Engelthaler et al. (1999) used data obtained by the remote sensing of the Earth to identify the high-risk areas for HPS in southwestern United States.

They based their work on the assumption that weather fluctuations (especially excessive precipitation), probably caused by El Niño, are associated with a change in vegetation, with fluctuations in the population of *P. maniculatus* and with an increased risk of HPS for humans.

Unexpected rains in 1991 and 1992 during the usually dry spring and summer, and the mild winter in 1992, are believed to have created suitable conditions for the increase in local rodent populations. This hypothesis is based on the following observations:

- The timing and abundance of precipitation in the southwestern United States tends to be influenced by El Niño/Southern Oscillation (ENSO);
- Some *P. maniculatus* populations increased dramatically in areas where precipitation was above average, but remained near normal levels where precipitation did not increase; and
- Excess proliferation of rodents changed in short distances. Households where people were suffering from HPS had higher rodent populations than neighbouring households where people were free of the disease, or than in randomly selected households located at least 25 km away.

The response of local populations to environmental fluctuations can substantially affect the risk for humans. Localizing HPS cases to those sites where the people met infected rodents, the authors studied the relationship between the environment and HPS risk. The following two sources were used:

- data on precipitation from 196 meteorological stations from March to June
- satellite imaging.

Three archived Landsat Thematic Mapper (TM) images, dated mid-July 1992, were selected and TM images of the reflected light in six bands (three in the visible and three in the infrared spectrum) were used. The resolution was 30 m. The images were imported into the Raster Geographic Information System. Three satellite images from mid-June 1995, when no ENSO phenomenon was observed, were selected for comparison.

Distribution of HPS cases in the HPS area was compared to the environmental factors measured with the TM imaging. The original model covered an area of 12 279 km^2. This area included 14 sites with HPS cases and 36 control sites. For validation analysis, 14 sites for HPS and 134 control sites were used in an area of 92 921 km^2 (Glass et al., 2000).

Climatic conditions are tightly associated with the biocenosis pattern in the respective landscape and with the conditions for the existence of natural foci of hantaviruses. There is little data in the literature that provides ecological evaluations of the natural foci of hantaviruses. Plank, Režucha and Rojković (1961) briefly characterize European biocenoses with the occurrence of HFRS. According to them, natural foci constitute a continuous strip of land from north-east Croatia, through Hungary and eastern Slovakia, up to Trans-Carpathian Ukraine. Sporadic cases also occurred in Bulgaria and the Danube Lowland. This is a largely hilly terrain with oak or beech forests. The occurrence of the disease has two peaks, one during early spring, and the second in late August and September.

Vasilenko et al. (1990) reported that 83.8% of all Bulgarian patients treated for HFRS lived at an altitude of more than 900 m above sea level (ASL), and 73.4% of them worked in the wood-logging and primary processing of wood pulp industries. In the former Yugoslavia, this disease is also found in forest complexes in submountainous elevations (Avsic-Zupanc et al., 1989). An HFRS focus was described from the area of Plitvice Lakes, where 11 out of 14 workers worked in wood logging and 13 of them even lived in cabins built in natural beech forests (Vesenjak-Hirjan et al., 1971). The most heavily affected area in Germany is a territory between the Main, Danube and Rhine rivers in altitudes ranging from 500 m to 700 m above sea level. The most important biocenosis in this respect is the beech forest (Pilaski et al., 1991). The majority of HFRS cases in France are located north of the river Seine, where the climate is more humid (Van Ypersele de Strihou, 1991).

Verhagen et al. (1987) described one of the most detailed ecological characteristics of the natural foci of Puumala virus in Belgium. Of 80 infected terrestrial mammals, 33 were caught in humid forests, 26 in wetlands (mostly in the contact zones with humid forests) and 21 in fields, moorlands and in scrublands.

Grešíková (1988 a) summarizes a geo-botanical evaluation of different biotopes of natural foci of HFRS in Slovakia. The bedrock is not homogeneous there. With respect to the climate, the localities can be classified from warm and dry areas with mild winters to mildly cold areas. Annual average temperatures range from 7.5 °C to 9.5 °C. Annual precipitation ranges from 566 mm to 740 mm. Grešíková also mentioned some similarities regarding vegetation. In the localities of east Slovakia, the original plant community was oak-hornbeam forest with relatively thermophilic plants in the undergrowth. Current vegetation there is young oak-hornbeam forest depleted by some original species in the undergrowth. In other localities, the original oak-hornbeam forest was destroyed and replaced by scrublands with blackthorn, or converted into grassland at other sites. West Slovak localities are situated in a territory influenced by large rivers in floodplain forests, which are markedly affected by human activities and by the quantity of synanthropic plant species. Suitable hiding places and a good supply of food for small rodents are important common factors in all these localities. Uncut undergrowth of high grasses or a layer of dead leaves that have been resisting decay for a long time characterizes these localities.

Unar, Daneš and Pejčoch (1996) documented the relationship between the climate and hantavirus circulation in nature in south Moravia. The analysis of the topographic and ecological situation in the natural foci of the hantavirus in south Moravia revealed that the dissemination of infection in the population of terrestrial mammals depends on the macroclimatic, local and microclimatic conditions of the respective locality. In the lowlands and highlands of the investigated areas, the localities with positive findings of hantavirus are particularly bound to the alluvia of the rivers, outside the reach of permanent swamping, and to the bases of adjacent slopes. In higher altitudes, where the macroclimate is generally colder and more humid, infected populations of rodents can also be found in the upper parts of the slopes and in terrain elevations, which are sunnier and somewhat drier. These findings are in accordance with other data from the relevant literature on Europe.

In southern Europe, the natural foci of hantaviruses are primarily situated in the forests of submountainous areas because they are lacking in warm and dry lowlands. In France, the foci are bound to the somewhat colder and more humid northern part of the country, while in Belgium, which has an ocean climate, the foci occur mostly in mesophilic sites. The rugged pattern of the territory enables a more precise determination of the range of altitudes and related macroclimatic conditions, in which hantaviral foci are located. The analysis of the plant species patterns of the individual biotopes provides evidence of the mesophilic character of the foci. A limiting influence of dry weather on the dissemination of hantaviruses, highlighted for example by French authors (Bowen et al., 1995), can also be shown in the Czech Republic.

In the Czech Republic, two phytocenological images from two neighbouring localities in the southwest part of the Třebíč district demonstrate the link between the natural foci of hantaviruses and mesophilic conditions. In the first image, hantavirus was found to circulate in the vole population in the Kostníky site, while the occurrence of hantavirus was not confirmed in the Police locality. This is despite the fact that the air distance between the two localities is less than 2 km, the difference in altitude is less than 10 m and the orientation and declination of both localities are identical. The important differences were the variance in terrain pattern and related soil conditions of the two localities. While the Kostníky site is situated on a mild gradual slope with heavy loamy soil, the Police locality is situated on the peak of a smaller terrain elevation with milder clay and sand soil that is more prone to drying. Hence, there are more mesophilic perennial species in the Kostníky locality (such as *Dactylis glomerata, Rumex crispus, Plantago major, Arctium tomentosum, Luzula campestris, Arrhenatherum elatius, Achillea millefolium, Carduus acanthoides, Trifolium pratense* and *Heracleum sphondylium*). In the second image, annual plants (sometimes even having an ephemeral nature) can be seen in fair numbers (for example, *Myosotis arvensis, Arabidopsis thaliana, Veronica polita, Erophila verna, Myosurus minimus* and *Galium tricorne*).

The connection of hantaviruses with the climate can even be traced in geologic history. The current distribution of distinct genospecies of hantaviruses reflects a complicated history of cospeciation and rodent migration (the last great migration of rodents resulted from the consequences of glaciation and the thawing of the northern hemisphere). The distribution patterns form a basis for the circulation of hantaviruses between different continents, for their coexistence in some geographic areas and for the geographic reassortment of genetic variants of hantavirus. Different strains of the respective hantavirus type show geographic clustering that reflects a somewhat complicated history in the shifting of the distribution area of the hosts. For example, the latest changes in the distribution area of C. glareolus, a host of the Puumala virus, resulted from the recolonization of Europe, and in particular Fennoscandia, after the last glacial period 10 000 years ago. These migrations affected the evolution of the Puumala virus.

The multiyear population density cycles of rodents have also had an influence on the dynamics of hantavirus circulation in nature. Although not well understood, these cycles are definitively related to the climate. For example, C. glareolus have 3 to 4 year population cycles in northern Europe and, as mentioned previously, in temperate Europe higher than usual summer temperature leads to mast years and increased populations of seed-eating forest rodents. HFRS outbreaks are closely associated with these increases in rodent densities.

Computer models suggest that in the year 2050, the Czech lowlands will have similar average temperatures as those experienced today in Toulouse or near Lyon. The number of tropical days will probably increase in south Moravia, with maximum temperatures sometimes reaching over 40 °C (Matyáš, 2002). These changes will affect the distribution area of natural hosts and vectors of some diseases. The anticipated changes are for the most part negative. Infections with natural foci, such as hantavirus infections, will be affected both directly and through their hosts. When studying climatic changes, the influence of the environment cannot be separated from other influences, including the influence of human activities. Therefore, Harvell et al. (2002) focussed on the influence of climatic changes on plant and animal pathogens. They confirmed that climate warming has significant influence on the sometimes catastrophic onset of new pathogens. The circumstances associated with the epidemic occurrence of the West Nile virus in North Africa have not yet been clarified.

Recently, there have been attempts to find an association between the HPS outbreak in 1993 and the phenomenon called El Niño (Escutenaire et al., 2001).

6.6.4 **Prevention of Hantavirus Infection**

Preventive measures are based on a knowledge of the biology of hantavirus-hosting rodents, and on a detailed surveillance of hantavirus infections. Procedures aimed at reducing contact with rodents and their excrement in human dwellings, during leisure time and at work, rodent control activities and the use of respirators in high-risk areas are recommended to minimize the risk of infection. Measures complying with the BSL-2 and BSL-3 levels are recommended for laboratory work. Closed areas are subject to the highest risk of hantavirus infection spreading (Mills et al., 2002).

Education of the population is an essential part of preventive measures in areas with endemic occurrence of pathogenic hantaviruses. The vaccination of humans against hantavirus infections is not wide-scale and remains in the investigation phase.

Hantavirus-related issues have not been elucidated sufficiently to date, despite the fact that research in this area has achieved considerable progress. The results of the latest investigation have not fully reached health care professionals and many hantaviruses remain unidentified. In addition, the ecology and phenology of hantaviruses in nature has not been well investigated to date. Special attention should be paid to estimating which areas are endemic with the occurrence of pathogenic hantaviruses. One cannot exclude the possibility that other genospecies of hantaviruses will be described in the future from those rodents in which they have not been demonstrated to date. Closer cooperation with agricultural experts, veterinary services and clinical facilities would be helpful. Examination methods should be improved to achieve a rapid diagnosis of hantavirus infections and introduced into routine practice. Safe and effective vaccine protection against hantavirus infection should be provided for selected groups of the population in endemic areas and for people exposed to occupational risks. Last but not least, sufficient awareness among health care professionals and among people exposed to risk should be ensured.

6.6.5 Conclusions

Although hantaviruses were not classified as an independent taxonomic category until 1978, knowledge of these viruses has increased rapidly, especially after the identification of HPS in the United States in 1993. Over 1700 scientific reports and papers have been published regarding hantaviruses to date, and the number is still growing. Although the most serious diseases caused by hantaviruses occur in Asia and the Americas, thousands of newly diseased people are also reported every year in Europe, particularly in the northern parts of the continent and in the Balkans. In Europe, diseases caused by the hantavirus have been known about since the early 20th century, although the causative agents were not identified at that time. Currently, a total of four hantavirus genotypes are known in Europe: Puumala, Dobrava, Saaremaa and Tula. In addition, the Topografov virus seems to be circulating among lemmings in northern Russia. Of these hantavirus genotypes, at least Puumala, Dobrava and Saaremaa are human pathogens.

Hantaviruses circulate among live, wild rodents in nature, and the spread of these viruses is directly and indirectly influenced by the climate. Climatic changes alter the ecology and phenology of hantaviruses in their natural reservoirs. There are many scientific works describing the influence of the nature of biocenoses on hantavirus foci, and the influence of climate on the distribution and evolution of not only modern hantaviruses, but also those of the geological past. The most important influences are believed to be from the results of the ice age and the subsequent thaw.

Accordingly, if we want to achieve a comprehensive understanding of hantaviruses, we should also examine the influence of the climate and weather on their existence and evolution.

References

Aberle SW et al (1999) Nephropathia epidemica and Puumala virus in Austria. European Journal of Clinical Microbiology and Infectious Diseases 18(7):467–472

Ahlm C et al (1994) Prevalence of serum IgG antibodies to Puumala virus in northern Sweden. Epidemiology & Infection 113(1):129–136

Ahlm C et al (1997) Prevalence of serum antibodies to hantaviruses in northern Sweden as measured by recombinant nucleocapsid proteins. Scandinavian Journal of Infectious Diseases 29(4):349–354

Ahlm C et al (1998) Prevalence of antibodies specific to Puumala virus among farmers in Sweden. Scandinavian Journal of Work, Environment and Health 24(2):104–108

Ahlm C et al (2000) Serologic evidence of Puumala virus infection in wild moose in northern Sweden. American Journal of Tropical Medicine and Hygiene 62(1):106–111

Alexeyev OA et al (1996 a) Hantaan and Puumala virus antibodies in blood donors in Samara, an HFRS-endemic region in European Russia. Lancet 347(9013):1483

Alexeyev OA et al (1996 b) Hantavirus antigen detection using human serum immunoglobulin M as the capturing antibody in an enzyme-linked immunosorbent assay. American Journal of Tropical Medicine and Hygiene 54(4):367–371

Anděra M, Beneš B (2001) Atlas rožšíření savců v České republice: předběžná verze. IV., Hlodavci (Rodentia), Křečkovití (Cricetidae), hrabošovití (Arvicolidae), plchovití (Gliridae) Část 1 [Atlas of the mammals of the Czech Republic. A provisional version. IV. Rodents (Rodentia) – Part 1. Hamsters (Cricetidae), voles (Arvicolidae), dormice (Gliridae)]. National Museum, Praha, p 154

Antoniadis A et al (1996) Direct genetic detection of Dobrava virus in Greek and Albanian patients with haemorrhagic fever with renal syndrome. Journal of Infectious Diseases 174(2):407–410

Asikainen K et al (2000) Molecular evolution of Puumala hantavirus in Fennoscandia: Phylogenetic analysis of strains from two recolonization routes, Karelia and Denmark. Journal of General Virology 81(12):2833–2841

Avsic-Zupanc T et al (1989) Evidence for Hantavirus disease in Slovenia, Yugoslavia. Acta Virologica 33(4):327–337

Avsic-Zupanc T et al (1992) Characterization of Dobrava virus: A hantavirus from Slovenia, Yugoslavia. Journal of Medical Virology 38(2):132–137

Avsic-Zupanc T et al (1994) Isolation of a strain of a Hantaan virus from a fatal case of hemorrhagic fever with renal syndrome in Slovenia. American Journal of Tropical Medicine and Hygiene 51(4):393–400

Avsic-Zupanc T et al (1999) Hemorrhagic fever with renal syndrome in the Dolenjska Region of Slovenia – a 10-year survey. Clinical Infectious Diseases 28(4):860–865

Avsic-Zupanc T et al (2000) Genetic analysis of wild-type Dobrava hantavirus in Slovenia: co-existence of two distinct genetic lineages within the same natural focus. Journal of General Virology 81(7):1747–1755

Bilčíková H et al (1989) Výskyt a klinický obraz hemoragickej horúčky s renálnym syndrómom (HFRS) na východnom Slovensku: západný a východný typ [Incidence and clinical picture of haemorrhagic fever with renal syndrome (HFRS) in East Slovakia: the western and eastern type]. Bratislavské Lekárske Listy [Bratislava Medical Journal] 90(11):852–856

Bowen MD et al (1995) Genetic characterization of a human isolate of Puumala hantavirus from France. Virus Research 38 (2/3):279–289

Bowen MD et al (1997) Puumala virus and two genetic variants of Tula virus are present in Austrian rodents. Journal of Medical Virology 53(2):174–181

Brummer-Korvenkontio M et al (1980) Nephropathia epidemica: detection of antigen in bank voles and serologic diagnosis of human infection. The Journal of Infectious Diseases 141(2):131–134

Brummer-Korvenkontio M et al (1999) Epidemiological study of nephropathia epidemica in Finland 1989–96. Scandinavian Journal of Infectious Diseases 31(5):427–435

Carey DE et al (1971) Thottapalayam virus: a presumptive arbovirus isolated from a shrew in India. The Indian Journal of Medical Research 59(11):1758–1760

Cho HW, Howard CR (1999) Antibody responses in humans to an inactivated hantavirus vaccine (Hantavax registered). Vaccine 17(20/21):2569–2575

Daneš L (1985) Hemoragická horečka s ledvinným syndromem (HHLS) [Hemorrhagic fever with renal syndrome (HFRS)]. Acta Hygienica Epidemiologica et Microbiologica 15(2):52–57

Daneš L (1987) Někeré nové poznatky o původcích hemorrhagické horečky s ledvinným syndromem (HHLS) [Some new knowledge about agens hemorrhagic fever with renal syndrome (HFRS)]. Acta Hygienica Epidemiologica et Microbiologica 17(2):24–27

Daneš L et al (1986) Hemorrhagic fever with renal syndrome in Czechoslovakia: Detection of antigen in small terrestrial mammals and specific serum antibodies in man. Journal of Hygiene, Epidemiology, Microbiology and Immunology 30(1):79–85

Daneš L et al (1991) Hantaviruses in small wild living mammals in Czechoslovakia. Results of a 1983–89 study. Journal of Hygiene, Epidemiology, Microbiology and Immunology 35(3):281–288

Daneš L et al (1992 a) Anti-hantavirus antibodies in human sera in Czechoslovakia. Journal of Hygiene, Epidemiology, Microbiology and Immunology 36(1):55–62

Daneš L et al (1992 b) Protilátky proti hantavirům u lovné zvěře a turu domácího v České republice [Antibodies against hantaviruses in game and domestic oxen in the Czech Republic]. Československá epidemiologie, mikrobiologie, imunologie [Czechoslovak Epidemiology, Microbiology, Immunology] 41(1):15–18

Diglisic G et al (1994) Isolation of a Puumala virus from Mus musculus captured in Yugoslavia and its association with severe HFRS. Journal of Infectious Diseases 169(1):204–207

Dungel J, Šebela M (1993) Savci střední Evropy: Ilustrovaná encyklopedie [The mammals of Central Europe. The Illustrated Encyclopaedia], Jota, Brno, p 158

Dušek J et al (2001) Interstitial nephritis in children due to hantavirus infection. Pediatric Nephrology 16(9):44

Dzagurova T et al (1995) Antigenic relationships of hantavirus strains analysed by monoclonal antibodies. Archives of Virology 140:1763–1773

Dzagurova T et al (1996) Antigennaya differenciaciya chantavirusov s pomoschyu monoklonalnych antitel [The antigenic differentiation of hantaviruses by using monoclonal antibodies]. Meditsinskaia Parazitologiia i Parazitarnye Bolezni 3:47–52

Elgh F (1996) Human antibody responses to Hantavirus recombinant proteins & Development of diagnostic methods [dissertation]. Umeå Universitet, Umeå, Sweden

Elgh F et al (1997) Serological diagnosis of hantavirus infections by an enzyme-linked immunosorbent assay based on the detection of immunoglobulin G and M responses to recombinant nucleocapsid proteins of five viral serotypes. Journal of Clinical Microbiology 35(5):1122–1130

Elgh F et al (1998) Development of humoral cross-reactivity to the nucleocapsid protein of heterologous hantaviruses in nephropathia epidemica. FEMS Immunology and Medical Microbiology 22(4):309–315

Engelthaler DM et al (1999) Climatic and environmental patterns associated with hantavirus pulmonary syndrome Four Corners Region United States. Emerging Infectious Diseases 5(1):87-94

Escutenaire S et al (2000a) Spatial and temporal dynamics of Puumala hantavirus infection in red bank vole (Clethrionomys glareolus) populations in Belgium. Virus Research 67(1):91–107

Escutenaire S et al (2000b) Evidence of Puumala hantavirus infection in red foxes (Vulpes vulpes) in Belgium. Veterinary Record 147:365–366

Escutenaire S, Pastoret PP (2000) Hantavirus infections. Revue Scientifique et Technique 19(1):64–78

Escutenaire S et al (2001) Genetic characterization of Puumala hantavirus strains from Belgium: Evidence for a distinct phylogenetic lineage. Virus Research 74(1/2):1–15

Gavrilovskaya IN et al (1999) Cellular entry of hantaviruses which cause haemorrhagic fever with renal syndrome is mediated by β_3 integrins. Journal of Virology 73(5): 3951–3959

Glass GE et al (2000) Using remotely sensed data to identify areas at risk for hantavirus pulmonary syndrome. Emergerging Infectious Diseases 6 (3):238–247

Gligic A et al (1992) Belgrade virus – a new Hantavirus causing severe haemorrhagic fever with renal syndrome in Yugoslavia. Journal of Infectious Diseases 166(1):113–120

Golovljova I et al (2000) Puumala and Dobrava hantaviruses causing hemorrhagic fever with renal syndrome in Estonia. European Journal of Clinical Microbiology & Infectious Diseases 19(12):968–969

Grešíková M et al (1984) Haemorrhagic fever virus with renal syndrome in small rodents in Czechoslovakia. Acta Virologica 28(5):416–421

Grešíková M et al (1986) Nález protilátok proti vírusu hemoragickej horúčky s renálnym syndromom (HFRS) [The occurrence of antibodies against the virus of haemorrhagic fever with renal syndrome (HFRS)]. Bratislavské Lekárske Listy [Bratislava Medical Journal] 85(6):655–660

Grešíková M et al (1988a) Hemoragická horečka s renálnym syndromom [Hemorrhagic fever with renal syndrome]. In: Veda, Bratislava

Grešíková M et al (1988b) Detection of the antigen and antibodies to the eastern subtype of haemorrhagic fever with renal syndrome virus in small rodents in Slovakia. Acta Virologica 32(2):164–167

Grešíková M et al (1988c) A simple method of preparing a complement-fixing antigen from the virus of haemorrhagic fever with renal syndrome (Western type). Acta Virology 32(3):272–275

Grešíková M et al (1994) Serological evidence of Hantaan virus in Slovakia during 1989–1991. Acta Virologica 38:295–296

Groen J et al (1991a) Identification of Hantavirus serotypes by testing of post-infection sera in immunofluorescence and enzyme-linked immunosorbent assays. Journal of Medical Virology 33(1):26–32

Groen J et al (1991b) Different hantavirus serotypes in western Europe. Lancet 337(8741):621–622

Groen J et al (1995a) Hantavirus infections in The Netherlands: epidemiology and disease. Epidemiology and Infection 114(2):373–383

Groen J et al (1995b) A macaque model for hantavirus infection. Journal of Infectious Diseases 172(7):38–44

Hamzic, S et al (2003) Serologic diagnosis of hemorrhagic fever with renal syndrome in Bosnia and Herzegovina in 2002. Acta medica Croatica 57(5):381–385

Harvell CD et al (2002) Climate warming and disease risks for terrestrial and marine biota. Science 296(5576):2158–2162

Heider H et al (2001) A chemiluminescence detection method of hantaviral antigens in neutralisation assays and inhibitor studies. Journal of Virological Methods 96(1):17–23

Heyman P et al (1999) A major outbreak of hantavirus infection in Belgium in 1995 and 1996. Epidemiology and Infection 122:447–453

Heyman P et al (2002) Tula hantavirus in Belgium. Epidemiology and Infection 128(2):251–256

Hjelle B et al (1995) Hantaviruses: clinical, microbiologic, and epidemiologic aspects. Critical Reviews in Clinical Laboratory Sciences 32(5):469–508

Hooper JW, Li D (2000) Vaccines against hantaviruses. Current Topics in Microbiology and Immunology 256:171–191

Hujakka H et al (2001) New immunochromatographic rapid test for diagnosis of acute Puumala virus infection. Journal of Clinical Microbiology 39(6):2146–2150

Hörling J et al (1992) Antibodies to Puumala virus in humans determined by neutralization test. Journal of Virological Methods 39:139–147

Hörling J et al (1996) Distribution and genetic heterogeneity of Puumala virus in Sweden. Journal of General Virology 77:2555–2562

Iakimenko VV et al (2000) O rasprostranenii chantavirusov v zapadnoy Sibiri [The spread of hantaviruses in western Siberia]. Meditsinskaia Parazitologiia i Parazitarnye Bolezni 3:2–28

Jonsson CB, Schmaljohn CS (2000) Replication of hantaviruses. Current Topics in Microbiology and Immunology 256:15–32

Kariwa H et al (1995) Evidence for the existence of Puumala-related virus among Clethrionomys rufocanus in Hokkaido, Japan. American Journal of Tropical Medicine and Hygiene 53(3):222–227

Kariwa H et al (1999) Genetic diversities of hantaviruses among rodents in Hokaido, Japan and far east Russia. Virus Research 59(2):219–228

Kobzík J et al (1992) Laboratorně potvrzené případy hemoragické horečky s ledvinovým syndromem v letech 1989–1990 na Břeclavsku [Laboratory-confirmed cases of haemorrhagic fever with renal syndrome which occurred in Breclav 1989–1990]. Československá Epidemiologie, Mikrobiologie, Imunologie [Czechoslovak Epidemiology, Microbiology, Immunology] 41(2):65–68

Kožuch O et al (1995) Mixed natural focus of tick-borne encephalitis, tularemia and haemorrhagic fever with renal syndrome in West Slovakia. Acta Virologica 39:95–98

Lee HW, Lee PW, Johnson KM (1978) Isolation of the etiologic agent of Korean hemorrhagic fever. Journal of Infectious Diseases 137(3):298–308

Lee HW et al (1981) Electron microscope appearance of Hantaan virus, the causative agent of Korean haemorrhagic fever. Lancet 1:1070–1072

Lee HW et al (1991) Hemorrhagic fever. An occupational risk for Osp technicians? Ryo Moon Gak Ltd, Seoul, Korea

Lee PW et al (1981) Propagation of Korean hemorrhagic fever virus in laboratory rats. Infection and Immunity 31(1):334–338

Leitmeyer K et al (2001) First molecular evidence for Puumala hantavirus in Slovakia. Virus Genes 23(2):165–169

Linderholm M, Elgh F (2000) Clinical characteristics of hantavirus infections on the Eurasian continent. Current Topics in Microbiology and Immunology 256:135–151

Lundkvist A et al (1993) The humoral response to Puumala virus infection (nephropathia epidemica) investigated by viral protein specific immunoassays. Archives of Virology 130(1/2):21–30

Lundkvist A et al (1997a) Dobrava hantavirus outbreak in Russia. Lancet 350(9080):781–782

Lundkvist A et al (1997b) Puumala and Dobrava viruses cause hemorrhagic fever with renal syndrome in Bosnia-Herzegovina: Evidence of highly cross-neutralizing antibody responses in early patient sera. Journal of Medical Virology 53(1):51–59

Mackow ER, Gavrilovskaya IN (2000) Cellular receptors and hantavirus pathogenesis. Current Topics in Microbiology and Immunology 256:91–115

Markotic A et al (2002) Characteristics of Puumala and Dobrava infections in Croatia. Journal of Medical Virology 66(4):542–551

Marshall E (1993) Hantavirus outbreak yields to PCR. Science 262(5135):832, 834–836

Martens H (2000) Serologische Untersuchungen zur Prevalenz und zum Verlauf von Hantavirus-Infektionen in Mecklenburg-Vorpommern. Gesundheitswesen 62(2):71–77

Matyáš J (2002) Česko čeká výrazné oteplení [Czech Republic waits the conspicuous warming] Lidové noviny. [People's Newspaper] 22 June: 21

McKenna P et al (1994) Serological evidence of hantavirus disease in Northern Ireland. Journal of Medical Virology. 43(1): 33–38

Mentel R et al (1999) Hantavirus Dobrava infection with pulmonary manifestation. Medical Microbiology and Immunology 188(1):51–53

Meyer BJ, Schmaljohn C (2000) The accumulation of terminally deleted RNAs may play a role in hantavirus persistence. Journal of Virology 74:1321–1331

Mustonen J et al (1998) Epidemiology of hantavirus infections in Europe. Nephrology, Dialysis, Transplantation: official publication of the European Dialysis and Transplant Association – European Renal Association 13(11):2729–2731

Myhrman G (1951) Nephropathia epidemica: a new infectious disease in Northern Scandinavia. Acta medica scandinavica 140(1):52–56

Nemirov K et al (1999) Isolation and characterization of Dobrava hantavirus carried by the striped field mouse (Apodemus agrarius) in Estonia. Journal of General Virology 80(2):371–379

Nichol ST (1996) Changing landscapes and emerging RNA viruses. RTG Symposium: The phylogeny of life and the accomplishments of phylogenetic biology. Abstracts of invited presentations (October)

Nichol ST (2000) Emerging viral diseases. Proceedings of the National Academy of Sciences of the United States of America 97(23):12411–12412

Niklasson B, Leduc JW (1987) Epidemiology of nephropathia epidemica in Sweden. Journal of Infectious Diseases 155:269–276

Nolte KB et al (1995) Hantavirus pulmonary syndrome in the United States: A pathological description of a disease caused by a new agent. Human Pathology 26(1):110–120

Nuti M et al (1992) Seroprevalence of antibodies to hantaviruses and leptospires in selected Italian population groups. European Journal of Epidemiology 8(1):98–102

Nuti M et al (1993) Infections in an alpine environment: antibodies to hantaviruses, leptospires, rickettsia, and Borrelia burgdorferi in defined Italian populations. American Journal of Tropical Medicine and Hygiene 48(1):20–25

Pacsa AS et al (2002) Hantavirus-specific antibodies in rodents and humans living in Kuwait. FEMS Immunology and Medical Microbiology 33(2):139–142

Papa A, Antoniadis A (2001) Hantavirus infections in Greece – an update. European Journal of Epidemiology 17(2):189–194

Papa A et al (1998) Retrospective serological and genetic study of the distribution of hantaviruses in Greece. Journal of Medical Virology 55(4):321–327

Papa A et al (2000). First case of Puumala virus infection in Greece. Infection 28(5): 334–335

Papa A et al (2001) Isolation of Dobrava virus from Apodemus flavicollis in Greece. Journal of Clinical Microbiology 39(6):2291–2293

Pejčoch M et al (1992) Prevalence nákaz drobných savců hantaviry v Jihomoravském kraji [Prevalence of infection of small mammals with hantaviruses in Southern Moravia]. Československá Epidemiologie, Mikrobiologie, Imunologie [Czechoslovak Epidemiology, Microbiology, Immunology] 41(2):92–100

Pejčoch M et al (1996) Importovaný případ hemoragické horečky s renálním syndromem [An imported case of hemorrhagic fever with renal syndrome]. Epidemiologie, Mikrobiologie, Imunologie [Epidemiology, Microbiology, Immunology (Prague)] 45(3):127–129

Pejčoch M et al (2003) Nálezy hantavirového antigenu u hlodavců v České republice [Hantavirus antigen in rodents in the Czech Republic]. Epidemiologie, Mikrobiologie, Immunologie [Epidemiology, Microbiology, Immunology (Prague)] 52(1):18–24

Pelikán J, Gaisler J, Rödl P (1979) Naši savci [Our Mammals]. Academia, Praha

Peters CJ, Simpson GL, Levy H (1999) Spectrum of hantavirus infection: Hemorrhagic fever with renal syndrome and hantavirus pulmonary syndrome. Annual Review of Medicine 50:531–545

Petrů K et al (1997) Hemoragická horečka s renálním syndromem [Hemorrhagic fever with renal syndrome]. Časopis Lékařů Českých [Journal of Czech Physicians] 136(23):739–740

Pilaski J, Zöller L, Blenk H (1986) Hämorrhagisches Fieber mit renalem Syndrom (HFRS): eine durch Nagetiere übertragene Nephropathie des Menschen. Wehrmedizinische Monatsschrift 10:435–444

Pilaski J et al (1991) Haemorrhagic fever with renal syndrome in Germany. Lancet 337(8733):111–112

Pilaski J et al (1994) Genetic identification of a new Puumala virus strain causing severe haemorrhagic fever with renal syndrome in Germany. Journal of Infectious Diseases 170:1456–1462

Plank I et al (1955) Prvé dva diagnostikované prípady hemoragickej nefroso-nefritídy na území našej republiky [First two cases of hemorrhagic nephroso-nephritis in Czechoslovakia]. Časopis Lékařů Českých [Journal of Czech Physicians] 94(40): 1078–1084

Plank J, Režucha M, Rojkovič D (1961) Výskyt hemorhagických horúčok v Európe [The incidence of hemorrhagic fever in Europe]. Almanac of the Regional Pathology of Eastern Slovakia 1:291–301

Plyusnin A et al (1994) Tula virus: a newly detected hantavirus carried by European common voles. Journal of Virology 68(12):7833–7839

Plyusnin A et al (1995) Genetic variation in Tula hantaviruses: sequence analysis of the S and M segments of strains from central Europe. Virus Research 39(2/3):237–250

Plyusnin A et al (1996) Hantaviruses: genome structure, expression and evolution. The Journal of General Virology 77:2677–2687

Plyusnin A et al (1996) Newly recognised hantavirus in Siberian lemmings. Lancet 347(9018):1835–1836

Plyusnin A et al (1997) Dobrava hantavirus in Estonia: does the virus exist throughout Europe? Lancet 349(9062):1369–1370

Plyusnin A et al (1999) Dobrava hantavirus in Russia. Lancet 353(9148):207

Poljak V (1956) Případ hemoragické nefroso-nefritidy u nás [The case of hemorrhagic nephroso-nephritis in Czechoslovakia]. Časopis Lékařů Českých [Journal of Czech Physicians] 95(10):265–267

Rollin PE et al (1995) Short report: isolation and partial characterization of a Puumala virus from a human case of nephropathia epidemica in France. American Journal of Tropical Medicine and Hygiene 52(6):577–578

Rosický B, Bárdoš V (1961) Preventívne opatrenia pri výskyte hemoragickej nefrózonefritídy [Direction for the prevention of hemorrhagic nephroso-nephritis]. Almanac of the Regional Pathology of Eastern Slovakia 1:302–311

Rosický B et al (1961). Fauna přírodního ohniska haemorrhagické nefrosonefritidy [The fauna of the natural focus of hemorrhagic nephrosonephritis]. Almanac of the Regional Pathology of Eastern Slovakia 1:9–29

Scharninghausen JJ et al (1999) Genetic evidence of Dobrava virus in Apodemus agrarius in Hungary. Emerging Infectious Diseases 5(3):468–470

Schmaljohn CS et al (1985) Antigenic and genetic properties of viruses linked to hemorrhagic fever with renal syndrome. Science 227(4690):1041–1044

Schultze D et al (2002) Tula virus infection associated with fever and exanthema after a wild rodent bite. European Journal of Clinical Microbiology and Infectious Diseases 21(4):304–306

Sibold C et al (1995) Genetic characterization of a new hantavirus detected in Microtus arvalis from Slovakia. Virus Genes 10(3):277–281

Sibold C et al (1999 a) Recombination in Tula hantavirus evolution: Analysis of genetic lineages from Slovakia. Journal of Virology 73(1):667–675

Sibold C et al (1999 b) Short report: Simultaneous occurrence of Dobrava, Puumala, and Tula hantaviruses in Slovakia. American Journal of Tropical Medicine and Hygiene 61(3):409–411

Sibold C et al (2001) Dobrava hantavirus causes hemorrhagic fever with renal syndrome in central Europe and is carried by two different Apodemus mice species. Journal of Medical Virology 63(2):158–167

Sironen T, Vaheri A, Plyusnin A (2001) Molecular evolution of Puumala Hantavirus. Journal of Virology 75(23):11803–11810

Sjölander KB et al (1997) Evaluation of serological methods for diagnosis of Puumala hantavirus infection (Nephropathia epidemica). Journal of Clinical Microbiology 35(12):3264–3268

Sjölander KB et al (2001) Serological divergence of Dobrava and Saaremaa hantaviruses: evidence for two distinct serotypes. Epidemiology and Infection 128(1):99–103

Snell NJC (2001) New treatments for viral respiratory tract infections – opportunities and problems. Journal of Antimicrobial Chemotherapy 47(3):251–259

Song JW, Gligic A, Yanagihara R (2002) Identification of Tula hantavirus in Pitymys subterraneus captured in the Cacak region of Serbia-Yugoslavia. International Journal of Infectious Diseases 6(1):31–36

Song JW et al (1998) Characterization of Tula hantavirus isolated from Microtus arvalis captured in Poland. In: Abstracts of the Fourth International Conference on HFRS and hantaviruses, Atlanta, Georgia, March 5–7, 78

Song JW et al (1999) Muju virus: a Puumala-related hantavirus harboured by Eothenomys regulus in Korea (abstract no. V11RS. 1) In: Abstracts of the XIth International Congress of Virology, Sydney, Australia, August 9–13, 290

Taller AM et al (1993) Belgrade virus, a cause of hemorrhagic fever with renal syndrome in the Balkans, is closely related to Dobrava virus of field mice. Journal of Infectious Diseases 168(3):750–753

Tang YW et al (1991) Distribution of Hantavirus serotypes Hantaan and Seoul causing hemorrhagic fever with renal syndrome and identification by hemagglutination inhibition assay. Journal of Clinical Microbiology 29(9):1924–1927

Tomiyama T et al (1990) Rapid serodiagnosis of hantavirus infections using high-density particle agglutination. Archives of Virology (Suppl 1):29–33

Tsai TF (1987) Hemorrhagic fever with renal syndrome: mode of transmission to humans. Laboratory Animal Science 37(4):428–430

Tsai TF et al (1982) Intracerebral inoculation of suckling mice with Hantaan virus. Lancet 2(8296):503–504

Tsai TF et al (1984) Hemagglutination-inhibiting antibody in hemorrhagic fever with renal syndrome. Journal of Infectious Diseases 150(6):895–898

Unar J, Daneš L, Pejčoch M (1996) The basic ecology and vegetation characteristic of the natural foci of hantaviruses in South Moravia. In: Zprávy Vlastivědného muzea v Olomouci [Reports of Museum of National History and Arts in Olomouc] 273: 19–55

Vacková M et al (2000) Problematika hantavirových nákaz v armádě České republiky [The problem of hantavirus infections in the Czech army]. Vojenské Zdravotnické Listy [Military Medical Journal] 69(3):93–97

Vacková M et al (2002) Febrilní stavy způsobené hantaviry [Febrile conditions caused by hantaviruses]. Praktický Lékař [General Practitioner (Prague)] 82(2):84–86

Van Loock F et al (1999) A case-control study after a hantavirus infection outbreak in the south of Belgium: Who is at risk? Clinical Infectious Diseases 28(4):834–839

Van Ypersele de Strihou C (1991) Clinical features of hemorrhagic fever with renal syndrome in Europe. Kidney International 40 (Suppl 35):80–83

Vapalahti O et al (1996) Isolation and characterization of Tula virus, a distinct serotype in the genus Hantavirus, family Bunyaviridae. The Journal of General Virology 77(12):3063–3067

Vapalahti O et al (1999) Puumala virus infections in Finland: Increased occupational risk for farmers. American Journal of Epidemiology 149(12):1142–1151

Vasilenko S et al (1990) Hemorrhagic fever with renal syndrome in Bulgaria: isolation of hantaviruses and epidemiologic considerations. Archives of Virology (Suppl 1): 63–67

Verhagen R et al (1987) Occurrence and distribution of Hantavirus in wild living mammals in Belgium. Acta Virologica 31(1):43–52

Vesenjak-Hirjan J et al (1971) An outbreak of hemorrhagic fever with a renal syndrome in the Plitvice lakes area. Folia Parasitologica 18:165–169

Wichmann D et al (2001) Hemorrhagic fever with renal syndrome: Diagnostic problems with a known disease. Journal of Clinical Microbiology 39(9):3414–3416

Xiao SY et al (1993) Dobrava virus as a new Hantavirus: evidenced by comparative sequence analysis. Journal of Medical Virology 39(2):152–155

Xiao SY et al (1994) Phylogenetic analyses of virus isolates in the genus Hantavirus, family Bunyaviridae. Virology 198(1):205–217

Yanagihara R et al (1990) Experimental infection of human vascular endothelial cells by pathogenic and nonpathogenic hantaviruses. Archives of Virology 111:281–286

Yashina LN et al (2000) Genetic diversity of hantaviruses associated with hemorrhagic fever with renal syndrome in the far east of Russia. Virus Research 70(1/2):31–44

Zöller L et al (1990) Hantavirus, Infektionen. Die gelben Hefte 30(1):9–18

Zöller L et al (1995) Seroprevalence of hantavirus antibodies in Germany as determined by a new recombinant enzyme immunoassay. European Journal of Clinical Microbiology and Infectious Diseases 14(4):305–313

Appendix: Distribution of Hantaviruses in the European Region

(Please note that in many of the studies listed below, reliable methods for detection and typing of hantavirus infections were not used)

Country	PUU	DOB	TUL	No specification	Notes	Source
Albania		+			Genetic detection in patients	Antoniadis et al., 1996
Austria	+				Nephropathia in Austria	Aberle et al., 1999
Austria	+		+		In Austrian rodents	Bowen et al., 1997
Belgium	+		+		In common vole	Heyman et al., 2002
Belgium		+			In bank vole and red fox	Escutenair et al., 2000 a, b
Belgium			+		Outbreak in 1995 and 1996	Heyman et al., 1999
Belgium		+			Occurrence	Heyman et al., 2002
Bosnia and Herzegovina	+	+			HFRS	Lundkvist et al., 1997 b
Croatia	+	+			Occurrence	Hamzic et al., 2003; Markotic et al., 2002
Czech Republic	+			+	In game and domestic oxen	Daneš et al., 1992 b
Czech Republic				+	HFRS in two people	Kobzik et al., 1992
Czech Republic				+	Hantavirus antigen in small rodents	Pejčoch et al., 1992
Czech Republic	+				Puumala nucleotide sequence in bank vole	
Czech Republic		+			In small rodents	Křivanec (unpublished)
Czech Republic			+		The first isolation from common voles	Plyusnin et al., 1995; Vapalahti et al., 1996
Denmark	+				Molecular evidence	Asikainen et al., 2000

Country	PUU	DOB	TUL	No specification	Notes	Source
Estonia		+			New distinct Saaremaa genotype	Sjölander et al., 2001
Finland	+				In bank voles and serologic findings in humans	Brummer-Korvenkontio et al., 1980; Vapalahti et al., 1999
France	+				Genetic characterization of a human isolate	Bowen et al., 1995; Rollin et al., 1995
France-Ardennes	+				Outbreak 1992–1994	Van Loock et al., 1999
Germany	+				In bank vole mainly in south-west Germany	Pilaski et al., 1994
Germany		+			Striped field mouse – north-east Germany	Wichmann et al., 2001
Germany				+	HFRS	Martens, 2000; Pilaski, Zöller and Blenk, 1986; Pilaski et al., 1991
Germany	+				Severe HFRS	Pilaski et al., 1994
Germany		+			HFRS and Apodemus mice	Sibold et al., 2001
Greece	+				Infections of humans	Papa et al., 1998, 2000; Papa and Antoniadis, 2001
Greece		+			Isolation from Apodemus flavicollis	Papa et al., 2001
Greece		+			Genetic detection in patients	Antoniadis et al., 1996
Hungary		+			In Apodemus agrarius	Scharninghausen et al., 1999
Italy	+				Seroprevalence in Italian populations	Nuti et al., 1992, 1993
Russian Karelia	+				Molecular evidence	Asikainen et al., 2000

Country	PUU	DOB	TUL	No specification	Notes	Source
Northern Ireland				+	Serological evidence	McKenna et al., 1994
Norway	+				Molecular evidence	Asikainen et al., 2000
Poland			+		Isolation from common vole	Song et al., 1998
Russia	+	+?			Antibodies in blood donors in Samara	Alexeyev et al., 1996
Russia		+			Outbreak in Russia	Lundkvist et al., 1997; Plyusnin et al., 1999
Russia				+	A newly detected hantavirus	Plyusnin et al., 1994
Serbia			+		In *Microtus subterraneus*	Song, Gligic and Yanagihara, 2002
Slovakia				+	Serological evidence in humans	Grešíková et al., 1994
Slovakia				+	Natural focus in west Slovakia	Kouch et al., 1995
Slovakia	+				Molecular evidence	Leitmeyer et al., 2001
Slovakia	+	+	+		Occurrence in Slovakia	Sibold et al., 1999 b
Slovakia			+		Genetic analysis	Sibold et al., 1995, 1999 a
Slovenia	+	+			HFRS in the Dolenjska region	Avsic-Zupanc et al., 1999
Slovenia		+			Genetic analysis	Avsic-Zupanc et al., 2000
Sweden	+				Serum IgG antibodies in humans	Ahlm et al., 1994, 1997, 1998; Hörling et al., 1996; Niklasson and Leduc, 1987
Sweden	+				Serologic evidence in wild moose	Ahlm et al., 2000
Switzerland			+		Human infection	Schultze et al., 2002
The Netherlands	+				Main reservoir – bank vole	Groen et al., 1995 a
The Netherlands	+				Epidemiology	Groen et al., 1991 b, 1995 a

7 Climate, Weather and Enteric Disease

R. Sari Kovats, Cristina Tirado

Contributing authors: Kristie L. Ebi, Shakoor Hajat, Sally Edwards, Annemarie Kaesbohrer

7.1 Introduction

This chapter reviews research on the effect of weather and climate on intestinal infectious diseases that are transmitted through food or water. Such diseases are the main causes of infectious diarrhoea and cause significant amounts of illness each year in Europe. Approximately 20% of the population in western Europe are affected by episodes of diarrhoea each year (IID study team, 2002; van Pelt et al., 2003). Such infections have a significant economic impact through treatment costs and loss of working time (Roberts et al., 2003). Some enteric pathogens can also have more serious complications in vulnerable individuals.

Laboratory studies have shown that the growth of some bacterial microorganisms increases with increasing temperature, up to a threshold value, providing other conditions are met. Epidemiological studies have also shown associations between environmental or outdoor temperature and related outcomes such as reported diarrhoeal episodes or hospital admissions (Singh et al., 2001; Kovats et al., 2004). For these reasons, increases in enteric disease have been identified as a potential health consequence of global climate change (Rose et al., 2001; Bentham, 2001; Charron et al., 2004).

Diseases associated with water are varied and cover multiple environmental pathways. Water-borne diseases are usually understood to be those diseases that are spread via water that is contaminated with faecal material and then ingested (so called faecal-oral diseases). Faecal-orally transmitted disease can also be affected by ambient temperatures. Diseases caused by pathogenic organisms (water-based diseases) that spend part of their life cycle in aquatic organisms are often associated with standing water and so are potentially affected by climate. In this chapter, we consider only enteric diseases that are transmitted through the faecal-oral route of transmission, and potentially affected by temperature and/or rainfall factors. Specifically, this chapter:

- reviews published literature on the effects of ambient temperature on enteric diseases, and the methods used to evaluate this relationship
- reviews current knowledge of the role of weather in water-borne disease transmission and outbreaks
- reviews assessments of the future impact of climate change on water- and food-borne disease and

discusses current adaptation strategies for the surveillance and control of food- and water-borne disease in the context of future climate change.

The impact of weather and climate variability on food- and water-borne infectious diseases has not been well addressed in climate change assessments, in contrast to the treatment of vector-borne diseases. This is, in part, because little epidemiological research has been undertaken on the role of weather in the transmission of food- and water-borne disease in Europe, or elsewhere. Outbreaks are investigated with the primary aim of identifying the causal agent. Much research is also focused on the microbiology of individual enteric pathogens. Climate is not a limiting factor in the geographical distribution of infectious intestinal disease, but is likely to play an important role in the seasonal distribution of human infections.

7.2   Food-borne Diseases

Food-borne illnesses are defined by the WHO Surveillance Programme for Control of Food-borne Infections and Intoxications in Europe as diseases, usually either infectious or toxic in nature, caused or thought to be caused by agents that enter the body through the ingestion of food or water. The definition of a food-borne "outbreak" varies among countries. A food-borne-disease outbreak is defined by the WHO Surveillance Programme for Control of Food-borne Infections and Intoxications in Europe as an incident in which two or more persons experience a similar illness resulting from the ingestion of a common food (Schmidt and Tirado, 2001). Infectious intestinal disease are not always self-limiting and can result in severe infections, particulary in those who have weak immune systems (Adak et al., 2002).

Many food-borne diseases show strong seasonal patterns that reflect their mode of transmission. Infections with salmonella peak in the summer months, whereas infections with campylobacter generally peak in the early spring time. The diarrhoeal diseases that are caused by viruses, such as rotavirus, tend to peak in the late winter months. However, a seasonal pattern per se does not provide strong evidence of a causal role of temperature or rainfall. Other factors vary seasonally that are also important determinants of transmission of campylobacter infections, such as travel or the growth kinetics of fly species (Nichols, 2005).

7.2.1 Temperature and Diarrhoeal Disease

Few studies have estimated the effect of temperature on routinely reported episodes of diarrhoeal disease, even though the seasonal patterns have been observed for many years. Temperature was found to be strongly associated with increased hospital admissions in children in Lima, Peru (Checkley et al., 2000; Speelman et al., 2000). A study in Fiji also found an association between monthly temperature and diarrhoeal episodes reported in the following month (Singh et al., 2001). Similar studies on nonspecific clinical diarrhoeal outcomes have not yet been undertaken in Europe.

The monthly incidence of "food poisoning" in England was significantly associated with the temperature above an identified "threshold", which was found to be approximately 7.5 °C (Bentham and Langford, 2001). The data on "food poisoning" was derived from GP notifications based on clinical diagnosis, and these data would have included a range of pathogens, including campylobacter that do not multiply at ambient temperatures, as well as nonbacterial causes of food poisoning. Further studies found that the effect of temperature was strongest for high temperatures in the previous 2 to 5 weeks to the person becoming ill (Atkinson and Maguire, 1998; Bentham and Langford, 2001). An important limitation of these studies is that the data were not specific to a single pathogen. Such studies provide little insight into causal mechanisms of disease transmission and, therefore, cannot provide useful information on potential interventions.

7.2.2 Temperature and Salmonellosis in Europe

Salmonella is one of the most important food-borne pathogens affecting European populations. *Salmonella* spp. accounts for around 70% of all outbreaks of food-borne disease reported to the WHO Surveillance Programme for Control of Food-borne Infections and Intoxications in Europe, among these approximately 30% are caused by S. *Enteritidis* (Tirado and Schmidt, 2001). There was a large increase of S. *Enteritidis* in several European countries between 1985 and 1992, since then Salmonellosis incidence has decreased in most countries due to the control measures implemented and to a greater awareness of the risk among the public (Tirado, 2005). In England and Wales, salmonella infection caused more deaths annually than any other food-borne pathogen (Adak et al., 2002).

The effect of temperature on the growth of salmonellas in food is well understood (Gomez et al., 1997). In the laboratory, the rate of multiplication of *Salmonella* spp. is directly related to temperature (Baird-Parker, 1990, 1994). Thus, in the absence of other controls, ambient (outdoor) temperature might be expected to influence the reproduction of salmonellas at various points along the food chain from farm to fork (Heyndrickx et al., 2002).

In Europe, "temperature misuse" is a contributory factor in 32% of the investigated outbreaks, many of them related to *Salmonella* spp. (Tirado and Schmidt, 2001). From these factors, inadequate refrigeration or cooling, inadequate cooking, reheating or hot-holding accounted for a similar percentage. This distribution varies according to geographical location in Europe. For example, in Mediterranean countries, the main contributing factor is inadequate refrigeration, whereas in Northern countries the main contributing factor is the inadequate cooking, reheating or hot holding (Tirado and Schmidt, 2001). The preparation of large quantities of foods too far in advance and stored inadequately are also important factors related to "temperature misuse".

Outdoor temperatures might also affect the exposure of individuals to salmonellas through seasonal changes in eating patterns (e.g. consumption of foods from buffets, barbecued foods and salads) and behaviour (e.g. outdoor recreational activities such as swimming or hiking that increase contact with sources of salmonella in the environment).

As part of the cCASHh project, the effect of temperature on salmonellosis in European countries was investigated (Kovats et al., 2004). Surveillance data were analysed for ten populations using a standardised regression analysis method to analyse weekly time series data. The objective was to quantify any short-term effects of temperature on disease while adjusting for potential confounders, such as seasonal changes. Indicator variables were used to control for the effect of public holidays (typically rates were low during holidays, and high following them). Ideally, time series analysis would use dates of the onset of illness but this information is not routinely available in surveillance. The onset of an illness in sporadic cases is self-reported and this was analysed when available.

Figure 1 describes the fitted relationship between salmonella cases (all types) and temperature (average of 0–9 weeks preceding case). The centre line is the estimated spline curve, and the upper and lower lines represent the 95% upper and lower confidence limits, respectively. For most countries, the relationship is approximately linear above a threshold temperature, or simply linear. For the Slovak Republic and Denmark, however, there was no clear association of case occurrence and temperature.

Figure 2 illustrates the per cent change in cases associated with temperature measured on each separate week before the onset of illness, up to a lag period of 9 weeks in England and Wales (all lags included in the model simultaneously). The greatest effect of temperature is one week before the onset of illness, with diminishing but positive effects up to 5 weeks. Data from other countries using estimated date of onset show broadly similar patterns. These results indicate a shorter lag period than that found by Langford and Bentham (1995).

Age-specific analyses were undertaken in England and Wales, Scotland, the Netherlands, Denmark and Switzerland, assuming the country-specific thresholds. The adult age group (15–64) appears to be the most sensitive to temperature effects on the incidence of salmonellosis. The differences between the age groups are not statistically significant at the 5% level except in England and Wales. The reasons why adults are more likely to get salmonella on a "hot day" are not clear. Food hygiene behaviour or food vehicles relevant to the effects of temperature on disease may be different in adults.

Infection with S. Enteritidis appeared to be more sensitive to the effects of environmental temperature than infection with S. Typhimurium. Lower

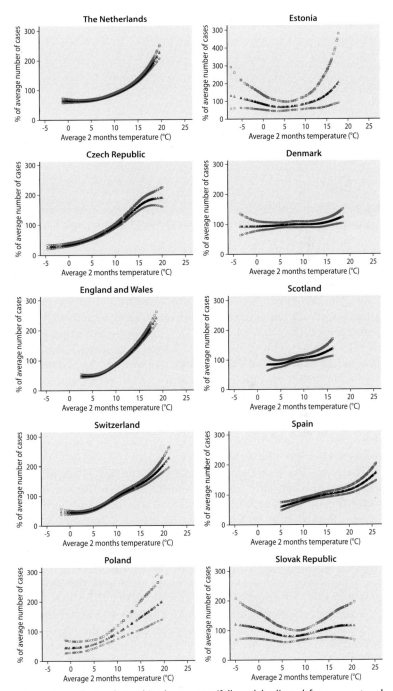

Fig. 1. Temperature-salmonellosis relationships by country (full model adjusted for season, trend and holidays), with temperature (°C) on the x-axis (0–9 week average), and salmonellosis cases on the y-axis as represented by percentage of the average number of cases

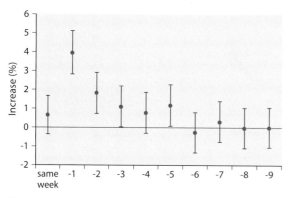

Fig. 2. The effect of temperature for each week lag between onset of illness and temperature exposure (% change), for temperatures above threshold (5 °C) in England and Wales

thresholds were also apparent for the *S. Enteritidis* series. In the Netherlands, the difference was statistically significant (increase in cases per degree increase in temperature: *S. Enteritidis* 12.6% (95% CI 11.1, 14.2), *S. Typhimurium* 6.1% (95% CI 5.0, 7.2)). If the same threshold is assumed within each country, the estimates are largely unchanged; however, the difference in England and Wales becomes statistically significant. Infection with *S. Typhimurium* is more common in rural areas and can be obtained through non-food contact (in the environment). *S. Enteritidis* is more strictly related to transmission via food. This further supports the hypothesis that temperature effects are more strongly mediated through the activities related to food preparation (and particularly egg handling behaviour) rather than other non-food sources.

There are likely to be some differences in the effect of temperature on outbreaks and sporadic cases of salmonellosis. Outbreaks are important because they affect more than one person and are also more likely to be reported. Surveillance of food-borne disease is discussed in more detail below. In some European countries, travel-associated cases make up a small but important proportion of all episodes of food-borne disease.

There are limitations in this study. First, not all cases in the community are represented in national surveillance data and the degree of underreporting varies by country (de Wit et al., 2000; IID study team, 2002) and by pathogen. However, except where changes to surveillance systems have occurred, there is no reason to believe that the degree of underreporting has varied over time. Second, it is estimated that cases reported to national surveillance are not more than 10% of all cases in the general population, and the extent of underreporting varies among countries (Tirado and Schmidt, 2001). It is unlikely that there are important differences in reporting during the year (e.g. cases of salmonellosis are not more likely to be reported and detected in a laboratory during hot weeks). The substantial

Table 1. Effect of temperature on salmonella or foodborne disease notifications: summary of published results

City/Country	Annual average count	Temp. range (°C)	Threshold (95% CI) (°C)	% Increase per degree	95% CI
Adelaide	397	9–26	No	4.9	3.4, 6.4
Perth	427	9–28	No	4.1	3.1, 5.2
Brisbane	504	14–26	No	11.0	7.7, 11.2
Melbourne	715	9–24	No	5.1	3.8, 6.5
Sydney	721	11–25	No	5.6 **	4.3, 7.0
Poland	21159	−5–20	6 (. , 7)	8.7	4.7, 12.9
Scotland	2108	−4–18	3 (., 12)	5.0	2.2, 7.9
Denmark	2657	−8–21	15 (., .)	0.3	−1.1, 1.8
England and Wales	24346	−2–22	5 (5, 6)	12.5	11.6, 13.4
Estonia	840	−10–21	13 (3, 14)	9.2	−0.9, 20.2
Netherlands	3285	−8–23	7 (7, 8)	8.8	8.0, 9.5
Czech Republic	40970	−11–25	−2 (−6, −1)	9.2	7.8, 10.7
Switzerland	4632	−5–24	3 (., 3)	9.1	7.9, 10.4
Slovak Republic	11816	−8–23	6 (., .)	2.5	−2.6, 7.8
Spain	5238	2–30	6 (., 8)	4.9	3.4, 6.4

** common threshold used (6 °C)
Sources: D'Souza et al., 2004; Kovats et al., 2004

underascertainment undoubtedly in the surveillance data series would not be expected to have caused bias in the observed associations.

The effect of temperature on salmonella is quite consistent across a range of different countries and cities (Table 1). Details of the thresholds, however, differ between countries and do not follow an obvious pattern, such as by latitude or mean summer temperature. For many countries, a threshold is not apparent and the relationship is approximately linear over the whole temperature range. Thresholds were not apparent for the effect of temperature on salmonellosis in five Australian cities (D'Souza et al., 2004).

7.2.3 Campylobacter

Campylobacter bacteria are important agents of enteritis and reported infections have been increasing in most European countries (Schmidt and Tirado, 2001). A range of mechanisms for Campylobacter transmission in Europe are reported in the literature: the consumption of contaminated foods, mostly chicken, and water and the consumption of insufficiently treated milk products. Other risk factors identified for Campylobacter infection include recent travel abroad and pets. Campylobacter infection is rarely transmitted from person to person (Dowell, 2001). Current knowledge of risk factors cannot explain all cases (the known attributable fraction is less than 50%). Risk assessment of *Campylobacter* spp. in broiler chickens has been undertaken by WHO and FAO in the framework of the Joint FAO/WHO Expert Consultation on Risk Assessment of Microbiological hazards in food (FAO and WHO, 2002).

Although it has been suggested that Campylobacter may be transmitted through the public water supply there is no epidemiological evidence that heavy rainfall events or run-off are associated with increases in the number of cases. The proportion of cases transmitted through the public water supply is unknown.

The effect of weather and climate was investigated in relation to reported cases of Campylobacteriosis in Europe. For the cCASHh project, data on reported campylobacteriosis were analysed for 15 populations in Europe, Canada, Australia and New Zealand. Weekly time series data were obtained for most countries. Most countries in Europe show an early spring peak (typically in April or May) in Campylobacter infection; however, not all countries follow this pattern (Fig. 3). The Czech Republic appears to have two peaks of infection in summer. Denmark, Switzerland and the Netherlands have late summer peaks with the peak of cases occurring after the peak of temperature. The seasonality is less pronounced in Australian cities than in New Zealand. For all Canada, the peak occurs in late June to early July, and lowest in February to March.

Within countries, there can be geographical variation in the seasonal patterns. In Ireland, the highest incidence rates are in the west of the country and the lowest rates are in the north-east and middle of the country. There is a sustained plateau in incidence seasonal patterns, from mid May until early August. However, the timing of the seasonal peak is earlier in the west of the country compared to the north-east (McKeown, 2003, personal communication). In Scotland, the prominence of the peak varied between regions (Miller et al., 2004). The authors noted that Lothian, with a mixed urban/rural population, had a more prominent peak than Greater Glasgow, which has a predominantly urban population. Differences in the timing of the peak of infection have also been observed between the North and South Island in New Zealand (Hearnden et al., 2003).

A statistically significant association was detected between both mean winter and spring temperature and timing of peak (Kovats et al., 2005). That is, campylobacter peaks earlier in warmer countries. There is no strong evidence that the year-to-year variation in the onset of the peak was related to seasonal temperatures. However, climate may explain some differences in the seasonal variation between countries (Kovats et al., 2005). Although there is an apparent relationship with temperature and cases in the time series studies (Tam et al., 2005), it should be interpreted with caution, as the affect is confined to temperatures between about 5 °C and 10–15 °C, corresponding to the spring months. If temperature was an important mechanism, cases would remain high throughout the summer months,

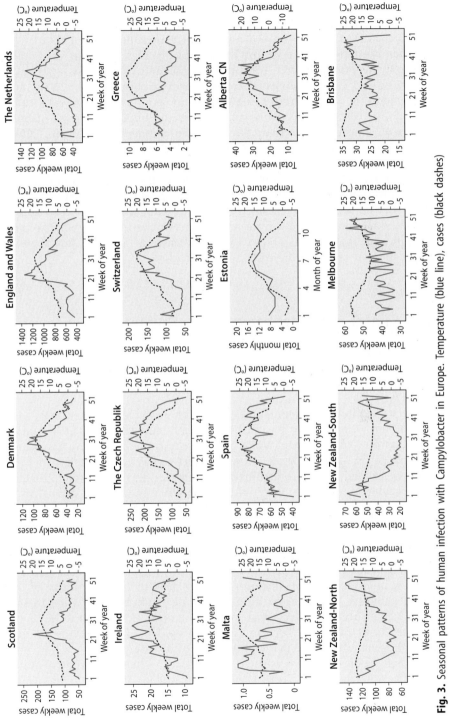

Fig. 3. Seasonal patterns of human infection with Campylobacter in Europe. Temperature (blue line), cases (black dashes)

as with salmonellosis. Laboratory studies show that Campylobacters only replicate in microaeorophilic environments and do not multiply at room temperature on food (Kapperud and Aasen, 1992; Altekruse et al., 1999).

It has been recently suggested the non-biting flies may play an important role in campylobacter transmission (Nichols, 2005). Fly activity is closely related to environmental temperatures (Goulson et al., 2005), and flies emerge in Spring around the same time as Campylobacteriosis cases begin to increase. Flies have been shown to be an important source of campylobacter infection of broiler flocks in the summer (Hald et al., 2004). The fly mechanism would also explain the lack of identification of single food vehicle or source. However, this hypothesis needs further investigation. The transmission of campylobacteriosis in humans is a complex ecological process with multiple hosts and routes (Skelly and Weinstein, 2003). More specific questions relating to the role of climate variability can be addressed once more detailed information on vehicles or serotyping become available. Campylobacters are so widely distributed in nature that there is no prospect of reducing the reservoir of bacteria.

7.3   Rainfall and Water-borne Disease Outbreaks

It has been suggested that climate change may increase water-borne disease due to an increase in extreme rainfall events (Rose et al., 2001; Charron et al., 2004). Specific mechanisms by which heavy rainfall may affect health include:

- heavy precipitation (sometimes but not necessarily associated with flooding) which causes the wash-in of contaminated rainwater or runoff in unprotected catchment areas, and thereby contaminated water reaches public water supply, private wells, or direct contact with humans
- heavy rainfall causes contamination of surface or coastal water (including recreational waters) where the sewers are used as storm drains
- heavy rainfall leads to failure in a water treatment plant leading to contamination of public water supply.

Microbiological contamination of the public supply is an important concern in Europe. However, few studies have investigated an association between rainfall and human disease (Table 3). Due to the complexity of the causal pathways on the route from surface or ground water to the household, it is difficult to detect what may be a small contribution to the overall burden of disease. There is also little epidemiological evidence that rainfall events play a role in sporadic cases of disease.

Several studies have investigated an association between drinking water turbidity and health (Schwartz and Levin, 1999; Aramini et al., 2000; Schwartz et al., 2000; Lim et al., 2002), and there is some indication that turbidity is a determinant of gastrointestinal illness in the general popula-

Table 2. Types of water-related diseases

Type	Description (examples from Europe)	Potentially affected by rainfall	
		Increased rainfall events	Decreased rainfall events
Faecal-oral (may be waterborne or water-washed)	Low infective dose: cholera, typhoid, Giardia, cryptosporidiosis. High infective dose: amoebic and bacillary dysentery, ascariasis, infectious hepatitis, paratyphoid, enteroviruses (some), and hookworm	Yes	Yes
Water-related insect vectors	Breeding in water: malaria, arboviral infections (e.g. West Nile fever)	+ see Chapter 6	+ see Chapter 6
Water-based	Penetrating skin: cercarial dermatitis. Ingested: none	No	No
Water-washed (strictly)	Skin and eye infections: trachoma, skin sepsis and ulcers, scabies, conjunctivitis. Other: insect and arachnid-borne typhus	No	Yes

Adapted from Cairncross and Feachem, 1993

Table 3. Summary of studies that investigate the association between rainfall and waterborne disease

Population	Exposure (Rainfall-related)	Outcome (Health-related)	Study design	Results	Reference
Germany, Rhineland	Extreme rainfall	Microbial load (faecal coliforms etc.)	Microbiological sampling of water courses	Microbiological load increased following extreme rainfall events	Kistemann et al., 2002
Delaware river basin, United States	Rainfall	Microbial load (Giardia and crypto)	Microbiological sampling of water courses	Microbiological load increased following extreme rainfall events	Atherholt et al., 1998
England and Wales	Monthly precipitation, runoff (3 rivers)	Sporadic cases of cryptosporidiosis	Multiple regression	Effect of temperature on crypto, and rainfall effect only in spring months	Lake et al., 2005
United States (whole country)	Extreme (90 centile) value of monthly rainfall	548 reported outbreaks of waterborne disease, 1948–1994	Cross-sectional analysis, stratified by ground-water/surface water contamination, and adjusted for season and hydrological region	51% outbreaks were preceded by an extreme rainfall event	Curriero et al., 2001
Mississipi, United States	Major floods (Midwest floods 2001)	Self-reported diarrhoea (daily)	Longitudinal survey (both pre- and post-flood)	Increase in gastro-intestinal symptoms in flooded persons. Rate ratio 1.29 (95% CI 1.06, 1.58)	Wade et al., 2004
Canada, 1975–2001	Extreme rainfall events	Waterborne disease outbreaks	Case-cross over design, using years without outbreak as controls	Positive association	Thomas et al. 2005

tion, at least in North America and Europe. Rainfall appears to increase concentrations of *Giardia* and *Cryptosporidium* in surface water through its effect on the amount of matter in the water. There may be a link between when *Cryptosporidium* and *Giardia* concentrations peak and when, for instance, turbidity reaches its highest level. Open finished water reservoirs are at risk for post-treatment faecal contamination by animals including those that shed species of *Giardia* and *Cryptosporidium* that are potentially pathogenic to humans. A study carried out in the United States found a statistically significant association between extreme rainfall events and monthly reports of outbreaks (Curriero et al., 2001).

In Europe, it is likely that extreme weather contributes to only a small proportion of water-borne disease episodes. When a link to the water supply has been established (by epidemiological survey), only a proportion of the reported outbreaks are related to weather events. Some outbreaks are related to maintenance failures, with rainfall as an additional causative factor, such as the *Cryptosporidium* outbreak in Milwaukee (MacKenzie et al., 1994). Many outbreaks in Europe have also been preceded by heavy rainfall events (e.g. Atherton et al., 1995; Miettinen et al., 2001). Table 3 summarizes the limited studies published so far that have investigated an association between rainfall extremes and water-borne disease.

Recreational waters (either inland surface waters or coastal waters) are also an important source of water-borne diseases in bathers (Zmirou et al., 2003). Transmission of water-borne disease from faecal contamination, as well as naturally occurring pathogens, may also be potentially affected by warmer temperatures as well as changes in rainfall and runoff (Hunter, 2003; Schijven and de Roda Husman, 2005).

7.3.1 Cryptosporidiosis

The most significant water-borne disease associated with the public water supply in Western Europe is cryptosporidiosis. Cryptosporidium is an intracellular parasite of the gastrointestinal and respiratory tracts of numerous animals. Cryptosporidium oocysts can survive several months in water at $48\,°C$ and are among the most chlorine-resistant pathogens. The sporocysts are resistant to most chemical disinfectants (including water treatment chemicals), but are susceptible to drying and ultraviolet sunlight. When contamination occurs, it has the potential to infect very large numbers of people (Meinhardt et al., 1996).

A recent study assessed the role of environmental factors in all cases of cryptosporidiosis by examining the association between all monthly reported cases, weather, and river flows in England and Wales between 1989 and 1996 (Lake et al., 2005). In April, cases of cryptosporidiosis were positively related to maximum river flow in the current month. Between May and July, cryptosporidiosis was positively linked to maximum river flows in the current and previous month. Cryptosporidiosis in August and September was negatively associated with rainfall in the previous three months and no associations were found between October and March. These relationships are consistent with an animal-to-human transmission pathway, and indicate that rainfall may play an important role in sporadic cases of the disease in the spring time, due to inadequate water treatment especially in northern parts of the country.

7.4 Drought, Water Supply and Health

Extreme weather could have a major impact on water resources where the infrastructure is poorly maintained and at risk of extreme rainfall events (floods and droughts). It has been suggested that reductions in rainfall may be associated with a decline (either short or long term) in the efficiency of local water supply or sewage systems. However, there is little direct evidence linking drought events with human health outcomes. All treatment systems are designed for a specific load. If the volume to be treated increases, the hydraulic load increases but the pollution load remains the same, leading to a reduction in treatment efficiency. In periods of extreme low flow, treatment efficiency is also compromised. This is illustrated by observations in the Rhine and Meuse rivers in the Netherlands during a particularly dry and warm year (Senhorst and Zwolsman, 2005). In 2003, a long drought period was associated with notable changes in water quality, including water temperature, dissolved oxygen and chloride concentrations.

Some areas in Europe may be vulnerable to drought events. Some populations in Eastern Europe have a sanitation and water infrastructure that is not robust enough to cope with current climate variability. For example, in 2003, a severe drought caused low levels in reservoirs supplying Vladivostok, Russia. The drought, in conjunction with poor infrastructure, led to severe rationing of the drinking water for over two months. Water supply was cut off in residential areas, but the supply was maintained to hospitals, schools and kindergartens. The inhabitants of Vladivostok (pop. 600 000) were supplied with water only once every two days for four hours at a time (IRC, 2003).

Limited information by country is available on access to "improved" water and sanitation in Europe. Countries in Western Europe have good access, whereas countries in Eastern Europe and Central Asia have limited piped water supplies provided by the state. Many such populations especially in rural areas have their own wells. A range of indicators of household water access and quality have been developed by the WHO, but reporting is very incomplete.

7.5   Assessment of Potential Impact of Climate Change

An increase in food-borne disease has been suggested as an outcome of climate change on human health (Rose et al., 2001; Hall et al., 2002). Although common infectious intestinal diseases are known to be temperature sensitive, quantitative projections have rarely been included in assessment of the impact of global climate change in human health. The United Kingdom National Assessment quantified future cases of "food poisoning" attributable to climate change, and estimated that cases may increase by up to 10 000 cases per year by the 2050s, although this is likely to be an overestimate (Department of Health, 2002). In Australia, it was estimated the climate warming may lead to an increase in hospital admissions for diarrhoea in children (McMichael et al., 2003). The Portuguese National Assessment did not undertake a quantitative risk assessment, but concluded that "if precipitation variability and temperature increase as the climate scenarios indicate, water- and food-borne transmission risk may increase as these climatic changes favour pathogen survival" (Casimiro and Calheiros, 2002, p. 245).

If a relationship between temperature and cases of salmonellosis has been quantified, then it is possible to estimate the number of cases attributable to "warm weather" in an average year. The population attributable fraction can be estimated as the proportion attributable to be hot weather, above the "threshold" temperature. Thus, it was estimated that temperature influences transmission of infection in about 35% of all cases of salmonellosis in England and Wales, Poland, the Netherlands, the Czech Republic, Switzerland and Spain (Kovats et al., 2004). It is reasonable to assume that the relationship with reported cases is applicable to all cases in the community. Given the approximately linear relationship between temperature and cases, it is relatively simple to estimate future cases assuming that climate alone changes in the future. However, the incidence of food-borne disease is likely to change in the future, and certainly over decadal time scales relevant to climate change. Recently many countries have observed marked declines in the reporting of salmonellosis coincident with falls in *S. Enteritidis* infection, following the introduction of various interventions to control the carriage of *S. Enteritidis* in poultry flocks (Tirado, 2005) (Fig. 4). The baseline incidence of infection is the most important determinant of the absolute effect of climate warming on food-borne disease.

7.5.1 Global Burden of Diarrhoeal Disease

The WHO Global Burden of Disease (GBD) estimated additional cases of diarrhoea due to increases in annual temperatures due to global climate change (Campbell-Lendrum et al., 2003). Based on two published studies described above (Checkley et al., 2000; Singh et al., 2001), a dose-response relationship of 5% increase in diarrhoea incidence per degree Centigrade

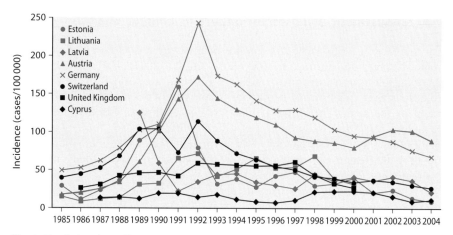

Fig. 4. Trends in salmonellosis in selected European countries, 1985–2004

increase in temperature was applied. The same dose-response was applied to both sexes and all age groups, and in all regions.

Adaptive capacity is assumed to be related to the level of sanitation or access to clean water. To allow for the modifying effect of development, countries were defined as 'developing' if they had (or are predicted to have, for future assessment years) per capita incomes lower than the richer of the two study countries (Fiji) in the year 2000 – approximately $ 6000/year in 1990 US dollars. A wide uncertainty range (0–10%) was placed on this value in extrapolating these relationships both geographically and into the future. For developed countries, in the absence of further information, WHO assumed a (probably conservative) increase of 0% in diarrhoea incidence per degree centigrade increase in temperature (uncertainty interval – 5 to 5%). The results suggest that richer countries in Western Europe (EUR A) will be able to avoid an additional burden due to global climate change (Table 4).

7.6 Adaptation: Current Strategies for Surveillance and Control

The most important mechanisms to prevent food- and water-borne disease are surveillance and monitoring, microbiological risk assessment, risk management and risk communication. The contamination of food products usually arises from improper handling, preparation or storage of food. Food safety is also an important economic concern. The control of food safety is an enormous challenge, both now and in a future world with climate change. New technologies will need to be developed, together with legislative measures and a greater reliance on voluntary compliance, as well

Table 4. Global burden of diarrhoeal disease due to global climate change globally and by region (compared to baseline climate in 1970s). (Campbell-Lendrum et al. 2003)

Region [a]	2000 DALYs (1000s)	Deaths (1000s)
Afr D	154	5
Afr E	260	8
Amr A	0	0
Amr B	0	0
Amr D	17	1
Emr B	14	0
Emr D	277	8
Eur A	0	0
Eur B	6	0
Eur C	3	0
Sear B	28	1
Sear D	612	22
Wpr A	0	0
Wpr B	89	2
World	1459	47

[a] WHO GBD World Regions

as education of consumers and professional food handlers (Kaferstein and Abdussalam, 1999).

7.6.1 Food-borne Disease Surveillance

The WHO Surveillance Programme for Control of Food-borne Diseases in Europe was launched almost 25 years ago as a result of the international awareness of the socioeconomical impacts of the increase of food-borne disease. This Programme is coordinated by the FAO/WHO Collaborating Centre in Berlin under the supervision of the WHO Food Safety Programme in Europe. The Programme is not mandatory and is based on surveillance activities at a national level. The main objective of the Programme is to provide support for the development of appropriate measures for the prevention and control of food-borne diseases in the region. The Programme aims to identify the epidemiology of food-borne diseases in the region, to disseminate relevant information on surveillance, and to collaborate with national authorities in identifying priorities for the reinforcement of the systems of prevention and control.

The Programme focuses on epidemiologically investigated outbreaks and the data and information requested from participant countries includes the number of affected people, causal agents, incriminated foods, places where the foods were contaminated or mishandled, places where foods were acquired or consumed and the contributing factors in the outbreaks. This kind of information is essential to establish preventive and control measures at National and Regional levels (Tirado and Schmidt, 2001). The Programme also provides surveillance data based on the statutory notification for hazard identification and for the analysis of trends in the European Region.

In the near future, the WHO Surveillance Programme is planning to collect quantitative and qualitative data from epidemiologically investigated outbreaks which will be essential for microbiological risk assessment. This is necessary to assist the work of the Codex Committee on Food Hygiene in the provision of data to the joint FAO/WHO expert consultation on risk assessment of microbiological hazards in food (Tirado and Schmidt, 2001).

WHO has a main role in assisting countries to build their capacity on surveillance, response, food contamination monitoring and food control. Coordinated international capacity building is essential for the effective prevention of food-borne diseases. The Global WHO Salmonella Surveillance (GSS) network is a partnership of several WHO Collaborating Centres including the Danish Veterinary Institute, the CDC, Institute Pasteur and Health Canada among others, who provide technical assistance to Salmonella Reference Laboratories in microbiological analyses and quality assurance and also provide training on epidemiological methods for surveillance of food-borne diseases. The WHO GSS has developed three levels of training courses on Integrated Laboratory Surveillance covering: *Salmonella* isolation, serotyping and antimicrobial susceptibility testing, *Campylobacter* isolation and identification, PDGE, PCR, Hip gene and integrated surveillance. The two first levels are for training of laboratory microbiologists from the clinical, veterinary and food microbiological Reference Laboratories. The third level course is designed to foster collaboration among the national networks of microbiologists and epidemiologists working with food-borne diseases. Curricula of future course levels are being developed and previews the training of risk assessors and managers.

In the European Union, Decision 2119/98/EC covers the Epidemiological Surveillance and Control of Communicable diseases. The European commission also co-ordinates several activities to improve food safety.

7.6.2 Zoonosis Monitoring in the European Union

In the European Union, there is a long tradition to control the traditional animal diseases. Zoonosis programes include routine testing of the relevant animal population and detailed control measures in case of a confirmed infection. In 1992, Council Directive 92/117/EEC was put in place, which made it obligatory for each Member State of the EU to monitor zoonoses and the agents thereof along the food chain and to control certain zoonoses. In this legislation, the main focus of attention was directed towards salmonellosis. For breeding flocks of poultry, detailed minimum measures were designed to stimulate the eradication of *S. Enteritidis* and *S. Typhimurium*.

In addition to the programmes that are compulsory on the EU or National level, voluntary activities are run in various countries to control for salmonella or campylobacter in the animal population, but no common approach is taken. Most attention is laid on salmonella control in laying hens for table egg production and broilers. Monitoring or control activities on campylobacter in broilers are already in place in several countries.

In 2003, the Council Directive 92/117/EEC was replaced by Directive 2003/99/EC on the monitoring of zoonoses and zoonotic agents. There, the number of diseases to be mandatory monitored has been extended to eight zoonoses and agents thereof. In a separate regulation, No. 2160/2003 on the control of Salmonella and other specified food-borne zoonotic agents, a progressive approach was fixed to control for all salmonella serovars with public health significance in poultry and pigs. Thus, the legal basis for monitoring and control of salmonella will be strengthened in the future. No common control strategy for campylobacter is currently in progress, but a harmonized monitoring scheme for Campylobacter in broilers, which should be applied on a voluntary basis, has been agreed on already.

7.6.3 Microbiological Risk Assessment

Microbiological Risk Assessments of Salmonella in Eggs and Broiler Chicken

The FAO and WHO undertook a risk assessment of Salmonella in eggs and broiler chickens in response to requests for expert advice on this issue from their member countries and from the Codex Alimentarius Commission (WHO and FAO, 2002). The risk assessment had several objectives:

- to develop a resource document of all currently available information relevant to the risk assessment of Salmonella in eggs and broiler chickens and also to identify the current gaps in the data that need to be filled to more completely address this issue
- to develop an example risk assessment framework and model for worldwide application
- to use this risk assessment work to consider the efficacy of some risk management interventions for addressing the problems associated with Salmonella in eggs and broiler chickens.

This Salmonella risk assessment provides information that should be useful in determining the impact intervention strategies may have on reducing cases of salmonellosis from contaminated eggs and poultry. In the risk assessment of Salmonella in broiler chickens, for example, it was determined that there is a relationship between reducing the prevalence of Salmonella on the broiler chickens and reducing the risk of illness per serving (WHO and FAO, 2002).

In the risk assessment of S. *Enteritidis* in eggs, reducing the prevalence of S. *Enteritidis* in poultry flocks was directly proportional to the reduction in risk to human health. The model can also be used to estimate the change in risk of human illness from changing storage times or temperature of eggs. The effects of time and temperature restrictions were evaluated assuming a flock prevalence of 25%. Keeping retail storage temperature at no more than 7.7 °C reduced risk of illness per serving by about 60% and when shelf-life to be reduced to 7 days, risk per serving would also be reduced by about 60% (FAO and WHO, 2002). Potential interventions evaluated included vaccination and refrigeration.

Comparison of the effectiveness of intervention measures, i.e. sensitivity analysis, cannot be done because this risk assessment is not conducted for a specific region or country, or for global settings. Data were collected from different countries for different input parameters. If those data were changed to reflect a specific national situation, the impact of a measure would also be changed. Therefore, caution is needed in interpreting the results of this risk assessment in Codex activities related to risk management.

Microbiological Risk Assessment of Campylobacter in Broiler Chickens

An understanding of thermophilic. *Campylobacter* spp., and specifically *Campylobacter jejuni* in broiler chickens is important from both the public health and trade perspectives. In order to achieve such an understanding, evaluation of this pathogen-commodity combination by quantitative risk assessment methodology was undertaken by the FAO and WHO (FAO and WHO, 2002). The first steps of this process, included hazard identification, hazard characterization and the conceptual model for exposure assessment. The final step, risk characterization, integrated the exposure and dose-response information in order to attempt to quantify the human-health risk attributable to pathogenic thermophilic *Campylobacter* spp. in broiler chickens.

This risk assessment attempted to understand how the incidence of human campylobacteriosis is influenced by various factors during chicken rearing, processing, distribution, retail storage, consumer handling, meal preparation and finally consumption. The benefit of this approach is that it enables consideration of the broadest range of intervention strategies. The FAO/WHO expert consultation found a linear relationship was found to exist between flock prevalence and probability of illness; however, it was recognized that there is a lack of systematic and fundamental investigation into the key processes throughout the production-to-consumption continuum that may lead to human infection as a result of chicken consumption (FAO and WHO, 2002). The FAO/WHO expert consultation concluded that risk characterization could provide risk managers with the ability to gain insight into the relative effectiveness of various mitigation strategies but recommended full exploration of the model to fully understand the scope and limitations.

7.6.4 Water-borne Disease Control and Surveillance

A considerable proportion of the population in the WHO European region does not have access to improved water and is at risk of water-borne infectious disease. New and emerging pathogens, such as *Giardia, Cryptosporidium* and chemicals, pose additional challenges in the short-term, while extreme weather events, such as floods and possibly increased water scarcity, pose challenges for the medium-term future. In the WHO European Region, almost 140 million (16%) do not have a household connection to a drinking-water supply; 85 million (10%) do not have improved sanitation; and over 41 million (5%) lack access to a safe drinking-water supply.

The WHO-UN ECE Protocol on Water and Health to the 1992 Convention on the Protection and Use of Transboundary Watercourses and International Lakes is now ratified by 16 European countries. The Protocol has identified cholera, bacillary dysentery, enterohaemorrhagic *Escherichia coli*, typhoid (and paratyphoid) and viral hepatitis A as priorities for action. Parties to the Protocol are mandated to review their systems for disease surveillance and outbreak detection, and implement the most appropriate measures to reduce disease, including vaccination or water treatment and distribution measures. Chemical contaminants of drinking-water and related diseases are also under review.

Successful implementation of the Protocol will contribute to achieving two Millennium Development Goals that relate to improving the water supply and sanitation and to reducing child mortality. The incidence of infectious diseases caused by poor-quality drinking-water is often highest in children aged 6–11 months. In the WHO European Region, this risk factor was estimated to cause over 13 000 deaths from diarrhoea among children aged 0–14 years (5.3% of all deaths in this age group) each year, with the countries of central and Eastern Europe and central Asia bearing the largest share of the burden (Ezzati et al., 2004).

The protocol asks countries to take all appropriate measures towards achieving:
- adequate supplies of wholesome drinking-water
- adequate sanitation of a standard which sufficiently protects human health and the environment
- effective protection of water resources used as sources of drinking water, and their related water ecosystems, from pollution from other causes
- adequate safeguards for human health against water-related diseases
- effective systems for monitoring and responding to outbreaks or incidents of water-related.

General improvements in access to water and sanitation will improve adaptive capacity to address the potential impact of climate change, and particularly reduce the impact of weather extremes (McMichael et al., 2003). Given the large burden of disease due to inadequate water and sanitation, it is an urgent requirement that access is improved.

Based on the epidemiological information so far, from monitoring and outbreak investigations, water supply managers are advised to sample surface water sources during and following rainfall events to determine maximum concentrations of *Giardia* and *Cryptosporidium*. Regular samples may be inadequate for representing the microbiological contamination of surface water (pre- or post-treatment). The procedures of raw water surveillance should include sampling during extreme runoff situations (Kistemann et al., 2002).

7.7 Conclusions

Diarrhoeal disease is one of the most important causes of ill-health in Europe. It is recognized to be highly sensitive to climate, showing strong seasonal variations in numerous sites. This is supported by regression analyses of the effects of temperature on reported cases of salmonellosis in a wide range of populations. The climate sensitivity of diarrhoeal disease is consistent with observations of the direct effects of meteorological variables on the individual pathogens.

However, the effects of weather on the transmission of pathogens are not generalizable, and depend on upon the local situation, and on the pathogen. The role of weather in triggering short-term increases in infection with campylobacter has yet to be resolved. There are various potential transmissions routes (water supplies, bird activity, fly activity and recreational contact) that could be affected by weather. However, the effect of short-term increases in temperature on campylobacter transmission is, at most, weak, in contrast to that consistently observed with salmonella transmission.

Diseases associated with water are varied and cover multiple environmental pathways. Some notable outbreaks of water-borne disease (such as cryptosporidiosis) have been associated with heavy rainfall events. At present, there is insufficient evidence to estimate the role of rainfall extremes in disease outbreaks in Europe, and it is not possible to assess whether climate change would have an overall impact on the burden of water-borne disease in Europe. The role of rainfall is likely to be a contributing factor to water treatment failures, and improved water treatment in the future is likely to diminish the effect of extreme weather.

Projected changes in climate for Europe indicate an increase in mean summer temperatures in the range of 1.5 °C to 2 °C by the 2050s. Any climate change-related increase in cases would, however, be dependent on the overall baseline incidence of the disease in the population. Reported salmonellosis is declining in many European countries due to active control measures. Food safety is an area of public health that is already subject to heavy regulation at the national and European levels. Legislation and enforcement, especially with regard to storage and refrigeration, is the most important strategy to address the potential effects climate change.

Acknowledgements

Thanks to Roger Aertgeerts, Hiroko Takasawa, and Sandy Cairncross for very helpful comments on early drafts. Special thanks to those who supplied surveillance data and helped with the analyses:

- J Cowden, C McGuigan: Scottish Centre for Infection and Environmental Health, Scotland
- C Gauci, Department of Public Health, Malta
- G Hernández Pezzi, Instituto de Salud Carlos III, Spain
- B Kriz, National Institute of Public Health, The Czech Republic
- K Kutsar, Health Protection Inspectorate, Estonia
- P McKeown, National Disease Surveillance Centre, Ireland
- K Mellou, Hellenic Centre of Infectious Disease Control, Greece
- S O'Brien, GK Adak, I Gillespie: Health Protection Agency, United Kingdom
- W van Pelt, National Institute of Public Health and the Environment, The Netherlands
- H Schmid, Federal Office of Public Health, Switzerland
- P Gerner-Smidt, K Moelbak, S Ethelberg and K Beicher, Statens Serum Institut, Denmark
- Z Kristufkova, State Health Institute of Slovak Republic, Slovak Republic
- W Magdzik, National Institute of Hygiene, Poland
- D Charron, M Fleury at Health Canada/University of Guelph, Canada and A Maarouf at Environment Canada
- Simon Hales, Wellington School of Medicine, and M Baker at Environmental Sciences Research, New Zealand
- RM D'Souza, G Hunt, National Centre for Epidemiology and Population Health, Australia, Communicable Diseases Network Australia, National Notifiable Diseases Surveillance System, and the Queensland Department of Natural Resources and Mines
- C Koppe at Deutscher Wetterdienst, Germany

References

Adak G, Long SM, O'Brien S (2002) Trends in indigenous food-borne disease and deaths, England and Wales, 1992–2000. Gut 51:832–841

Altekruse SF, Stern NJ, Fields PI, Swerdlow DL (1999) Campylobacter jejuni – An emerging food-borne pathogen. Emerging Infectious Diseases 5:28–35

Aramini J, Wilson J, Allen B, Sears W, Holt J (2000) Drinking water quality and health care utilization for gastrointestinal illness in Greater Vancouver. Division of Enteric, Food-borne and Water-borne Diseases, Health Canada, Guelph

Atherholt TB, Lechevalier MW, Norton WD, Rosen JS (1998) Effect of rainfall on Giardia and cryptosporidiosis. J Am Water Works Assoc 90:66–80

Atherton F, Newman C, Casemore DP (1995) An outbreak of water-borne cryptosporidiosis associated with a public water supply in the UK. Epidemiology and Infection 115:123–131

Atkinson P, Maguire H (1998) Is food poisoning a clinical or laboratory diagnosis? A survey of local authority practices in the South Thames region. Communicable Disease and Public Health 1:161–164

Baird-Parker AC (1990) Food-borne salmonellosis. Lancet 336:1231–1235

Baird-Parker AC (1994) 1993 Fred Griffith Review Lecture. Foods and microbiological risks. Microbiology 140 (Pt 4):687–695

Bentham G (2001) Food poisoning and climate change, In: Department of Health (ed) Health effects of climate change in the UK. Department of Health, London, pp 81–85

Bentham G, Langford IH (1995) Climate change and the incidence of food poisoning in England and Wales. Int J Biometeorol 39:81–86

Bentham G, Langford IH (2001) Environmental temperatures and the incidence of food poisoning in England and Wales. Int J Biometeorol 45:22–26

Cairncross AM, Feachem RG (1993) Environmental Health Engineering in the Tropics: an introductory text. John Wiley & Sons, Chichester

Campbell-Lendrum D, Pruss-Ustun A, Corvalan C (2003) How much disease could climate change cause? In: McMichael AJ, Campbell-Lendrum D, Corvalan C, Ebi KL, Githeko AK, Scheraga JS, Woodward A (eds) Climate Change and Health: Risks and Responses. World Health Organization, Geneva

Casimiro E, Calheiros JM (2002) Human Health. In: Santos FD, Forbes K, Moita R (eds) Climate Change in Portugal: Scenarios, Impacts, and Adaptation Measures – SIAM project. Gradiva, Lisbon, pp 241–300

Charron D. Thomas M, Waltner-Toews D, Aramini J, Edge T, Kent R, Maarouf A, Wilson J (2004) Vulnerability of water-borne diseases to climate change in Canada: a review. J Toxicol Environ Health A 67:1667–1677

Checkley W, Epstein LD, Gilman RH, Figueroa D, Cama R, Patz JA, Black RE (2000) Effects of El Nino and ambient temperature on hospital admissions for diarrhoel diseases in Peruvian children. Lancet 355:442–450

Curriero F, Patz JA, Rose JB, Lele S (2001) The association between extreme precipitation and water-borne disease outbreaks in the United States, 1948–1994. American Journal of Public Health 91;1194–1199

D'Souza RM, Becker N, Hall G, Moodie KB (2004) Does ambient temperature affect food-borne disease? Epidemiology 15:86–92

de Wit MA, Hoogenboom-Verdegaal AM, Goosen ES, Sprenger MJ, Borgdorff MW (2000) A population-based longitudinal study on the incidence and disease burden of gastroenteritis and Campylobacter and Salmonella infection in four regions of The Netherlands: Eur.J Epidemiol 16:713–718

Department of Health (2002) Department of Health Effects of Climate Change in the UK. London, Department of Health

Dowell SF (2001) Seasonal variation in host susceptibility and cycles of certain infectious diseases. Emerging Infectious Diseases 7:369–374

Ezzati M, Lopez AD, Rodgers A, Murray CJ (eds) (2004) Comparative Quantification of Health Risks: Global and Regional Burden of Disease due to Selected Major Risk Factors, Vols 1 and 2. World Health Organization, Geneva

FAO, WHO (2002) Report of the Joint FAO/WHO Expert Consultation on Risk Assessment of Microbiological Hazards in Foods; Hazard identification, exposure assessment and hazard characterization of Campylobacter spp. in broiler chickens and Vibrio spp. in seafood. WHO, Geneva

Gomez T, Motarjemi Y, Miyagawa S, Kaferstein F, Stohr K (1997) Food-borne salmonellosis: World Health Stat Q 50:81–89

Goulson D, Derwent LC, Hanley M, Dunn DW, Abolins SR (2005) Predicting calyptrate fly populations from the weather, and the likely consequences of climate change. Unpublished work

Hald B, Skovgard H, Bang DD, Pedersen K, Dybdahl J, Jespersen JB, Madsen M (2004) Flies and Campylobacter infection of broiler flocks. Emerging Infectious Diseases 10:1490–1492

Hall GV, D'Souza RM, Kirk MD (2002) Food-borne disease in the new millennium: out of the frying pan and into the fire. Medical Journal of Australia 177:614–618

Hearnden M, Skelly C, Eyles R, Weinstein P (2003) The regionality of campylobacteriosis seasonality in New Zealand: International Journal of Environmental Health Research 13:337–348

Heyndrickx M, Vanderkerchove D, Herman L, Rollier I, Grijspeerdt K, DeZutter I (2002) Routes for salmonella contamination of poultry meat: epidemiological study from hatchery to slaughterhouse. Epidemiology and Infection 129:253–265

Hunter PR (2003) Climate change and water-borne and vectorborne disease. J Appl Microbiol 94 (Suppl 1):37–46

IID study team (2002) A report of the study of infectious intestinal disease in England. The Stationery Office, London

IRC (2003) Russia, Vladivostok: drought leads to water cuts. International Water and Sanitation Centre (IRC), http://www.irc.nl/content/view/full/4163, accessed 28 Sept 2003

Kaferstein F, Abdussalam M (1999) Food safety in the 21st century. Bulletin of the World Health Organization 77:347–351

Kapperud G, Aasen S (1992) Descriptive epidemiology of infections due to thermotolerant Campylobacter spp. in Norway, 1979–1988. APMIS 100:883–890

Kistemann T, Classen T, Koch C, Dangendorf F, Fischeder R, Gebel J, Vacata V, Exner A (2002) Microbial Load of Drinking Water Reservoir Tributaries during Extreme Rainfall and Runoff. Appl Environ Microbiol 68:2188–2197

Kovats RS et al (2005) Climate variability and Campylobacter infection: an international study: Int J Biometeorol 49:207–214

Kovats RS, Edwards S, Hajat S, Armstrong B, Ebi KL, Menne B (2004) The effect of temperature on food poisoning: time series analysis in 10 European countries. Epidemiology and Infection 132:443–453

Lake I, Bentham G, Kovats RS, Nichols G (2005) Effects of weather and river flow on cryptosporidiosis. Water and Health (in Press)

Lim G, Aramini J, Fleury M, Ibarra R, Meyers R (2002) Investigating the relationship between drinking water and gastro-enteritis in Edmonton, 1993–1998. Division of Enteric, Food-borne and Water-borne Diseases, Health Canada, Guelph

MacKenzie WR et al (1994) A massive outbreak in Milwaukee of Cryptospiridium infection transmitted through the public water supply. New England Journal of Medicine 331(3):161–167

McMichael AJ, Campbell-Lendrum D, Corvalan C, Githeko AK, Ebi KL, Scheraga JS, Woodward A (eds) (2003) Climate Change and Health: Risks and Responses. WHO, Geneva

McMichael AJ, Woodruff RE, Whetton P, Hennessy K, Nicholls N, Hales S, Woodward A, Kjellstrom T (2003) Human Health and Climate Change in Oceania: Risk Assessment 2002. Commonwealth of Australia, Department of Health and Ageing, Canberra

Meinhardt PL, Casemore DP, Miller KB (1996) Epidemiologic aspects of human cryptosporidiosis and the role of water-borne transmission. Epidemiological Reviews 18(2):118–136

Miettinen IT, Zacheus O, von Bonsdorff CH, Vartiainen T (2001) Water-borne epidemics in Finland in 1998–1999. Water Science and Technology 43:67–71

Miller G, Dunn GM, Smith-Palmer A, Ogden ID, Strachen NJ (2004) Human campylobacteriosis in Scotland: seasonality, regional trends and bursts of infection. Epidemiology and Infection 132:585–593

Nichols G (2005) Fly transmission of Campylobacter. Emerging Infectious Diseases

Roberts JA, Cumberland P, Sockett P, Wheeler JG, Rodrigues LC, Roderick PJ (2003) The study of infectious intestinal disease in England: socio-economic impact. Epidemiology and Infection 130:1–11

Rose JB, Epstein PR, Lipp EK, Sherman BH, Bernard SM, Patz JA (2001) Climate variability and change in the United States: potential impacts on water- and food-borne diseases caused by microbiologic agents. Environmental Health Perspectives 109:211–221

Schijven JF, de Roda Husman AM (2005) Effect of climate changes on water-borne disease in the Netherlands. Water Science and Technology 51:79–87

Schmidt K, Tirado C (2001) Seventh Report on Surveillance of Food-borne Diseases in Europe 1993–1998. Federal Institute for Health Protection of Consumers and Veterinary Medicine/WHO, European Centre on Environment and Health, Berlin/Rome

Schwartz J, Levin R (1999) Drinking water turbidity and health. Epidemiology 10:86–89

Schwartz J, Levin R, Goldstein BD (2000) Drinking water turbidity and gastro-intestinal illness among the elderly of Philadelphia. Journal of Epidemiology and Community Health 54:45–51

Senhorst HAJ, Zwolsman JJG (2005) Climate change and effects on water quality: a first impression. Water Science and Technology 51:53–59

Singh RBK, Hales S, deWet N, Raj R, Hearnden M, Weinstein P (2001) The influence of climate variation and change on diarrhoeal disease in the pacific islands. Environmental Health Perspectives 109:155–159

Skelly C, Weinstein P (2003) Pathogen survival trajectories: an eco-environmental approach to the modelling of human campylobacteriosis ecology. Environmental Health Perspectives 111:19–28

Speelman EC, Checkley W, Gilman RH, Patz JA, Calderon M, Manga S (2000) Cholera incidence and El Nino related higher ambient temperature: JAMA 283:3072–3074

Tam CC, Rodrigues LC, O'Brien S, Hajat S (2005) Temperature dependence of reported Campylobacter infection in England, 1989–1999. Epidemiology and Infection (in press)

Thomas K et al. (2005) Water-borne disease outbreaks in Canada [abstract]. Epidemiology 16:592–593

Tirado C (2005) WHO Surveillance Programme for control of Food-borne Diseases in Europe. Trends on food-borne diseases during the last 25 years. WHO, Copenhagen (forthcoming)

Tirado C, Schmidt K (2001) WHO Surveillance Programme for Control of Food-borne Infections and Intoxications: preliminary results and trends across Greater Europe. Journal of Infection 43:80–84 van Pelt W, de Wit MA, Wannet WJ, Ligtvoet EJ, Widdowson MA, van Duynhoven YT (2003) Laboratory surveillance of bacterial gastroenteric pathogens in The Netherlands, 1991–2001. Epidemiology and Infection 130:431–441

Wade TJ, Sandu SK, Levy D, Lee S, LeChevallier MW, Katz L, Colford JM (2004) Did a severe flood in the midwest cause an increase in the incidence of gastro-intestingal symptoms? Am J Epidemiol 159:398–405

WHO, FAO (2002) Risk assessments of Salmonella in eggs and broiler chicken. Microbiological risk assessment series 1. WHO/FAO, Copenhagen

Zmirou D, Pena I, Ledrans M, Letertre A (2003) Risks associated with the microbiological quality of bodies of fresh and marine water used for recreational purposes: summary estimates based on published epidemiological studies. Archives of Environmental Health 58:703–711

Policy Implications of Climate Change-related Health Risks in Europe

Kristie L. Ebi, Ian Burton, Bettina Menne

Introduction

As detailed throughout this volume, weather and climate are projected to have effects on human health in Europe, including adverse health outcomes due to thermal extremes, extreme weather events (particularly floods), vector-borne diseases, and foodborne diseases. As events demonstrated over the past few years, climate variability is negatively affecting population health now. Policy-makers are beginning to respond to the risks that climate change poses for Europe to prevent further morbidity and mortality. The policy responses will need to address impacts from both increasing short-term weather variability and gradual long-term changes in temperature, precipitation and other weather variables due to climate change. This chapter reviews the policy process, presents background information on adaptation to climate change, then suggests a risk assessment framework for thinking about the policy implications of climate change-related health risks in Europe. In the penultimate section, the paper turns to current international and regional responses at the European and global levels, including examples of current policies.

Policy Process

WHO has defined public health as 'the art of applying science in the context of politics so as to reduce inequalities in health while ensuring the best health for the greatest number' (Yach, 1996). Health policy refers to the objectives and policies adopted by public authorities from the community and regional level to state and national governments, such as Ministries of Health, and to international institutions such as the European Union. Policy-making is a process through which these authorities select desirable objectives, and identify and implement measures to achieve them. The process is composed of three main elements:

- the selection and acceptance of broad policy objectives
- the identification, design, and implementation of specific measures to achieve those objectives and
- decisions about the setting or level at which the measures will be applied. Settings are often relatively easy to adjust compared with changing measures or policy objectives. Levels are periodically adjusted, sometimes in response to assessments of their effectiveness in achieving policy objec-

tives, or in response to other considerations such as financial constraints, or a combination of these and other factors.

For example, a policy objective may be to maintain and, where practicable, to improve the level of food safety. The chosen measures might include a programme of public education, inspections of food handling establishments, and regulatory standards. The implementation setting or level could include the amount of funding allocated to the education programme, the frequency of inspection of restaurants and other food handling establishments, and the degree of stringency in penalties and sanctions applied in cases of default or failure to meet the standards established. Box 1 discusses measures for control of foodborne diseases, particularly salmonella.

Box 1: Measures to control foodborne diseases in europe

The European Union (EU) has a long tradition of controlling the transmission of food-borne diseases and infections via programes that include routine testing of the relevant animal populations and detailed control measures in case of a confirmed infection. Council Directive 92/117/EEC made it obligatory for each EU Member State to monitor and control zoonoses along the food chain. A major focus of attention has been on salmonellosis. Control measures taken at the poultry and broiler level aim to eradicate the two main strains of salmonella, including

- restrictions on flock movements;
- treatment or culling of affected animals;
- treatment of hatching eggs;
- heat-treating feeding supplies;
- disinfecting buildings; and
- regular monitoring, with eventual slaughter of broilers. Monitoring or control activities for campylobacter in broilers are in place in four countries.

The level of implementation of control activities varies by country. For example, the Swedish salmonella control programme was initiated in 1953 following a large outbreak. Extensive measures were taken to prevent the recurrence of salmonella, especially in domestic animals, food, and feed. The interventions were successful, particularly in cattle and broiler flocks, and the amount of salmonella decreased to less than 1%. The goal is to have Swedish food of animal origin completely salmonella free. Each flock is tested before slaughter and any report of salmonella is notifiable. Infected flocks are culled, the farm is put under restrictions, and the poultry house disinfected. Few other European countries have this level of control. Although the number of salmonella cases in Europe is declining, the total number of cases remains high. Thus, strengthening implementation of salmonella control programs is important.

The CD 92/117/EEC extended monitoring of zoonosis now to eight zoonoses and agents thereof. Additional regulation is under consideration for control for all salmonella serovars with public health significance in poultry and pigs. No common control strategy for campylobacter is under consideration.

Policy development is rarely orderly and linear. In practice, the three steps continually interact in response to changing societal needs and circumstances, and in response to the ways in which the needs and circumstances are perceived and communicated (Scheepens, 1994; Schein, 1995; Scheraga et al., 2003). Policy change often begins with reassessment of the settings when adverse impacts occur. Under circumstances where changing the settings does not or cannot produce the wanted results, new measures may be identified and implemented. In other situations, it becomes necessary to reformulate the policy objectives themselves to respond to new circumstances, new information (including about climate-related risks), and changing human perceptions, values and priorities. Policy change comes about most frequently by modification of existing policies in response to newly perceived or enhanced risks. For example, in Europe the increase in the number of overweight and obese individuals is leading to proposals to ensure that urban planning and transport arrangements are more favourable and supportive to pedestrians and cyclists.

The policy process is driven by strategic considerations at the level of decision-makers, and by a myriad of social, behavioural, economic, environmental, scientific and technological variables that work from the bottom-up to influence policy.

Adaptation

In the language of the United Nations Framework Convention on Climate Change, adaptation refers to the actions taken to prevent or reduce projected impacts, including the health impacts of climate change. Mitigation refers primarily to actions taken to control emissions of greenhouse gases. Although public health has more than 150 years dealing with a variety of risks to population health, including implementing policies and measures to cope with climate variability, there is room for improvement as evidenced in the various preceding chapters of this book. The present state of public health reflects the success or otherwise of a range of policies and measures, including those designed to address climate-related health outcomes.

The following questions are now beginning to be addressed to determine whether current policies are likely to be sufficient to cope with current and projected health impacts of climate change (Smit et al., 2001).

▪ Adaptation to what? This refers to the stimuli, stresses or exposure to which populations need to be adapted. The policies and measures implemented to address a possible change in the range of tick-borne encephalitis will be different from the policies and measures implemented to address increases in the frequency and intensity of flooding events. For this reason, some have questioned the need for generic climate change adaptation strategies, and consider that adaptation policy and measures are best designed for specific exposures. On the other hand, the synergies across different risk domains, and the need to increase public awareness point to a need to consider adaptation to climate change using an integrated and multi-sectoral approach.

- Who or what adapts? This refers to how adaptation occurs. Adaptations can occur at various levels to achieve a policy objective, such as to reduce vulnerability to heat events. Individuals may adapt by changing behaviour, such as activity levels or seeking cooler environments. Communities and nations may adapt by implementing heat event warning systems, by establishing building codes and landscape designs to address urban heat island issues, or by providing cooler buildings or shelters in which more susceptible people may seek refuge. Policies and measures need to clarify who is expected to take action and how those actions will be facilitated.

- Are additional interventions needed? This asks whether current interventions are as effective as possible in reducing health impacts. If not, to the extent that the burden of climate-sensitive diseases falls short of acceptable standards, there is an adaptation deficit. For example, the 2003 heat-wave in Europe resulted in significant loss of life. It is clear in hindsight that precautionary adaptation measures could have substantially reduced the loss of life. In the light of this experience, many European countries are introducing or strengthening their public health precautions to cope with future heat-waves and expanding their emergency response capacity.

- How and when is climate change projected to change health outcomes? Where are the most vulnerable populations? When should additional policies and measures be implemented? The burden of the 2003 European heat-wave revealed that public health programmes and infrastructure were not sufficiently prepared to cope with the extent and severity of extreme weather events that are likely to be associated with future weather (Stott et al., 2004). Combined with an increasing population of older adults, this suggests that current policies and measures need to take climate change projections into account. While this is starting to be done in the case of heat-wave risks and floods, it is more difficult to gain the attention of decision-makers and policy-makers and the public at risk in the absence of recent adverse experience. It is frequently observed that anticipatory adaptation measures are difficult to implement on the basis of scientific forecasts alone, and without the confirmation of some adverse experience.

The choice of which policies and measures to implement and the timing of their implementation is determined by the responsible authorities and civil society, particularly those likely to be affected by the policy choice (Scheraga et al., 2003). Effective risk management can facilitate adaptation to climate change.

Risk Management

When it becomes apparent that specific measures and their settings are not adequately reducing a risk, or when a new risk arises, scientists, experts, and managers are frequently called upon to conduct a risk assessment and provide recommendations on alternative response options to improve health risk management. However, note that some policy changes needed to reduce current and projected climate change-related health risks will take place outside the health sector, such as in water, agriculture and other sectors.

Because there is no such thing as absolute safety in any sphere of life, risk assessors and policy-makers seek to understand "how safe is safe enough?" The answer is, of course, that it depends upon the risk criteria that have been established or are the social norm in a given society or jurisdiction. There are several approaches for evaluating whether the risks associated with an exposure or activity are acceptable, including (1) estimates of how much increased (or decreased) mortality or morbidity will occur, (2) an assessment of comparative risks, and (3) a benefit-risk assessment. A comparative risk assessment determines whether alternatives have comparable levels of risk. A benefit-risk assessment determines the costs and benefits of risk reduction. In addition, there is the concept of balanced risk in situations where it is possible to choose among alternative risks.

In the cases of foodborne or vector-borne diseases, it is important to know how much the risks have increased or are projected to do so, and how much of any such increase can be attributed to climate change. Existing policies, measures, and settings may be sufficient to cope with small increases in risk. More information and advice is needed to assist decision-makers to assess the need to prepare for the management of the larger risks that are projected to occur with climate change.

Comparative risk refers to the notion that risk management policies, measures and settings can be deployed such that in any society exposure to risks is approximately equal across populations, regions and policy domains. Such a fine balance has proved practically impossible to achieve in practice, and risks are commonly much higher in some policy domains than others. One assumption underlying this approach is that policy-makers should not invest in policies, measures and settings for risks that are low compared with other risks. The health sector regularly evaluates the risks of various diseases and makes funding allocations based on which diseases pose larger threats to a community.

One characteristic of this approach that emerges over time is that successful interventions often suffer substantial budget cuts precisely because they are successful in reducing the burden of a disease. An example is the disease prevention and control programmes established in the Americas during the first half of the 20th century that effectively controlled vectors for yellow fever, dengue fever, and malaria, thus, significantly reducing the burden of these diseases (Gubler and Wilson, 2005). Success resulted in complacency, which in turn led to policy changes that resulted in dramatically decreased resources (and deterioration) in the public health infra-structure needed to deal with vector-borne diseases, and in support for re-search on new and more effective vector control strategies. The explanation seems to be that budget cuts are applied not only to the programmes or actions that are introduced to cope with the risk, but also to the compo-nents of the programmes that are needed for long-term surveillance and maintenance. These public health changes, combined with population growth, modern transportation, increased movement of people, animals, and commodities, and other demographic and societal changes over the past 50 years, have resulted in a resurgence of these diseases.

Benefit-risk assessments compare the benefits to be gained from a par-ticular intervention with the amount of risk reduction to be achieved. One underlying assumption is that society should not invest in policies, mea-sures and settings for which there will be little social gain. This is particu-larly relevant for risks that have been reduced to a fairly low level. Given that risks cannot be reduced to zero, policy-makers need to decide whether the effort required for further reduction in risk is an appropriate allocation of scarce public health resources. When considering the projected health impacts of climate change, benefit-risk assessments need to determine the benefits of future risk reductions.

Lessons learned from the 2002 flood and the 2003 heat events (discussed in Box 2) demonstrate that additional adaptation is required to reduce the burden of climate-sensitive diseases in Europe. In addition, policies and measures have not been evaluated to determine if they are as effective as would be desirable for addressing projected climate change. Research such as the cCASHh project is determining how much risk levels may be ele-vated by changing climatic conditions, and how the magnitude of the cli-mate-related risks compare with each other. Additional research is needed to determine how much reduction in the burden of climate-sensitive dis-ease might be achieved by new policies and/or stronger measures.

▌ **Box 2: Lessons from the 2003 heat event in France**

A severe heat event in August 2003, resulted in an estimated 14 800 excess deaths in France. Meteo France issued warnings to the media, but these were not passed on be-cause authorities did not understand the potential scale of the impact (Senat, 2004; Michelon et al., 2005). Deaths were not detected in real time because data from emer-gency and medical services and from death certificates were not used for health sur-veillance. An inquiry by the General Directorate of Health (DGS) concluded that the 2003 heat event was unforeseen, was only detected belatedly and that it highlighted deficiencies in the French public health system, including a limited number of experts; lack of preparation for a heat event, including defining responsibilities across public or-ganizations and developing effective mechanisms for information exchange; health au-thorities and crematoria/cemeteries overwhelmed by the influx of patients and bodies; few nursing homes equipped with air-conditioning; and a large number of elderly peo-ple living alone without a support system and without proper guidelines for appropriate responses to a heat event (Michelon et al., 2005). These conclusions indicated the need for strong international surveillance of emerging risks, for more research into the health consequences of heat events, and for better coordination between the expert agencies and research bodies. One issue that arose was that the ministry with primary responsibility for managing a heat event varied by how heat events were classified. If a heat event was classified as a natural catastrophe, then the ministry of the interior would have primary responsibility. But if a heat event was classified as a health catastrophe, then the ministry of health would have primary responsibility.

Since 2003, the French government formulated short- and medium-term actions to reduce future health impacts from heat events, including sponsoring research on at the risk factors associated with heat event-related mortality, implementing a health and environmental surveillance programme, and developing national and local action plans for heat events. If effective, these actions should reduce future vulnerability to heat events.

Community, State, and National Government Policy Responses to Address Climate Variability and Change in Europe

The design and implementation of adaptation policies and measures primarily focuses on existing climate-related risks to human health because most of the health outcomes projected to increase with climate change are also current problems. Likely responses will range from

- modification of existing prevention strategies including the introduction of new measures or higher settings of existing measures,
- translation of policies and knowledge from other countries or regions to address changes in the geographic range of disease,
- reinstitution of effective surveillance, maintenance and prevention programmes that have been neglected or abandoned, or
- development of new policies to address new threats (Ebi et al., 2005).

Most policy responses to climate-related health risks are likely to be incremental changes in current disease management programs to address gradual increases in the rate and range of diseases; these programs are organized into the broad categories of adverse health outcomes due to thermal extremes and to extreme weather events, vector-borne diseases, diseases associated with air pollutants and aeroallergens, and water- and foodborne diseases. Policy-makers assess the adequacy of existing measures and determine how, to what extent, and when the measures should be improved. For example, as discussed in Box 1, all European countries have programmes to limit salmonella contamination of food. Because the risk of salmonella may increase as warmer conditions provide more favourable conditions for the growth and spread of the bacteria, policy responses should focus on enhancing current programes and on improving measures to encourage more people to follow proper food handling guidelines. In another example, flooding risks will increase if projections are realized of more extreme storms and longer and more intense rainfall events. One policy response is to enhance programmes in place to address current flooding risks. For vector-, tick- and rodent-borne diseases, current surveillance programmes may need to be modified to detect early changes in the distribution of disease, such as tick-borne encephalitis (Lindgren, 2000).

Another category of policy response concerns the geographic spread of emerging or re-emerging diseases, primarily vector-, tick- and rodent-borne diseases. Because climate is one of the constraints on the range of many mosquitoes, ticks, and rodents, increasing average temperatures may allow temperature-sensitive vectors to migrate into places previously too cool to support reproduction and survival, carrying diseases with them. When designing responses to address the emergence or re-emergence of vector-borne diseases, policy development can be informed by lessons learned in other regions where the diseases are epidemic. In addition, regional and international cooperation is often needed to ensure the effectiveness of programmes because low levels of adaptation in one country could adversely affect other countries (Ebi et al. 2005). For some diseases, surveillance programmes may be coupled with an early warning system to provide timely interventions to reduce the magnitude or extent of a disease outbreak.

From a policy perspective, it is important to understand the various drivers of disease expansion and retreat. A variety of recent modelling efforts have shown that, assuming no future human-imposed constraints on malaria transmission, changes in temperature and precipitation could alter the geographic distribution and intensity of malaria transmission, with previously unsuitable areas of dense human population becoming suitable for transmission (e.g., van Lieshout et al., 2005). Projected changes include an expansion in latitude and altitude, and, in some regions, a longer season during which malaria may be present. Such changes could dramatically increase the number of people at risk for malaria.

The potential for malaria and other "tropical" diseases to invade southern Europe is commonly cited as an example of the territorial expansion of risk. However, many of these diseases existed in Europe in the past and have been essentially eliminated by public health programmes. For example, in the early part of the 20th century, malaria was endemic in many parts of southern Europe (Kuhn et al., 2004). The prevalence of malaria was reduced primarily via improved land drainage, better quality of housing construction, and higher levels of socioeconomic development, including better education and nutrition. Any role that climate played in malaria reduction would have been small. Note that this does not provide assurance that climate will not play a larger role in determining the future range and intensity of malaria transmission.

A third category is to determine if there are effective prevention problems that could be reinstated, such as mandatory vector-control programmes. Policy responses also will need to be developed for new health risks. AIDS, SARS, Creutzfeld Jakob, and Ebola are examples of diseases that were first observed relatively recently. Although their emergence was not connected with climate variability and change, these diseases are examples of public health risks that had varying degrees of impact and effectiveness of policy response. The strength and duration of the 2003 heat-wave also could be categorized as a new threat because it was outside the range of recent historical experience (Stott et al., 2004). A win-win strategy is the implementation of heat event warning systems in regions where heat events have rarely occurred, such as northern Europe. The systems are relatively inexpensive to implement and have been shown to reduce mortality in the event of a heat event if followed by appropriate actions (Ebi et al., 2004; Palecki et al., 2001).

International Policy Responses to Address Climate Variability and Change

States and international organizations have developed international agreements to protect populations from the adverse health consequences of exposure to a variety of pollutants that cross national boundaries. For example, the UN Framework Convention on Climate Change (UNFCCC) is designed to protect human health and societies from potentially catastrophic threats, such as sea level rise. Key issues that create difficulties for the creation of international agreements on climate change include that the global climate is a common resource that does not fall within the jurisdiction of any one state; each country has jurisdiction over their rate and amount of greenhouse gas emissions, but those emissions are creating potential public health problems in other countries; all countries are contributing to the atmospheric concentrations of greenhouse gases; and reduction of greenhouse gas emissions from any one country alone cannot be sufficient to prevent climate change, or even materially slow it down (Fidler, 2001). Therefore, many states must work together to reduce their greenhouse gas emissions in order to prevent further degradation of the global climate. Achieving this objective involves political, economic, and legal complications. The approach adopted in the UNFCCC has been to recognize that countries (Parties to the Convention) have "common but differentiated responsibilities". This refers to the fact that while global climate change is a commonly shared problem, responsibilities differ according to such factors as the wealth of countries, and their current and historic per capita greenhouse gas emissions. Accordingly in the Kyoto Protocol to the Convention, the developed countries have, with a few exceptions, agreed to cut their emissions by agreed percentages by 2008–2012. The Convention also recognizes the need to adopt adaptation policies and measures, and it is agreed that the developed countries will provide assistance in meeting the costs of adaptation to climate change by the most vulnerable countries. A number of voluntary funds have been established to help meet the costs of adaptation for developing countries; these are managed by the Global Environmental Facility (GEF), the financial instrument of the Convention. Many developing countries have now started to develop adaptation programmes, and international development agencies in Europe and elsewhere are beginning to take climate change risks into account in their assistance programmes.

▌ **The Policy Situation in the EU.** Article 152 of the Treaty establishing the European Community provides that a high level of human health protection should be ensured in the definition and implementation of all Community policy and activities. The vision of health is broad and includes all determinants that may impact health: the environment, social and economic conditions, lifestyle factors, and personal behaviour. The Commission and international organizations, particularly the European Environment Agency and the World Health Organization, cooperate with the Commission in maintaining and improving population health. WHO plays an important role in the selection of methods for determining priorities for action.

There are a number of actions aimed at addressing the burden of climate-sensitive diseases. For example, the Action Programme (2003–2008): Decision 1786/2002 states "actions should be guided by the need to increase life expectancy without disability or sickness, promote quality of life and minimize the economic and social consequences of ill health, thus, reducing health inequalities, while taking into account the regional approach to health issues. Priority should be given to health-promoting actions that address the major burdens of disease." This can be used to address increased mortality during heat events for, by example, promoting educational activities among older adults. Another objective of the Action Programme is for the Commission, in collaboration with Member States and international organizations, to improve public health information and knowledge by developing and operating a monitoring system to establish Community-level quantitative and qualitative health indicators. The system is collecting, analysing, and disseminating comparable and compatible age- and gender-specific health information, including socioeconomic, personal, and biological factors, health behaviours, and living, working, and environmental conditions, paying special attention to inequalities in health. An important component for addressing health risks related to climate change is the development of an information system for the early warning, detection, and surveillance of health threats. This could be used to address newly emerging diseases and diseases that might change their geographic range because of climate change.

On 9 June 2004, the Commission adopted the EU Environment & Health Action Plan 2004–2010. The action plan, which is based on the general orientations set out in the Commission's June 2003 Communication, to reduce the disease burden caused by environmental factors in the EU, to identify and prevent new health threats caused by environmental factors, and to strengthen EU capacity for policy-making in this area (COM (2003) 338 final, Brussels 11.6.2003). The Action Plan's added value is to improve the coordination between the health, environment and research sectors in order to provide a systematic review of the effect of our policy in improving health-related results so that it can be adjusted as needed. Responsibility for making progress in this complex area will be shared between Member States, which have responsibility for implementing monitoring and risk management measures, research and education and training; stakeholder groups such as industry and civil society, which play a key role in translating information about identified threats into preventive action and innovative responses; and the Commission, which will continue to engage with all main players and promote co-operation at the EU level, within its areas of competence, and liaise with the European Environment Agency, the European Food Safety Agency and other relevant bodies. Action 8 addresses the establishment of methodological systems for improved risk assessment and a system for early identification of emerging issues such as the effects of climate change on health. This process is being implemented through the EU Research Framework Programmes.

In 2005, there was a Communication from the Commission to the Council, the European Parliament, the European Economic and Social Committee, and the Committee of the Regions on 'Winning the Battle Against Global Climate Change' (COM (2005) 35, Brussels 9.2.2005). The adaptation challenge notes that few Member States have examined the need to reduce vulnerability and to increase their resilience to the effects of climate change. Adaptation will require additional research to predict the impacts at regional levels in order to enable local and regional public and private sector actors to develop cost-effective adaptation options. Another important aspect of adaptation is the prediction of more frequent and more damaging natural disasters; the Commission is already involved in an EU-wide early warning system for floods and forest fires to assist in preventing further damage from natural disasters. One conclusion from the Communication is that more resources need to be allocated to adapt effectively to climate change.

Discussion

Climate change is projected to have far-reaching effects on human health; the extent of the impacts will be partially determined by the efficiency and effectiveness of policies to increase the ability of individuals and societies to cope with what the future does bring. Policy-makers have begun to understand how weather and climate are projected to change, the potential health impacts of climate change, the effectiveness of current policies in reducing the impacts of weather and climate on health, and the range of choices available for enhancement of current or development of new policies, measures and settings. The degree of policy response is informed by the current burden of climate-sensitive diseases, the effectiveness of current interventions to protect the population from weather- and climate-related hazards, and projections of how the burden of disease could change as the climate changes. Understanding the rate and intensity of climate change could help differentiate situations where increases in mean temperature (or another weather factor) could lead to gradual increases in disease rates, from situations where thresholds could be crossed either because a disease was close to its boundary conditions or because there was a sudden and/or large change in weather.

The cCASHh project has provided timely and critical information on the potential health impacts of climate change during the 2002 floods and the 2003 heat-wave. This information is being used to design new policies and to improve current measures to address morbidity and mortality due to flooding and heat-waves. It is not so apparent that improvements are taking place at a sufficient rate in those risk domains where no recent disasters or emergencies have occurred. Further analyses are needed of policies and measures designed to reduce the burden of other climate-sensitive diseases, including vector-borne diseases, food- and waterborne diseases, and diseases due to air pollution and aeroallergens.

The evidence assembled and analysed in the chapters of this book show that current policies, measures and settings are probably not sufficient to prevent some health impacts of climate change. The shape and character of the health risks from climate change in Europe have been clearly identified, and further quantitative work is now needed on comparative risk and the benefits of new and enhanced public health policies.

References

Daniel M, Danielova V et al (2004) An attempt to elucidate the increased incidence of tick-borne encephalitis and its spread to higher altitudes in the Czech Republic. Int J Med Microbiol 293 (Suppl 37):55–62

Danielova V, Daniel M et al (2004) Prevalence of Borrelia burgdorferi sensu lato genospecies in host-seeking Ixodes ricinus ticks in selected South Bohemian locations (Czech Republic). Cent Eur J Public Health 12(3):151–161

Ebi KL, Teisberg TJ, Kalkstein LS, Robinson L, Weiher RF (2004) Heat watch/warning systems save lives: estimated costs and benefits for Philadelphia 1995-1998. Bulletin of the American Meteorological Society (BAMS) 85(8):1067–1073

Ebi KL, Smith J, Burton I, Scheraga J (2005) Some lessons learned from public health on the process of adaptation. Mitigation and Adaptation Strategies for Global Change (in press)

Fidler DP (2001) Challenges to humanity's health: the contributions of international environmental law to national and global public health. Environmental Law Institute 31ELR:10048–10078

Fuessel H, Klein R (2004) Conceptual Frameworks of Adaptation to Climate Change and their applicability to human health. Potsdam Institute for Climate Impact Research (PIK), Potsdam

Gubler DJ, Wilson ML (2005) The global resurgence of vector-borne diseases: lessons learned from successful and failed adaptation. In: Ebi KL, Smith J, Burton I (eds) Integration of Public Health with Adaptation to Climate Change: Lessons Learned and New Directions. Taylor & Francis, London, pp 44–59

Hajat S, Ebi KL et al (2003) The human health consequences of flooding in Europe and the implications for public health: a review of the evidence. Applied Environmental Science and Public Health 1(1):13–21

IPCC (2001) Climate Change 2001. Impacts, Adaptations and Vulnerability. Contribution of Working Group II to the Third Assessment Report of the Intergovernmental Panel on Climate Change. Cambridge University Press, New York

Karl TR, Trenberth KE (2003) Modern global climate change. Science 302(5651):1719–1723

Kirch W, Menne B: Extreme weather events and public health responses." Springer, Berlin Heidelberg New York

Koppe C, Jendritzky G et al (2004) Heat-waves: impacts and responses. World Health Organization, Copenhagen

Koppe C, Jendritzky G et al (2004) Heat-waves: Impacts and Responses. Health and Global Environmental Change Series. No 2. WROf Europe, World Health Organization, Copenhagen

Kovats RS, Edwards SJ et al (2004) The effect of temperature on food poisoning: a time-series analysis of salmonellosis in ten European countries. Epidemiol Infect 132(3):443–453

Kovats RS, Hajat S, et al. (2004) Contrasting patterns of mortality and hospital admissions during heatwaves in London, UK. Occup Environ Med 61(11)

Kuhn K, Campbell-Lendrum D, Haines A, Cox J (2004) Using climate to predict infectious diseases outbreaks: a review. WHO/SDE/OEH/04.01

Lindgren E, Talleklint L (2000) Impact of climatic change on the northern latitude limit and population density of the disease-transmitting European tick Ixodes ricinus. Environmental Health Perspectives 108:119–123

Louviere J, Hensher D et al (2000) Stated Choice Methods: Analysis and Applications. Cambridge University Press, Cambridge

Michelon T, Magne P, Simon-Delavalle F (2005) Lessons from the 2003 heat-wave in France and action taken to limit the effects of future heat-wave. In: Extreme Weather Events and Public Health Responses. Springer, Berlin, pp 131–140

Paldy ABJ, Vamos A, Kovats RS, Hajat S (2005) The effect of temperature and heat-waves on daily mortality in Budapest. In: Kirch W, Menne B, Bertollini R (eds) Extreme weather events and public health responses. Springer

Palecki MA, Changnon SA, Kunkel KE (2001) The nature and impacts of the July 1999 heatwave in the midwestern United States: learning from the lessons of 1995. Bulletin of the American Meteorological Society (BAMS) 82:1353–1367

Reiter P (2001) Climate change and mosquito-borne disease. Environmental Health Perspectives 109:41–61

Scheraga J, Ebi K, Moreno AR, Furlow JS (2003) From science to policy: developing responses to climate change. In: McMichael AJ, Campbell-Lendrum D, Corvalan CF, Ebi KL, Githeko A, Scheraga JD, Woodward A (eds) Climate change and human health: risks and responses. WHO/WMO/UNEP

Scheraga JS, Ebi KL et al (2003) From science to policy: developing responses to climate change. In: McMichael A, Campbell-Lendrum D, Corvalan C et al. eds. Climate change and health: risks and responses. WHO, Geneva, pp 237–266

Senat (2004) La France et les Francais face a la canicule: les lecons d'une crise. Rapport d'information no 195 (2003–2004) de Mme Letard, MM Flandre, S Lepeltier, fait au nom de la mission commune d'information du Senat, depose le 3 Fevrier 2004. Paris, France

Smit B, Pilifosova O et al. (2001) Adaptation to climate change in the context of sustainable development and equity. In: McCarthy JJ et al (eds) Climate change 2001. Impacts, adaptation and vulnerability. Contribution of Working Group II to the

Third Assessment Report of the Intergovernmental Panel on Climate Change. Cambridge University Press, Cambridge, pp 877–912

Stott PA, Stone DA, Allen MR (2004) Human contribution to the European heatwave of 2003. Nature 432:610–614

Van Lieshout M., Kovats RS, Livermore MTJ, Martens P (2004) Climate change and malaria: analysis of the SRES climate and socio-economic scenarios. Global Environmental Change 14:87–99

WHO (2001) Floods: Climate Change and Adaptation Strategies for Human Health. WHO, Copenhagen, EUR/01/503 6813

WHO (2004) Report of WHO meeting on Climate and Foodborne Disease. Rome, Italy, 27–28 February 2003) In preparation. Rome, WHO

WHO, EEA (2004) Report of the meeting "Extreme weather and climate events and public health responses", Bratislava, Slovakia, 9–10 February 2004. World Health Organization, Copenhagen

WHO Regional Office for Europe (2004) Declaration of the 4th Ministerial Confernce on Environment and Health, Budapest, Hungary, 23–25 June 2004 (EUR/04/5046267/6): 11

Yach D (1996) Redefining the scope of public health beyond the year 2000. Current Issues in Public Health 2:247–252

Yohe G, Ebi K (2005) Approaching adaptation: parallels and contrasts between the climate and health communities. A Public Health Perspective on Adaptation to Climate Change. K. Ebi and I. Burton

8.1 Preparedness for Extreme Weather among National Health Ministries of WHO's European Region

Tom Kosatsky, Bettina Menne

Summary

The clear impacts on population health and the high costs on health services of recent floods, windstorms and heat-waves in Europe have motivated the current assessment of preparedness for extreme weather events among National Health Ministries by the WHO Regional Office for European, Global Change and Health Programme.

> **! Goal**: To inform policy making and programme development in the area of climate change adaptation through a review of current European Region legislation and practice, with emphasis on the role of national public health authorities. Specifically, to describe the role of Health Ministries of the 52 WHO European Region Member States with respect to extreme weather events (floods, heat-waves, cold spells and wind storms) in terms of their mandates, programmes, capacities and limitations.

During June 2004, we wrote to the Chief Medical Officers or Directors of Public Health for the 52 Ministries of the European region asking that a staff member responsible for or "most knowledgeable about" prevention of adverse health impacts related to extreme weather be mandated to respond to our survey of Health Ministry preparedness. The questionnaire included both closed and open questions, covered both preparedness for and response to extreme weather, and was provided in English, French, German and Russian versions (Appendix 3).

As of 20 December 2004, 19 of 52 (37%) of the Ministries had responded. Several others, not included here, had acknowledged receipt and requested more time to complete the survey. Completed surveys were received from 9 of 25 (36%) EU States and 5 of 12 (42%) Newly Independent States (NIS). Of the 19 respondents, 16 (84%) related that their Ministries had specific legislation empowering them to act during emergencies. In 7 cases, this legislation was said to specify flood response, with windstorm response in 5 cases. There was less frequent specific mention of heat-waves (3 cases) and cold spells (1 case). In contrast, 6 Ministries had specific powers in the event of chemical spills, and 5 with respect to earthquakes.

Fifteen (79%) Ministries responded that they had a plan for mobilization and response during emergencies. Eleven of these plans had sections devoted to earthquakes and 10 to chemical spills; fewer mentioned extreme weather events: floods (9), windstorms (3), heat-waves (3), cold spells (1). Eight Ministries stated that their response plan had been revised within the last 5 years, of which 5 had been revised in 2003.

Three (16%) Ministries had collaborative agreements with national weather services: agreements covered the transmission of weather information, the drafting of joint communiqués, and responsibilities for risk management. Nine (47%) Ministries had cooperative agreements with the national civil protection agency on extreme weather/natural disasters, and 4 (21%) with the national armed forces.

Ten (53%) of the Ministries had programmes in place to monitor health during natural disasters, which in most cases included surveillance of medical visits, emergency services, hospitalizations, and recent deaths. However, in fewer than half of the 10 Ministries with programmes to monitor the population's health were extreme weather events included.

Other than national responsibilities in the area of extreme weather, in 16 (84%) countries there were also regional and in 18 cases (95%) also municipal mandates.

Asked about the attention that national media accorded extreme weather, 11 (58%) responded "much" and 6 (32%) "moderate".

Interpretation of responses is limited by the many nations that did not answer, by contextual issues concerning the different histories and mandates of public health in the various European Region member states, and by difficulties in locating a single best respondent for the complex issue of preparedness for extreme weather.

It does appear that most Health Ministries have become involved in the issue of response to extreme weather, that they participate in inter-Ministerial programmes, and that some have instituted strategies to identify and to rectify population vulnerabilities before such events occur. However, specific legislation referring to the health consequences of extreme weather is still the rarity in the European region, and Ministerial programmes designed to mitigate natural disasters often do not include extreme weather events. Few Ministries cooperate formally with their national meteorological agencies.

8.1.1 Background

The WHO European Region has experienced an unprecedented rate of warming in the recent past. During the period 1976–1999, the mean daily maximum temperature in most areas during the summer months has increased by more than 0.3 °C per decade (Koppe, 2004).

Increasing variability in the European climate has lead many experts to predict that extreme weather events including floods, heat-waves, cold spells and windstorms will become both more frequent and more severe (Navarra, 2004).

In the summer of 2003, heat-waves struck large areas of Western Europe and caused many unanticipated deaths in several countries. According to provisional data provided by national authorities, there were more than 14 800 excess deaths in France alone (Kovats, 2004).

Flooding is the most common natural disaster in the European Region. From January to December 2002, the Region suffered an estimated 15 major floods, which killed approximately 250 people and affected as many as 1 million. An international disaster database recorded 238 floods in the Region between 1975 and 2001. During the last decade, floods killed 1 940 people and made 417 000 homeless (Hajat, 2003).

Cold spells are a fact of winter life in many Northern and Central European countries and throughout Central Asia. In the United Kingdom, substandard housing has been associated with excess winter deaths (Wilkinson, 2004). Deaths due to hypothermia occur among the marginalized and among those isolated by storms or power failures (Hassi, 2004).

Wind storms, both tropical storms and tornadoes, have struck the WHO European Region. Much of the impact of tropical storms is due to flooding when the "storm surge" hits land. Windstorms are also associated with flying debris. Living in low-lying and high-density areas, and substandard and poorly anchored housing are risk factors for wind-related injury and death (Lillibridge, 1997).

In recent years, preventing and responding to the effects on human health of extreme weather has become a public health action priority. Health Ministries have responded by invoking the three classic public health strategies: disease prevention; wellness promotion; and health protection (Detels and Breslow, 2002). Reducing the transmission of infectious liver disease during floods through the maintenance of high levels of immunization against hepatitis A, and limiting foodborne illness during heat-waves through the implementation of high standards of hygiene, are examples of prevention-based public health preparedness. As a fit population is better able to withstand heat and cold stress, and an informed population better able to adopt protective measures, wellness promotion also has a place in preparedness for extreme weather. Health protection measures such as the provision of air conditioning, clean water, and shelter for the homeless might be included in programmes to reduce the sanitary impacts of heat-waves, floods, wind storms and cold spells.

Besides the classic strategies of prevention, promotion and protection, such integrative actions as vigilance (information gathering to signal early evidence of stress and/or impact on health) and resource coordination have been implicated in public health responses to extreme weather.

While each type of extreme weather event and each locale require a tailored response, public health is most effective where specific programmes form part of a well-developed network and approach. Thus, where a strong system of disease surveillance is already in place, public health authorities can more easily incorporate special measures to detect emerging health impacts related to new environmental stressors. Likewise, where the home care network is already active, it can more easily incorporate extraordinary measures to protect the most vulnerable during heat-waves and floods.

One might argue that public health preparedness for extreme weather should be based on the assessment of vulnerabilities coupled with the mobilization of resources from within and outside the health sector. The assessment of vulnerabilities includes both environmental factors (examples: residential zones most liable to be flooded, urban areas where heat is trapped) and social and behavioural factors (examples: marginalized or non-autonomous populations, persons with low levels of immunity to infectious disease). Given the complexity of responses, resources must be mobilized, both from within and outside the health sector, to mitigate these vulnerabilities. In order that these resources truly strengthen population resilience in the domain of health (as opposed to other worthwhile objectives), public health authorities must both assume a coordinating role and act as primary advocates for health.

Public health authorities can be involved in both preparation for, and response to, extreme weather. Both mandates involve the building and activation of partnerships, with weather services, civil protection agencies, civic authorities, health service providers and voluntary agencies. Issues of responsibilities, leadership and coordination must be defined.

Finally, public health authorities can be involved in influencing those environmental stressors that contribute to global climate change and with it, extreme weather. Specifically, in promoting a healthier world, public health authorities can advocate for the great variety of measures which diminish vulnerabilities, an example at the global level being the reduction of greenhouse gas emissions, or, at the local level, reduction of the urban heat island, or limitation of residential encroachment on the flood plain.

The WHO Regional Office for Europe is working to place climate change higher on the global public health agenda. As a first step, there is a need to document policy frameworks, programmes and lead agencies in relation to actual public health responses to extreme weather. The theoretical basis for public health intervention has been developed in two recent publications on heat-waves:

▌ Methods of assessing human health vulnerability and public health adaptation to climate change (Kovats et al 2003),
▌ Heat-waves: risks and responses (Koppe et al 2003),

The current survey, funded as part of the cCASHh project (European Commission grant No. EVK2-CT-2000-00070), has been designed to allow WHO to develop national and regional profiles of extreme weather prevention activities and policies. It aims to identify areas where national authorities are active and areas where more efforts are needed. As a result WHO will be in a better position to support Ministries of Health by providing technical assistance (e.g. capacity building, methods analyses) that better respond to national needs for the implementation and support of programmes to prevent the adverse effects on health of extreme weather.

Goal: To inform policy making and programme development in the area of climate change adaptation through a review of current European legislation and practice, with emphasis on the role of national public health authorities. Specifically, to describe the role of European Health Ministries with respect to extreme weather events (floods, heat-waves, cold spells and windstorms) in terms of their mandates, programmes, capacities and limitations.

8.1.2 Pertinent Literature

Two recent enquiries are particularly relevant. During the summer of 2002, the U.S. Center for Disease Control (CDC) reviewed emergency response plans for 18 US City Health Departments (Bernard and McGeehin, 2004). Of the 18, six had no plan which encompassed a heat-wave response, two included heat-waves in an overall disaster response plan, and 10 had stand-alone heat-wave plans. Of the latter 10, for seven, the plans were judged comprehensive, on the basis that: lead and participating agencies were identified; a consistent, standardized warning system was activated on the basis of weather forecasts; a communication plan was in place; the response plan targeted high risk groups; and mechanisms for information collection, evaluation and revision were in place. While the CDC review covered the emergency response function, it excluded issues relating to the involvement of city health departments in the promotion of medium and long-term adaptation measures.

The WHO European Region's Division of Country Support conducted, during 2004, a survey of disaster preparedness in 12 (mainly Central and Eastern European and Central Asian) countries (G Rockenshaub, WHO European Region, personal communication). Among issues queried were:

- the legal and administrative basis of disaster response
- the nature of national disaster plans, including roles and responsibilities and chain of command
- co-ordination of the national response, specific role of health authorities
- training
- monitoring and evaluation
  - indicators of potential human exposure
  - simulation exercises
  - periodic reviews
  - evaluation and response
- risk assessment
  - inclusion of risks to health institutions
- risk reduction, mitigation
- epidemiologic surveillance
- budget level, procedures for financial management.

8.1.3 **Methods**

In June 2004, the WHO European Center for Environment and Health (Rome) initiated a survey of Health Ministries in the 52 Member States of the European Region in order to describe their activities with respect to the prevention of impacts on health of extreme weather. Our review has been designed to assess the legislative basis for Ministerial activities, the history of their involvement in adaptation to extreme weather, who their partners are, what their programmes include, how these relate to other Ministerial mandates, and what are the limits on the level and nature of their efforts.

Our study questionnaire (Annex 3) was designed to cover the spectrum of powers and responses with respect to extreme weather events. While emphasis was placed on floods and heat-waves, windstorms and cold spells were also queried, and additional questions concerned chemical spills and earthquakes in order to assess the attention paid extreme weather events relative to environmental health emergencies not related to weather. While priority was given to planning and preventive activities, we also queried public health responses in the event of extreme weather. In order to assess the division of responsibilities within nations, we asked about whether regions and cities had mandates and programmes beyond those of their National Ministries. In order to validate and contextualize responses, we asked for written documentation of legal mandates and programme activities where available.

The questionnaire was pre-tested with colleagues at the WHO European Centre for Environment and Health and subsequently with French, Slovakian, and Canadian environmental public health authorities.

Respondents were identified by first asking the Chief Medical Officer or Director of Public Health in each Ministry of the 52 Member States of the WHO European region to mandate a staff member responsible for or "most knowledgeable about" prevention of adverse health impacts related to extreme weather to respond to our questionnaire. The questionnaire itself, written in English, was translated into Russian, German, and French. It included both open and closed questions, the answers to which were later coded based on response similarities. Respondents were also asked to provide copies of their Ministerial organizational charts, relevant legislation, and programme descriptions, and public communications concerning heat-waves, cold spells, floods and windstorms. The transmittal letter was also sent to the WHO country liaison officers appointed to Central and Eastern European and Central Asian countries (Appendix 1).

Ministerial transmittal letters and survey forms were distributed during June and July 2004. Completed survey forms were scanned for missing pages and questions left unanswered. When this occurred, we requested completion of the missing sections. Occasionally answers appeared inconsistent, as when a general question was answered "no", but follow-up questions, contingent on the general question receiving a positive response, were answered anyway. Given that these responses may have been the result of an ambiguous question or of a situation where the respondent's Ministry had particular legislation, programmes or activities, not considered when our questions were formulated, we accepted the responses as provided. Occasionally, inferences were required when a written response was received which did not fall into one of the suggested categories: these were coded as "uncertain".

For purposes of sub-regional assessment, analyses were also conducted for the 25 European Union (EU) and the 12 Newly Independent States (NIS).

8.1.4 Results

As of 20 December 2004, 19 of the 52 (37%) Ministries of the WHO European Region had responded (Appendices 1–3). Several others, not included here, had acknowledged receipt and requested more time to complete the survey. While we are aware that certain countries (among them Lithuania and Norway) have introduced legislation and programmes after submitting their questionnaire responses, they have not been taken into account in the present report. The degree of elaboration varied: for some countries, mainly those with little history of catastrophic extreme weather, responses were brief. Others provided comprehensive responses, and included legislation, reports and programme outlines. Some of the comments provided are cited in the paragraphs which follow. Completed surveys were received from 9 of 25 (36%) EU states, and 5 of 12 (42%) Newly Independent States.

Ministries were asked about the experience of their countries with extreme weather:

Major floods since 1994		
All (of 19)	EU (9)	NIS (5)
12	7	3

Major windstorms since 1994		
All (of 19)	EU (9)	NIS (5)
11	5	3

Major heat-waves since 1994		
All (of 19)	EU (9)	NIS (5)
11	6	4

Major cold spells since 1994		
All (of 19)	EU (9)	NIS (5)
7	4	2

Of the 19 countries that responded, 16 (84%) stated that their National Health Ministry had specific legislation authorizing it to act during emergencies. In 7 cases, this legislation was said to include specific mention of flood response, with windstorm response mentioned in 5 cases. There was less frequent specification of heat-waves (3 cases) and cold spells (1 case). In contrast, 6 Ministries had specific powers with respect to chemical spills, and 4 with respect to earthquakes. While most States were said to have a single law authorizing emergency response, examples being Malta's Civil Protection Act (1999), Lithuania's Order of the Ministry of Health Number V-529 on Emergency Situation Management (2004), or Iceland's Civil Protection Act, in other cases empowering legislation was said to derive from a variety of laws and cabinet orders, such as (among others) Poland's Act of 18 April 2002 on State of Environmental Disaster, Act of 6 September 2001 on Infectious Diseases and Infections, and the Regulation of the Council of Ministers of 11 June 2004 Regarding Detailed Range of Operations of the Ministry of Health (Appendices 2 and 3).

▌ **Box 1. Is there specific legislation that gives your Ministry authority to act during emergencies?**
Sveinn Magnússon, Head of Department,
Ministry of Health and Social Security, Iceland

"In Iceland, the forces of nature are very strong. Volcanic eruptions happen every 4–5 years, the last one last month. Earthquakes >3 are common. Strong winds >50 m/s happen many times every year. Flooding of rivers caused by snow and ice melting and flooding of coastal areas caused by high sea level happen yearly. Avalanches are a constant threat, both because of snow and because of gravel. Cold periods are common, this year the coldest days in Reykjavík in November in 100 years. Heat-waves are no problem in Iceland!
All these events have taken their toll and have taught us to organize the civil protection of all kinds in the best possible way; this is done under a special act of Civil Protection (Lög um almannavarnir nr 94/1962 with subsequent additions – not available in English) led by the Minister of Justice. Other Ministries are involved accordingly, like the Ministries of Health, Transportation and others.
All the actors in the civil protection are organized in one response unit in one building, where police, rescue services, coast guard and communication services are the main actors and train regularly together with personnel from the Ministry of Health and others."

Fifteen of 19 responding Ministries (79%) stated that they had a plan for mobilization and response during emergencies. The only Ministry to respond that did not have a plan told us that one was currently being drafted. Three Ministries responded that they had plans for specific environmental emergencies, but left blank the question concerning a generic emergency mobilization and response plan. Of the 15 Ministries with generic emergency plans, 10 included sections devoted to earthquakes and 10 to chemical spills; fewer mentioned extreme weather events: floods (9), windstorms (3), heat-waves (3), cold spells (1). Eight Ministries stated that their response plan had been revised within the last 5 years, of which 5 were revised in 2003.

Box 2. Does your Ministry have a plan (whether or not you have specific legislation) for mobilization and response during emergencies?
Edward Wlodarczyk, Director
Department of Defence Affair, Ministry of Health, Poland

"The Ministry of Health has special procedures for actions during particular emergencies. These procedures are being set and actualised according to the existing legislation. They regard, among others, mobilization and response in case of floods, chemical spills, radiation accidents."

Box 3. Does your Ministry have a plan for mobilization and response during emergencies?
Filomena Araújo, Coordinator
Environment and Health Division, General Direction of Health, Portugal

"The Ministry of Health has a Contingency Plan for heat-waves. Each hospital and health centre has its own emergency plan for logistics questions related to preparedness to emergencies. Also, INEM, the National Institute for Medical Emergencies, has its own plan for emergencies."

Three of the 19 Ministries which responded (16%) had collaborative agreements with national weather services: agreements covered the transmission of weather information, the drafting of joint communiqués, and responsibilities for risk management. In one case, a respondent noted that despite the absence of a formal agreement, "we actively talk to each other." In another, it was noted that the (National) Institute of Meteorology is obliged to cooperate with the Governmental Crisis Coordination Group, of which the Health Ministry is a member. Eight (44%) Ministries (including 3 of 9 EU Ministries, and 3 of 5 Ministries of the Newly Independent States) had cooperative agreements on extreme weather/natural disasters with the national civil protection agency, and 4 (22%, of which 2 were in the NIS) with the national armed forces.

Ten (53%, including 6 EU and 3 NIS) of the 19 Ministries had programmes in place to monitor health during natural disasters, which in most cases included surveillance of medical visits, emergency services, hospitalizations and recent deaths. However, in fewer than half of the 10 Ministries with emergency health monitoring programmes were extreme weather events covered.

Other than national responsibilities in the area of extreme weather, in 16 countries (84%) there were also regional and in 17 countries (89%) also municipal mandates.

Asked about the attention that national media accorded extreme weather, 11 (58%) responded "much" and 6 (32%) "moderate". Of the five respondents from the Newly Independent States, four (80%) responded "much". One Ministry noted that the media were an important tool for the transmission of weather warnings and advice as to protective measures.

Details were asked about prevention plans for floods and heat-waves. Eight (42%) Ministries said they had flood prevention plans and 5 (26%) had prevention plans for heat-wave impacts. Four of 9 (44%) responding EU Ministries had plans for floods and for heat-wave impact mitigation. Items commonly included in the flood impact prevention plans were assessment of environmental (flood plains) and social (ill or disadvantaged populations) vulnerabilities, communications strategies, and zoning regulations; only one Ministry stated that financial incentives were offered to citizens to increase their ability to resist the effects of floods. Four Ministries had heat watch warning systems where weather forecasts trigger advice and/or the implementation of emergency measures. Three Ministries mentioned that financial incentives were available to help citizens decrease their vulnerability to heat.

Two Ministries (11%) had a plan for the prevention of the health impacts of cold.

Box 4. In what ways could your Ministry play a larger role in the prevention of adverse health impacts from flooding?
Hristina Mileva, Chief State Expert
Health Promotion and Disease Prevention Section, Ministry of Health, Bulgaria

"Through establishment of a particular body for management of health services in cases of extreme weather events, accidents, and other emergency cases."

Box 5. In what ways could your Ministry play a larger role in the prevention of adverse health impacts from flooding?
Mikko Paunio, Senior Medical Officer
Unit of Health Protection, Department of Health, Finland

"If it occurs, local water authorities might seek for governmental advice or have a possibility to consult national public health epidemiology specialists."

8.1.5 Discussion and Conclusion

Interpretation of responses is limited by the many Ministries which did not answer, by contextual issues concerning the different histories and mandates of public health in the various European Member States, and by difficulties in locating a single best respondent for the complex issue of describing preparedness for extreme weather. It does appear that most Health Ministries have become involved in the issue of response to extreme weather, that they participate in inter-Ministerial programmes for disaster response, and that some have instituted strategies to identify and to rectify population vulnerabilities before such events occur. However, specific legislation referring to the health consequences of extreme weather is still the rarity in the European region, and Ministerial programmes designed to mitigate natural disasters often do not cover heat-waves and cold spells.

An alternative to the current mail-out survey might be to first identify Ministerial focal points and then to query these mandated persons about the activities of their government. Another alternative, used with success in 2004 by the European Region's Division of Country Support, would be to have questionnaires completed on-site during a training exercise.

Language, culture, the division of powers between federal and state authorities, and the organization of Health Ministries are more difficult issues. In some states, the federal role is limited, while regional authorities have extensive powers and responsibilities. Security concerns may have limited our ability to appreciate programmes in some states: we speculate that the response by two Ministries of the Newly Independent States, which informed us that disaster preparedness is under the authority of the "Second Department", may be indicative of the latter.

The issue of validity of responses goes beyond the question of rates of response. Typically, a variety of units and individuals within a Ministry are involved in extreme weather preparedness and response. In some States, an affiliated national institute of public health assumes Ministerial functions. We attempted to give voice to the various actors involved by first approaching national directors of public health and chief medical officers, who presumably see the larger picture of Ministerial activities, and asking them to mandate a person or persons capable of responding to the various issues addressed in the survey. In several cases, more than one individual contributed to the Ministerial response. A more proactive approach would be to develop the survey on the basis of case studies, with the various actors implicated in extreme weather event preparedness interviewed both individually and in a group encounter.

Despite these obstacles, we submit that the current survey has added knowledge to an area where, as yet, little work has been done. Both the survey and the collection of legislation and programmes it has identified provide examples of the wide variety of public health responses to extreme weather throughout the European Region: we expect the current project to help make these models known among the 52 Member States.

We found, in common with the WHO Division of Country Support survey, that there is a wide variation in preparedness and response activities among the 52 European Region Member States (G Rockenshaub, personal communication). While Ministerial activities in this area are mandated by current legislation in almost every country, the level of support and of policy development varies greatly. Furthermore, disaster preparedness is not always integrated into other Ministerial mandates and activities.

There appears to be more activity in the area of earthquakes than for extreme weather events: among extreme weather events, floods are most likely to be subjects of a prevention plan. It is to be noted, however, that several Ministries have recently put heat-wave prevention and response plans into place. Particularly impressive is the comprehensive nature and preventive orientation of many of these plans, which include strategies for the diagnosis of vulnerabilities, for health monitoring, for population advice, and for financial incentives to encourage vulnerability reduction. Also impressive is the high degree of integration of activities among Health and related Ministries existing in many countries. An important gap is the lack of formal collaborative agreements with the national weather services.

Clearly much has been done and much more remains to be done in raising the profile of extreme weather events within the WHO European Region.

References

Bernard SM, McGeehin MA (2004) Municipal heat wave response plans. American Journal of Public Health 94(9):1520–1522

Detels R, Breslow R (2002) Current scope and concerns in public health. In: Detels R, McEwen J, Beaglehole R, Tanaka H (eds) Oxford Textbook of Public Health. Oxford University Press, Oxford

Hajat S, Ebi K, Kovats S, Menne B, Edwards S, Haines A (2003) The human health consequences of flooding in Europe and the implications for public health: a review of the evidence. App Environmental Science and Public Health 1(1):13–21

Hassi J (2005) Cold extremes and impacts on health. In: Kirch W, Menne B, Bertollini R (eds) Extreme Weather Events and Public Health Responses. Springer, Heidelberg

Koppe C, Kovats S, Jendritzky G, Menne B (2004) Heat waves: risks and responses. World Health Organization, Copenhagen

Kovats S, Wolf T, Menne B (2004) Heatwave of August 2003 in Europe: provisional estimates of the impact on mortality. Eurosurveillance Weekly, 8(11). http://www.eurosurveillance.org/ew/2004/040311.asp

Lillibridge SR (1997) Tornadoes. In: Noji E (ed) The Public Health Consequences of Disasters. New York, Oxford Press, 1997, pp 228–244

Navarra A (2005) The climate dilemma. In: Kirch W, Menne B, Bertollini R (eds) Extreme Weather Events and Public Health Responses. Springer, Heidelberg

Wilkinson P, Pattenden S, Armstrong B, Fletcher A, Kovats RS, Mangtani P, McMichael AJ (2004) Vulnerability to winter mortality in elderly people in Britain: population based study. British Medical Journal 329(7467):647

Appendix 1: Member States of the WHO European Region (survey respondents in bold)

EU Members

Austria	**Latvia**
Belgium	**Lithuania**
Cyprus	Luxembourg
Czech Republic	**Malta**
Denmark	The Netherlands
Estonia	**Poland**
Finland	**Portugal**
France	**Slovak Republic**
Germany	Slovenia
Greece	**Spain**
Hungary	**Sweden**
Ireland	United Kingdom
Italy	

Newly Independent States

Armenia	**Republic of Moldova**
Azerbaijan	Russia
Belarus	Tajikistan
Georgia	**Turkmenistan**
Kazakhstan	Ukraine
Kyrgyzstan	**Uzbekistan**

Other European Region Member States

Andorra	Monaco
Albania	**Norway**
Bosnia and Herzegovina	Romania
Bulgaria	San Marino
Croatia	Serbia and Montenegro
Former Yugoslav Republic of Macedonia	Switzerland
Iceland	**Turkey**
Israel	

Appendix 2: Responses to selected questions about preparedness and response to heat, by country

Country	Last heat-wave	National programme to monitor heat health effects	Mobilization and response plan for heat	National heat prevention plan	Cooperative agreement between meteorological services and MOH
Azerbaijan	2003	Yes	No	No	No
Bulgaria	2002	No	No	No	No
Finland	No heat-waves	No	No	No	Informal
Iceland	No heat-waves	No	No	No	No information
Israel	Every year	Yes: deaths, medical reports, hospital admissions and ambulance calls	No	Yes, advice to vulnerable, no Heat Health Warning System	No
Kyrgyzstan	1997	No	Yes	No	No
Latvia	2003	Yes: medical reports, hospital admissions and ambulance calls	Out of the competence of the MOH	Plans under development	No
Lithuania	2002	Not for heat	No	No	No
Malta	2003	No	No	No	No
Moldova	2001 and 2002	No	No	No	No
Norway	No heat-waves	Municipalities responsible for all risks	No	No	No

Appendix 2 (continued)

Country	Last heat-wave	National programme to monitor heat health effects	Mobilization and response plan for heat	National heat prevention plan	Cooperative agreement between meteorological services and MOH
Poland	No information	No	No	No	Only with Governmental Crisis Coordination Group
Portugal	2003	Yes	Yes	Yes, Heat Health Warning System, rapid mortality surveillance, education, advice to emergency services	Yes
Slovak Republic	2003	Yes	Yes	Yes, with addressing vulnerable populations, Heat Health Warning System, education, advice to doctors	No
Spain	2003	Yes	Yes	Yes, addressing vulnerable populations, Heat Health Warning System, education, advice to doctors	Yes
Sweden	Not relevant	Not specific	No	No	No
Turkey	No information	No	Yes	Yes, assessment of susceptible persons	No
Turkmenistan	Every year	No information	No information	No	No
Uzbekistan	No information	No	No information	No	Yes

Appendix 3: Responses to selected questions about preparedness and response to floods, by country

Country	Last flood	National programme to monitor flood health effects	Mobilization and response plan for floods	National flood prevention plan	Cooperative agreement between meteorological services and MOH	Comments
▨ Azerbaijan	2003	Yes: deaths, medical reports, hospital admissions and ambulance calls	Yes	No	No	
▨ Bulgaria	2003	No	No	Yes	No	Plan includes: assessment of susceptible areas and populations and media advice
▨ Finland	2004	No	No	Yes	Informal	Plan includes: assessment of susceptible areas media advice, some building regulations and some flood resistance measures
▨ Iceland	2004	No	No	Act of civil protection under the Ministry of Justice	–	–

Appendix 3 (continued)

Country	Last flood	National programme to monitor flood health effects	Mobilization and response plan for floods	National flood prevention plan	Cooperative agreement between meteorological services and MOH	Comments
Israel	1992	Yes: deaths, medical reports, hospital admissions	No	–	No	
Kyrgystan	1998	No	Yes	No	No	
Latvia	1980/81	Yes: medical reports, hospital admissions and ambulance calls	Out of the competence of the MOH	Plans under development	No	
Lithuania	1958	Yes for flood	Yes but not for EWE	Yes	No	Including zoning regulation and housing flood resistance measures
Malta	2003	No	Yes	No	No	
Moldova	2002	Yes: deaths, medical reports, hospital admissions	No	Yes	No	Media announcements
Norway	1995	Municipalities and Health enterprises	No	No	No	

Appendix 3 (continued)

Country	Last flood	National programme to monitor flood health effects	Mobilization and response plan for floods	National flood prevention plan	Cooperative agreement between meteorological services and MOH	Comments
Poland	2001, 2000 and 1997	No	No	Ministry of Interior and the National Civil protection and at province level	Only with Governmental Crises Coordination Group	
Portugal	–	Yes	No	No	Yes	
Slovak Republic	2003	Yes: deaths	Yes	Yes: susceptible zone and people, media advice, zoning regulations and flood resistance measures for buildings	No	
Spain	–	Yes: medical reports, hospital admissions and ambulance calls	Yes	–	–	
Sweden	2004	Yes, not specific	No	No	No	
Turkey	2004	No	No	No	No	
Turkmenistan	No information	No information	Yes	No, although flood prevention and mitigation activities are in place	No	
Uzbekistan	No inform ation	No	Yes	Yes	Yes	

8.2 Indicators of Adaptive Capacity: Evidence from a Conjoint Choice Exercise

ANNA ALBERINI, ALINE CHIABAI

8.2.1 Introduction

Considerable attention has been focussed, as of late, on the role of adaptation to judging the economic implications of climate change and climate variability (IPCC, 2001), and on the characteristics of systems, such as communities and regions, that influence their propensity or ability to adapt (or their priorities for adaptation measures). Adaptive capacity is defined as the "potential, capability, or ability of a system to adapt to climate change stimuli or their effects of impacts" (IPCC, 2001), implying that, at least in principle, adaptation has the potential to reduce the damages of climate change, or to increase its benefits.

Yohe and Tol (2002) propose a formal model where vulnerability (i.e., the losses caused by climate change) is a function of exposure and sensitivity, which depend on adaptation. Adaptation is, in turn, a function of adaptive capacity, which is influenced by

- available technological options,
- resources,
- the structure of critical institution and decision-making authorities,
- the stock of human capital,
- the stock of social capital including the definition of property rights,
- the system's access to risk spreading processes,
- information management and the credibility of information supplied by decision-makers, and
- the public's perceptions about risks and exposure.

Yohe and Tol estimate an empirical model of vulnerability using a country panel dataset documenting the occurrence and effects of natural disasters, such as floods, earthquakes etc. Their regression model relates alternative measures of vulnerability (number of disaster events, fraction of population affected by disaster events, annual damage and number of deaths) to the country's per capita income, the Gini inequality coefficient, and population density. They conclude that these vulnerability measures are negatively related to income, and positively related to inequality and population density.

Yohe and Tol's model confirms the expected relationship between resources and vulnerability, but its stylized nature does not allow one to disentangle institutional factors, the effects of the health stock in the population, and the existence of adequate infrastructure and information channels.

In this paper, we propose a different approach to overcome these limitations of previous studies and identify the role of factors generally linked with adaptive capacity. We survey public health experts, climatologists, and emergency response officials to find out which factors they believe to be the most important determinants of adaptive capacity. Our survey questionnaire relies on conjoint choice questions.

In a typical conjoint choice exercise (Louviere et al., 2000), respondents are presented with a set of K hypothetical alternatives (the so-called "choice set") representing situations, goods or public policies. The alternatives are described by a vector of attributes, and differ from one another in the level of two or more attributes. Respondents are asked to indicate which of the K alternatives they deem the most attractive, K being at least two. It is assumed that in choosing the most preferred alternative, the respondent will trade off the attributes of the alternatives.

We adopt this approach in our survey about adaptive capacity, where the alternatives described to the respondents in the questionnaire are hypothetical countries that could be located anywhere in the world. Our attributes are a country's resources and distribution of income, the distribution of age in the population, measures of health status and quality and type of health care, and access to information and technology. In our choice questions, we present respondents with pairs of hypothetical countries ($K=2$), and we ask them to tell us which of the two countries they believe to have the higher adaptive capacity.

The purpose of our study is two-fold. First, we wish to infer which social, economic, health, and technological factors experts and decision-makers associate with a higher or lower degree of adaptive capacity. Second, we wish to find out how they trade off such factors against one another in assessing adaptive capacity. For example, when it comes to adaptive capacity, can a more egalitarian distribution of income make up for lower income per capita? Or, how much wealthier does a country need to be to make up for the absence of a universal health care system?

Our questionnaire was administered to a sample of public health officials and climate change experts intercepted at professional conferences and intergovernmental meetings, resulting in 100 completed questionnaires. We fit probit equations to the responses to the choice questions, finding that per capita income, inequality, universal health care coverage and access to information are judged to be important determinants of adaptive capacity. By contrast, our respondents do not consider the distribution of age the population, life expectancy or the number of physicians per 100 000 to influence adaptive capacity. The effect of a more equitable distribution of income is deemed equal to about US $ 4600 worth of per capita income, while universal coverage in health care and high access to information are equivalent to US $ 12 000–14 000 in per capita income.

The remainder of this paper is organized as follows. In section 8.2.2, we describe the structure of the questionnaire, focussing on the conjoint choice questions in section 8.2.3. In section 8.2.4, we provide a theoretical framework to interpret the responses to the conjoint choice questions, and describe our statistical models of the responses. In section 8.2.5, we present our data. Section 8.2.6 reports and discusses the results of our statistical models. We offer concluding remarks in Section 8.2.7.

8.2.2 Structure of the Questionnaire

Our questionnaire (reported in Annex 4) starts with explaining the topic of the study to the respondents, thanking them in advance for their participation in the survey and assuring them that their responses to the questions are confidential. Instructions for returning the completed questionnaire are provided on the cover page. For good measure, they are also repeated at the end of the questionnaire. The remainder of the questionnaire is divided into five sections.

Section A provides a brief description of global climate change and its effects. Respondents are told that there may be numerous effects of climate change on human health, but that in this survey we would like them to focus on three, namely (i) deaths and injuries associated with floods and landslides caused or aggravated by sustained rains or extreme weather events, (ii) cardiovascular or respiratory illnesses and deaths during heat waves and (iii) vector-borne infectious diseases.

In section B, we ask respondents how important (i), (ii) and (iii) are to the respondent's organization and to the respondent himself (or herself). We then briefly introduce the concept of adaptive capacity, explaining that country and local governments may implement adaptation policies to reduce (i), (ii) and (iii), and announce that the next section of the survey will ask questions about adaptive capacity.

Section C contains the conjoint choice questions. Respondents are explicitly told that they will be asked to consider pairs of hypothetical countries that could be located anywhere in the world.[1] These countries, the instructions continue, are described by seven attributes, and differ in the levels of at least two of these attributes. For example, respondents are told, one of the two countries (A or B) may have higher income, but lower life expectancy, than the other. Based on the description, the respondent will have to indicate which of the two countries has higher adaptive capacity.

Respondents face a total of four choice questions. The first two refer to countries which have relatively high population densities (400 people per square kilometre), a mild climate, significant amounts of coastline and mountains, a relatively high degree of deforestation, and are moderately

---

[1] While cCASHh wishes to examine adaptation to climate change with specific reference to Europe, our survey respondents include nationals of many countries. We, therefore, reasoned that respondents from countries on other continents may feel more comfortable answering questions about countries that are not necessarily located in Europe

subject to floods and landslides. The last two choice questions refer to pairs of countries with a cold climate, relatively low population density, little deforestation, significant amounts of coastline and mountains, and have rarely experienced extreme events in the past.

Section D asks questions about the professional background of the respondents, and section E concludes with debriefing questions that inquire about the clarity of the questions and of the choice exercises.

8.2.3 Choice Questions

While researchers have discussed the potential role of many social, economic, and institutional factors in determining a community's adaptive capacity to climate change, we rely on a relatively small number of attributes in our conjoint choice tasks due to sample size considerations and to limit the cognitive burden imposed on the respondent.

Our stylized, hypothetical countries are defined by a total of seven attributes:

- per capita income (in US dollars),
- a qualitative description of the level of inequality in the distribution of income ("high" or "low"),
- the proportion of the country's population of age 65 and older,
- life expectancy at birth,
- physicians per 100 000 residents,
- the type of health care system (universal coverage, or based on private health insurance), and
- access to information via television, radio, newspaper, and internet ("high" or "low").

We arrived at this set of attributes after reviewing the literature, consulting with the other members of the research team, and developing a first list of candidate descriptors, which was pared down after pre-testing earlier versions of the questionnaire at the WHO and cCASHh coordination meetings held in Freiburg in May 2003, in Prague in June 2003 and in Potsdam in July 2003.

Our experimental design calls for three possible levels of income per capita: US $ 13 000, US $ 20 000 and US $ 27 000. These were selected because they reflect the per capita incomes of accession countries like the Czech Republic (US $ 13 991) and of certain European Union countries. For example, in 2000 Spain's per capita income was US $ 19 472, Italy's was US $ 23 626 and Belgium's was US $ 27 178. We use two possible levels for the percentage of the elderly in the population: 12%, which corresponds to a relatively "young" country (like the Republic of Ireland), and 18%, which corresponds to a relatively "old" country in Europe (Italy) in 2000.

Regarding the number of physicians per 100 000, which is a measure of access to health care widely used in public health, statistics compiled by

WHO suggest that there is much variability across countries in this index.[2] In the end, we selected three possible levels: 250, 300 and 400 per 100 000. Life expectancy at birth is high in Western European countries, so we focus on 70 and 79 years. The former is life expectancy at birth in Eastern European countries like Rumania and Bulgaria, while the latter is the approximate figure for Italy, France and Sweden, among other Western European countries.

The remaining three attributes (health care system, inequality in the distribution of income, and access to information) are qualitative and are later converted into 0/1 dummy variables for the purpose of statistical modelling.

In our conjoint choice questions, each choice set consists of two artificially created, hypothetical countries. To create these pairs, we first created the full factorial design, i.e. all possible combinations of the levels of the attributes.[3] Next, we randomly selected two country profiles, but discarded pairs containing dominated alternatives.[4] This was repeated for a total of four conjoint choice questions per respondent, making sure that each set of four pairs did not contain duplicate pairs. We created 32 different versions of the questionnaire following this approach. Respondents were randomly assigned to a questionnaire version.

The first two conjoint choice questions refer to countries A and B, and C and D, respectively, which are portrayed as enjoying a mild climate, but a relatively high propensity to extreme events, and high deforestation. The countries in the remaining two pairs (E and F, and G and H, respectively) have colder climates, are relatively unlikely to experience extreme events, and have had little deforestation.

8.2.4 The Model

This section provides a theoretical framework to motivate our statistical models of the responses to the choice questions. This framework is similar to the random utility model generally employed by economists and market analysts with conjoint choice surveys where the alternatives are private goods, environmental goods or natural resource management plans (e.g. Adamowicz et al., 1994; Boxall et al., 1996; Hanley et al., 2001), and other public policies (Alberini et al., 2005).

---

[2] For example, there are 164 physicians per 100 000 in the UK, 345 in Bulgaria, and 554 in Italy

[3] This is comprised of $2^5 \times 3^2 = 32 \times 9 = 288$ possible combinations

[4] An alternative in a pair is dominated if it is obviously worse than the other. In deciding whether there is a dominating alternative, we reasoned that countries with higher income should be judged to have higher adaptive capacity, and so should countries with lower inequality in the distribution of income, longer life expectancy, more numerous physicians per capita, and better access to information

We assume that when answering the conjoint choice questions, individuals select the alternative with the higher level of adaptive capacity out of the two in the choice set.

We further assume that the level of adaptive capacity individual i associated with alternative j, A_{ij}, breaks down into two components: a deterministic component, which is a linear function of the attributes of the alternative via a vector of unknown, fixed coefficients, β, and a stochastic error term. Formally,

$$A_{ij} = \bar{A}_{ij} + \varepsilon_{ij} \tag{1}$$

where $\bar{A}_{ij} = \mathbf{x}_{ij}\beta$, $\mathbf{x}_{ij}$ is the $1 \times p$ vector of attributes describing alternative j (j = 1, 2) to individual i, β is a $p \times 1$ vector of coefficients, and ε is the error term. The error term captures individual- and alternative-specific factors that affect the choice and are known to the respondents, but not to the researcher.

Since we observe a discrete choice out of two possible alternatives, the appropriate statistical model is a binomial model that describes the probability that the respondent selects, say, option 1 between alternatives 1 and 2 in the choice set. Selecting 1 means that this country is deemed to have a greater adaptive capacity, and hence that A_{i1} is greater than A_{i2}:

$$\Pr(1) = \Pr(A_{i1} > A_{i2}) \tag{2}$$

which in turn implies that the probability of choosing country 1 is

$$\Pr(1) = \Pr((\varepsilon_{i2} - \varepsilon_{i1}) < (\bar{A}_{i1} - \bar{A}_{i2})) = \Pr(\eta_i < (\mathbf{x}_{i1} - \mathbf{x}_{i2})\beta) \tag{3}$$

where $\eta_i = \varepsilon_{i2} - \varepsilon_{i1}$. If we assume that η_i is normally distributed with mean zero and variance 1, probability (3) is equal to:

$$\Pr(1) = \Phi(\eta_i < (\mathbf{x}_{i1} - \mathbf{x}_{i2})\beta) \tag{4}$$

where $\Phi(\cdot)$ is the standard normal cdf. Equation (4) is, therefore, the contribution to the likelihood in a probit model where the dependent variable is a dummy indicator taking on a value of one if the respondent chooses country 1, and zero otherwise. The independent variables are the differences in the level of the attributes between country 1 and 2, i.e. $(\mathbf{x}_{i1}-\mathbf{x}_{i2})$.

Probit equation (4) may be amended to include variables capturing individual characteristics of the respondent,[5] which means that the probability of picking country 1 over country 2 is

$$\Pr(1) = \Phi((\mathbf{x}_{i1} - \mathbf{x}_{i2})\beta + \mathbf{z}_i\gamma) \tag{5}$$

It is also possible to include interaction terms between the individual characteristics of the respondents and $(\mathbf{x}_{i1} - \mathbf{x}_{i2})$ to allow the same attribute to appeal to a different extent to different individuals.

---

[5] This means that in equation (1) $\mathbf{x}_{ij}$ contains the individual characteristics of the respondents, which are multiplied by alternative-specific coefficients

Finally, since respondents engage in a total of four choice tasks, it is necessary to spell out our assumptions about the possible correlation between the errors η_{im}, where m denotes the choice task ($m = 1, \ldots, 4$), within the same individual. The simplest model treats these errors as mutually independent, so that the log likelihood function of the data is:

$$\sum_i \sum_{m=1}^{4} \sum_{j=1}^{2} I_{ijm} \cdot \log \Pr(j \text{ in task } m) \tag{6}$$

Alternatively, a random-effects probit (see Greene, 2003) can be specified to account for the presence of unobserved heterogeneity, i.e. unobserved factors that influence choice and are common to all of the responses contributed by the same individual. The random-effects probit assumes that $\eta_{im} = v_i + \xi_{im}$, where v_i is the individual-specific effect, which remains unchanged over all of the error terms of the same respondent (but varies across individuals), while ξ_{im} is a completely random error term. Both are assumed to be normally distributed, have mean zero, and be independent of one another. These assumptions imply that the pairwise correlation between any two η_{im} within individual i is the same.

Once the β coefficients are estimated by the method of maximum likelihood, we wish to check their statistical significance using asymptotic t tests (for individual coefficients) and likelihood ratio tests (for subsets of the entire vector of coefficients). This will allow us to conclude whether the socioeconomic variables we have used to describe adaptive capacity to our respondents *are* truly judged by them to be important determinants of adaptive capacity.

If so, we can use the estimated coefficients to examine how a change in the level of an attribute changes the probability that one country is deemed to be the one with the higher adaptive capacity out of a pair. This is accomplished by examining the impact of changing attribute l on the probability of choosing country j.

Finally, we compute the rate of substitution between any two attributes. For example, if we wish to know what increase in GDP per capita is necessary to offset a one-year loss in life expectancy at birth, while keeping adaptive capacity the same, we compute the ratio between their respective estimated β coefficients.

Implicit in the independent probit model (equation (6)) and in the random-effects probit discussed above is the assumption that the coefficients of the attributes do not change between the first two and the last two pairs of countries, which share certain common characteristics meant to capture climate and propensity to experience extreme events, and differ from the other pairs. Our first order of business is, therefore, to test empirically whether this is the case. Our probit models rely on continuous variables for income, physicians per capita, and life expectancy, and 0/1 dummy indicators for the others.

8.2.5 **The Data**

Our questionnaire was self-administered by a sample of public health officials and climate change experts intercepted at random at professional conferences and intergovernmental meetings from October 2003 to August 2004.[6] Study participants were offered the option to complete the questionnaire on the premises, or to return it by fax or mail. We received a total of 100 completed questionnaires.

We cannot make any claims that our sample is representative of the population of these professionals, so our first order of business is to describe the individual characteristics of our respondents. Males account for over two-thirds of our sample (67.35% of the respondents), and the age of the respondents ranged from 24 to 70 years, for an average of 48. Our respondents were from a total of 29 countries. Sixty-seven of them were from Western Europe (e.g. the United Kingdom, Ireland, Northern Ireland, Italy, France, Norway, Austria and Germany), fifteen were from former Eastern block countries (including Russia, Azerbaijan, Lithuania, Poland, Hungary, Croatia and Slovenia), five were from the United States, and the remainder came from Thailand, Turkey, Brazil, Japan, Congo, Israel and Nigeria.

Table 1a reports information about their professional background, showing that the medical, public health/epidemiology, engineering, and economics or business fields are roughly equally represented in our sample. Almost one-third of the respondents, however, are from another professional background.

In Table 1b, we show the composition of our sample by type of organization. Public health organizations are well represented in our sample (22.4% of the respondents), as are universities or other research institutions, which account for 38% of the sample. About 19 of our respondents work for government agencies, and the remainder of the sample is comprised of persons who work for health care organizations (both public and private), emergency response agencies, and other organizations.

Table 2 displays the frequencies of the responses about concern for the effects of climate change on human health within the respondent's organization. Table 3 refers to the respondent's professional concern about these issues.

These tables show that roughly one-third or more of the respondents stated that their organization was highly concerned about the three effects of climate change covered in this survey. Similar percentages selected the highest category of professional concern for these effects.

---

[6] These conferences include the 2003 International Healthy Cities Conference, held in Belfast, Northern Ireland, 19–22 October 2003; the World Climate Change Conference, held in Moscow, 29 September–3 October 2003; the 2003 IPCC Conference, held in Orlando, Florida, 21–24 September 2003. Additional participants were recruited among the participants of the MIT Global Climate Change Forum XXI, held in Cambridge, MA, 8–10 October 2003

Table 1 a. Professional background of the respondents

	Percentage (frequency)
Medical	19.15% (18)
Public health or epidemiology	15.96% (15)
Engineering	12.77% (12)
Economics or business administration	19.15% (18)
Other	32.98% (31)

Table 1 b. Type of organization where the respondent works

Type of organization	Percentage of the sample
Public health organization	22.4
Private or public health care organization	7.5
Emergency response agency	2.1
University or research institution	38.3
Another government organization	19.2
Nongovernment, non-profit organization	2.1
Private company	4.2
Another type of organization	4.2

Table 2. Organization's concern about the effects of climate change. Percentage selecting each response category

	Very concerned	Somewhat concerned	Not concerned at all	No position/ outside the organization's mission
Deaths and injuries due to floods, landslides and mudslides	31.00% (31)	45.00% (45)	10.00% (10)	14.00% (14)
Cardiovascular and respiratory illnesses due to heatwaves	43.00% (43)	34.00% (34)	12.00% (12)	11.00% (11)
Increased cases of vector-borne diseases	34.00% (34)	35.00% (35)	17.00% (17)	14.00% (14)

Table 3. Respondent's professional concern about the effects of climate change. Percent selecting each response category

	Very concerned	Somewhat concerned	Not concerned at all	No position/ outside of professional duties
▌Deaths and injuries due to floods, landslides and mudslides	27.55% (27)	37.76% (37)	12.24% (12)	22.45% (22)
▌Cardiovascular and respiratory illnesses due to heat-waves	33.67% (33)	36.73% (36)	7.14% (7)	22.45% (22)
▌Increased cases of vector-borne diseases	30.61% (30)	32.65% (32)	9.18% (9)	27.55% (27)

Table 4. Responses to choice question 1

▌Country A	38.38% (38)
▌Country B	61.62% (61)

Table 5. Responses to choice question 2

▌Country C	45.92% (45)
▌Country D	53.06% (52)
▌Other*	1.02% (1)

* The category "Other" represents only one respondent, who explained that there are other important variables which were not considered in the questionnaire, like personal capacity to react, health status of population etc.

Regarding the responses to the choice questions, Tables 4–7 display the relative and absolute frequencies for each hypothetical country. The responses were nicely split between the two hypothetical countries in each pair, confirming that there were no obvious choices, and suggesting that people engaged in trading off attributes against one another.

Further inspection of our data reveals that most of our study participants (92%) found the description of the consequences of climate change adequate, and only 15.5% noted that the information on climate change provided in the questionnaire was new to them. Roughly 89% of the respondents found the concept of adaptive capacity clearly explained, and a similar fraction of the sample (88%) found the text and table presentation of the hypothetical countries clear.

Table 6. Responses to choice question 3

Country E	46.88% (45)
Country F	52.08% (50)
Other*	1.04% (1)

* The category "Other" represents only one respondent, who explained that there are other important variables which were not considered in the questionnaire, like personal capacity to react, health status of population etc.

Table 7. Responses to choice question 4

Country G	50.52% (48)
Country H	49.47% (47)

About 66% of the respondents stated that they took into account all of the three major effects of climate change on human health when answering the choice questions. Almost 19% of the sample said that they had thought exclusively of the extreme weather events, and 5.3% told us that they had considered only thermal stresses. Vector-borne diseases were cited as the only reason for the responses to the choice question by 2.1% of our sample, and, finally, 7.4% said that their choice responses were motivated by effects of climate change.

8.2.6 Results

In this section we report the results of several variants on probit model (equation (6)). Our first order of business is to check whether people's β coefficients were different across the first two and the second two pairs of countries. To test for structural change, we use a likelihood ratio (LR) test based on a probit model where the right-hand side variables are the differences in the level of the attributes across the two countries in each pair. No individual characteristics of the respondents or interaction terms are included in this specification. Under the null of no structural change, the LR statistic is distributed as a chi square with 8 degrees of freedom.

The log likelihood function for the pooled data model (restricted likelihood) is –214.93. When the same probit equation is fit to the responses from the first two choice questions (196 observations), we obtain a log likelihood function equal to –104.84. A probit model of the responses to the last two choice questions (190 observations) produces a log likelihood function of –107.02. The likelihood ratio statistic is, therefore, 6.14 (p-value = 0.63), so the test fails to reject the null at the conventional levels. This means that it is reasonable to pool the data and fit probit models with one vector of coefficients β for the responses to all of the four choice tasks. We report four such models in Table 8. In model A, expected adaptive capacity depends only on country attributes. Model B adds individual characteristics of the respondents, and model C introduces interaction terms.

The results of model A show that adaptive capacity is judged to increase with income per capacity, and to be lower when inequality is high. The coefficients on the respective variables are large and significant at the 1% level or better. They imply, for example, that, all else the same, raising per capita income by US $ 5000 increases the likelihood that a country is selected as the higher adaptive capacity country to 68%.[7] A more equitable distribution of income is worth roughly US $ 4600 in per capita income.

By contrast, the age distribution and life expectancy at birth of the population are not significant predictors of the probability of choosing a country, even though the signs of the coefficients on these variables (negative and positive, respectively) are consistent with our expectations. Perhaps our respondents thought that these aspects of vulnerability of the population would be offset by sufficient resources, and would, thus, be only of secondary importance relative to per capita income.

Regarding the type and quality of the health care system, our respondents indicate that they associate a universal coverage system with a higher degree of adaptive capacity, as is implied by the positive sign of the coefficient on this dummy. The effect is strongly statistically significant (t statistic: 5.89). The magnitude of the coefficient on this dummy implies that removing universal health care and replacing it with a private health insurance system requires a change in per capita income of about US $ 12 800 to keep adaptive capacity the same. All else the same, the probability of selecting a country drops to only 0.23 if universal health care is removed.[8] Quality of health care as measured in physicians per capita, however, is not significantly associated with the likelihood of selecting a country over another.

High access to information via newspaper, television, radio and internet is associated with a higher adaptive capacity. Our respondents indicate that they considered moving from low to high access to information equivalent to a change in per capita income of US $ 14 107.

These results are robust to the inclusion of individual characteristics of the respondents, such as a gender dummy and dummies for the professional background of the respondent, as we do in model B. The LR test of the null hypothesis that the coefficients on the individual characteristics of the respondents are jointly equal to zero is equal to 5.18, failing to reject the null at the conventional levels.[9]

---

[7] This figure represents a 36% increase over 0.5, the likelihood of selecting either one of two completely identical countries

[8] This is a 54% decline from 0.5, the equal chance of selecting either one of two perfectly identical countries

[9] Under the null hypothesis, this LR statistic is distributed as a chi square with 4 degrees of freedom

Table 8. Probit model results. N = 100 respondents, total number of observations 386

Variable	Model A		Model B		Model C		Model D	
	Coefficient	t-Ratio	Coefficient	t-Ratio	Coefficient	t-Ratio	Coefficient	t-Ratio
ONE	−0.1046	−1.2558	−0.2422	−1.7434	−0.1134	−1.3542	−0.1023	−0.9298
INCOME	5.59E-05	5.3674	5.73E-05	5.3927	5.34E-05	4.8651	6.07E-05	5.1523
HIGHINEQ	−0.25516	−2.3957	−0.2665	−2.4847	−0.2522	−2.3607	−0.2829	−2.2073
PCT65	−0.01916	−1.0670	−0.0197	−1.0913	−0.0185	−1.0269	−0.0220	−1.2027
LIFEEXP	0.0021	0.1463	0.0029	0.2034	−0.0107	−0.6910	0.0028	0.1932
DOCTORS	0.0013	1.1888	0.0014	1.3087	1.11E-03	0.9891	1.40E-03	1.1184
UNIVERSA	0.7173	5.8917	0.7241	5.8637	0.7271	5.9405	0.7454	5.4462
HIGHINFO	0.7886	6.9080	0.7903	6.8299	0.7741	6.7256	0.8480	6.8674
MALE			0.1504	0.9771				
MEDICAL			0.2392	1.2644				
PUBLICHE			0.0570	0.2778				
ENGINEER			−0.0797	−0.3567				
INCOME1					2.10E-05	0.8915		
DOCTOR1					0.00012	0.0520		
LIFEEXP1					0.0719	2.0739		

Table 8 (continued)

Variable	Model A		Model B		Model C		Model D	
	Coefficient	t-Ratio	Coefficient	t-Ratio	Coefficient	t-Ratio	Coefficient	t-Ratio
▌ Correlation rho between error terms in random effects model							1.09E-01	1.1964
▌ Log likelihood	−214.93		−212.34		−212.32		−214.01	

INCOME per capita income in US dollars; *HIGHINEQ* high inequality in the distribution of income (dummy); *PCT65* percentage of the population older than 65; *LIFEEXP* life expectancy at birth; *DOCTORS* number of physicians per 100 000; *UNIVERSA* universal health care system coverage (dummy); *HIGHINFO* access to information via newspaper, television, radio, internet (dummy); *MALE* male (dummy); *MEDICAL* medical field (dummy); *PUBLICHE* public health or epidemiology field (dummy); *ENGINEER* engineering field (dummy); *INCOME1* income×engineer; *DOCTOR1* life expectancy×public health officials; *LIFEEXP1* physicians per 100 000×medical doctor

Model C drops the individual characteristics but includes interactions terms between country attributes and individual characteristics of the re-spondents ((income×engineer), (life expectancy × public health officials) and (physicians per 100 000×medical doctor)) to see if respondents tends to weight more heavily country attributes more affine to their professional background. We find modest evidence of this effect: The coefficient on the interaction between physicians per 100 000 residents and a dummy denot-ing whether the respondent is a medical doctor by training is significant at the 5% level, but the coefficients on the other interaction terms are in-significant.

Finally, we evidence no significant evidence of random effects (model D). The coefficient of correlation between the error terms underlying the choice responses within an individual is pegged at 0.11, and is not statisti-cally significant. A likelihood ratio test (1.84, p value 0.17) confirms that the model can be simplified back to the independent probit.

8.2.7 Conclusions

We have surveyed public health officials and climate change experts about the determinants of adaptive capacity to selected effects of climate change on human health. We have used conjoint choice questions that ask individ-uals to indicate which of two hypothetical countries has the higher adap-tive capacity, where a country is described by its per capita income, high or low level of adaptive capacity, various measures of the age distribution in the population, health stock and type and quality of the health care sys-tem, and high or low access to information.

The survey results imply that the resources available to a country and the level of inequality in the distribution of income are judged to be im-portant determinants of the distribution of income, as are the type of health care system coverage (universal coverage or a system based on pri-vate insurance) and access to information. A more equitable distribution of income is judged equivalent to US $ 4600 in per capita income, while uni-versal health care coverage and high access to information are judged equivalent to US $ 12000–US $ 14000 in per capita income. These results complement earlier empirical work by Yohe and Tol (2002) seeking to de-termine vulnerability to certain effects of climate change.

References

Adamowicz WL, Louvbiere J, Williams M (1994) Combining Revealed and Stated Preference Methods of Valuing Environmental Amenities. Journal of Environmental Economics and Management 26(3):271–292

Alberini A, Longo A, Tonin S, Trombetta F, Turvani M (2005) The Role of Liability, Regulation and Economic Incentives in Brownfield Remediation: Evidence from Surveys of Developers. Regional Science and Urban Economics 35(4):327–351

Boxall PC, Adamowicz WL, Swait J, Williams M, Louviere J (1996) A Comparison of Stated Preference Methods for Environmental Valuation. Ecological Economics 18(3):243–253

Hanley N, Mourato S, Wright RE (2001) Choice Modelling Approaches: A Superior Alternative for Environmental Valuation? Journal of Economic Surveys 15(3):435–462

Intergovernmental Panel on Climate Change (IPCC) (2001) IPCC, 2000. Impacts, Adaptation and Vulnerability. The Contribution of Working Group II to the Third Scientific Assessment of the Intergovernmental Panel on Climate Change. Cambridge University Press, Cambridge

Louviere J, Hensher DA, Swait JD (2000) Stated Choice Methods: Analysis and Applications. Cambridge University Press, Cambridge, UK

Train KE (1999) Mixed logit models for recreation demand. In: Herriges JA, Kling CL (eds) Valuing recreation and the environment. Revealed preference methods in theory and practice. Edward Elgar Publishing, Cheltenham, UK

Yohe G, Tol RSJ (2002) Indicators for social and economic coping capacity – moving towards a working definition of adaptive capacity. Global Environmental Change 12:25–40

9 Valuing the Mortality Effects of Heat-waves

Anna Alberini, Aline Chiabai, Giuseppe Nocella

9.1 Introduction and Background

Europe has experienced an unprecedented rate of warming in recent decades. In the period 1976–1999, the average annual number of periods of extreme warmth increased twice as fast as the corresponding reduction in the number of periods of extreme cold. In most of Europe, the increase in the mean daily maximum temperature during the summer months was greater than 0.3 °C per decade during the period 1976–1999. The frequency of very hot days in central England has increased since the 1960s, with extremely hot summers in 1976, 1983, 1990 and 1995. Sustained hot periods have become more frequent, particularly in May and July (Alberini and Menne, 2004).

An unprecedented heat-wave affected the European Region in summer 2003. A heat-wave struck France in early August 2003 after a warm June with temperatures 4 to 5 °C above seasonal averages, and a warmer-than-average last two weeks of July. The period from 4–12 August 2003 broke all historical records for Paris since 1873, in terms of minimum, maximum and average temperatures, and in terms of duration (two weeks).

This unprecedented heat-wave, which was associated with high levels of air pollution, was accompanied by an excess of mortality that started early and rose quickly.

Specifically, there were 300 deaths in excess of the average on 4 August, 1200 on 8 August, and 2200 on 12 August 2003. The total mortality excess between 1 and 20 August is pegged at 14 802 cases above the average daily mortality figures in the 2000–2002 period. The observed excess of mortality affected primarily the elderly (75 years-old and older), but was also severe for the 45–74 year-olds. Most of these excess deaths were attributed to cardio- and peripheral vascular, cerebrovascular, and respiratory causes.

While almost all of France was struck by the excess mortality, its intensity varied significantly from one region to another. For example, the excess mortality amounted to 20% of deaths for all causes in Languedoc-Roussillon (South), and 130% in the Ile-de-France (Paris and Suburbs). The excess mortality clearly increased with the duration of extreme temperatures. The mortality rate was the highest in nursing homes where the observed number of deaths was twice the expected number (INVS, 2003).

Higher excess mortality rates were also observed for other European countries, such as in Italy, where an increase of 15% in mortality for all causes was observed in comparison to the same period during the years 2001 and 2002.[1] Portugal observed a 26% increase of mortality in August 2003, compared to the average of the previous five years. Spain observed a 6% increase in total mortality.

If the mortality impacts of climate change are so serious, Europe would presumably stand to gain from implementing adaptation policies to offset these effects. Economists would recommend that, when setting these policies, at least some consideration be given to their costs and benefits. The purpose of this chapter is to present techniques for estimating the mortality benefits of adaptation to the human health effects of climate change, and to discuss the challenges and difficulties associated with these techniques. We focus on the economic valuation of these benefits, which means that the benefits are monetized for the purpose of comparing them with the costs (or with other categories of benefits). We also present the design, administration and results of an original survey of Italian residents conducted for the purpose to find out their willingness to pay for reductions in their risk of dying for cardiovascular and respiratory causes. The Value of a Statistical Life (VSL) obtained through this study – € 0.9 to € 3.7 million, depending on whether we use median or mean VSL – can be used to value the mortality benefits of adaptation to heat-waves.

The remainder of this chapter is organized as follows. Section 9.2 discusses non-market valuation techniques and introduces the concept of Value of a Statistical Life. Section 9.3 presents revealed-preference approaches to estimating the mortality benefits of adaptation policies to heat-waves, and Section 9.4 focusses on the method of contingent valuation.

In Section 9.5, we discuss factors that may influence the VSL. In Section 9.6 we present and discuss the results of an original contingent valuation survey conducted in Italy under the cCASHh research program to elicit WTP for reductions in the risk of dying for cardiovascular and respiratory causes, while in Section 9.7 conclusions are given.

---

[1] The authorities estimated that 34071 people over the age of 65 died during the period of 16 July to 15 August, the height of the heat-wave. That is 4175 more than the same time of the previous year. See http://www.cbsnews.com/stories/2003/09/11/world/main572686.shtml

9.2   Estimating the Mortality Benefits of Adaptation to the Climate Change

9.2.1   Revealed and Stated Preference Approaches

Adaptation to climate change – and more specifically to its adverse health effects – has received considerable attention in public health and policy circles as of late. Adaptation policies may help reduce or offset (i) the mortality effects of extreme weather events brought about by climate change, (ii) the surge in cardiovascular, cerebrovascular and respiratory illnesses and deaths associated with heat-waves, and (iii) the additional cases of vector-borne, food-borne and waterborne illnesses, malnutrition and psychosocial diseases associated with different climate patterns. In this chapter, attention is restricted to the excess mortality brought about by heat-waves, and to the monetized benefits of policies that avoid such loss of life.

To estimate the benefits of adaptation policies that save lives, it is necessary to place a monetary value on reductions in the risk of dying. It is, however, difficult to value reductions in mortality risks, because this good is not typically traded in regular markets. To circumvent this problem, economists have devised a number of techniques that are generally known as non-market valuation. Non-market valuation methods can be classified into two categories: revealed-preference and stated-preference methods. The purpose of these methods is to estimate an individual's – and hence, society's – willingness to pay to reduce one's risk of dying, which is the theoretically correct measure of its value.[2]

Revealed preference studies infer the value of reducing the risk of dying by observing individual behaviours and expenditures on related activities and goods. *Stated* preference studies are survey-based, and rely on what individuals *say* they would do under specified circumstances. Contingent valuation studies, for example, directly ask individuals to report their willingness to pay (WTP) to reduce their risk of dying. The circumstances under which the risk of dying is reduced are hypothetical and spelled out to the

---

[2] Willingness to pay is defined as the maximum amount that can be subtracted from an individual's income to keep his or her expected utility unchanged. Individuals are assumed to derive well-being, or utility, from the consumption of goods. Let $U(y)$ denote the utility function expressing the level of well-being produced by the level of consumption y when the individual is alive. Furthermore let R denote the risk of dying in the current period, and $V(y)$ the utility of consumption when dead. Expected utility is expressed as $EU=(1-R)\cdot U(y)+R\cdot V(y)$. This expression is simplified to $EU=(1-R)\cdot U(y)$ if it is further assumed that the utility of income is zero when the individual is dead

respondent in the course of the survey, and no actual transaction takes place.[3]

9.2.2 The Value of a Statistical Life

The Value of a Statistical Life (VSL) is a key input into the calculation of the benefits of adaptation policies that save lives. The mortality benefits are computed as VSL×L, where L is the expected number of lives saved by the policy.

The VSL is the marginal value of a reduction in the risk of dying and is, therefore, defined as the rate at which the people are prepared to trade off income for risk reduction:

$$VSL = \frac{\partial WTP}{\partial R} \tag{1}$$

where R is the risk of dying.[4] The VSL can equivalently be described as the total WTP by a group of N people experiencing a uniform reduction of 1/N in their risk of dying. To illustrate, consider a group of 10 000 individuals, and assume that each of them is willing to pay € 30 to reduce his or her own risk of dying by 1 in 10 000. The VSL implied by this WTP is € 30/0.0001, or € 300 000.

The concept of VSL is generally deemed as the appropriate construct for ex ante policy analyses, when the identities of the people whose lives are saved by the policy are not known yet. As shown in the above mentioned example, in practice VSL is computed by first estimating WTP for a specified risk reduction ΔR, and then by dividing WTP by ΔR.

9.3   Revealed Preference Approaches

9.3.1 Compensating Wage Studies

A number of alternative approaches have been used to estimate WTP for a specified risk reduction, and hence the VSL. The traditional approach holds that the value of a life is the present value of the stream of income

---

[3] The method of contingent valuation can be and has been used to place a value on public goods, environmental quality, as well as private goods, including episodes of illness and private mortality risk reductions. A recent bibliography (Carson et al., 2002) documents over 5000 papers and articles studying or reporting on applications of the method of contingent valuation

[4] In an expected utility framework with expected utility EU=(1–R) · U(y), the VSL can be expressed as VSL = U(y)/[(1–R) · U'(y)]

generated by a person over his or her remaining lifetime. This approach has been criticized because it places a very low value, or no value at all, on individuals that are not gainfully employed, such as homemakers, and on retired persons, even if these people *are* willing to pay for reduction in their own risk of dying.

Alternative estimates of the VSL can be derived from compensating wage differentials in labor markets. The rationale of compensating wage studies is that workers must be offered higher wages for them to accept jobs with greater risks of dying, and that employers are willing to do so to the extent that this is cheaper than installing safety equipment in the workplace.

In a typical compensating wage study, data are gathered on the wage rate, education, experience, occupation, and other individual characteristics of workers and workplace characteristics. These data are then used to run a regression relating the wage rate to the risks of fatal and non-fatal injuries, while controlling for education and experience of the worker, and other job and worker characteristics thought to influence wages. Viscusi (1993) argues that the correct specification of the wage regression is:

$$w_i = \beta_0 + \mathbf{x}_i\beta_1 + p_i\beta_2 + q_i\beta_3 + (q_i \times WC_i)\beta_4 + \varepsilon_i \tag{2}$$

where w is worker i's wage rate, and $\mathbf{x}$ is a vector of individual, workplace and occupational characteristics, such as experience, education, age, gender, marital status, union status of the worker, industry dummies, occupation dummies and geographical dummies. The variable p measures the risk of dying on the job, while q is the risk of nonfatal injuries. The βs are unknown coefficients, and the VSL can be inferred from β_1.[5,6]

One key issue when estimating equation (2) in a compensating wage study is how to measure p and q. Researchers have typically measured p and q using official statistics, but *perceived* risks – not objective risks – are what should truly drive wages. We are only aware of two studies, however, that attempted to elicit perceived risks and attempted to regress wages on perceived, rather than objective, risks (Gegax et al., 1988; Lanoie et al., 1995).

Estimates of VSL based on compensating wage studies are rife with econometric identification problems. For example, Leigh (1995) shows that if the regression equation attempts to control for inter-industry wage dif-

---

[5] In empirical work, the logarithmic transformation of the wage rate often replaces w as the dependent variable in the regression. The wage rate, w, and fatality risk, p, are usually measured on an annual basis.

[6] Viscusi argues that q must be included in the compensating wage equation. Since p and q are generally highly correlated, failure to do so would result in biased estimates of the β_1 coefficient, and hence of the VSL. Viscusi further argues that the expected worker compensation (the term $q \times WC$) should also be included in the equation.

ferentials, the coefficient on fatal risks becomes insignificant, rendering the researcher unable to estimate the VSL.

Recent compensating wage studies based on European labor markets (Siebert and Wei, 1994; Sandy and Elliott, 1996; Arabsheibani and Martin, 2000; Sandy et al., 2001), peg the VSL in the range between € 4.3 million and € 74.4 million. A meta-analysis by CSERGE (1999) generates a range of VSL figures between € 2.9 million and € 100 million, resulting in a weighted average equal to € 6.5 million (all 2000 €).

Clearly, when using the VSL estimated from labor market studies, one is implicitly assuming that the tradeoffs between risk and income observed in labor markets can be applied in other contexts, such as environmental or climate change policies. However, there is no particular reason to believe that the VSL observed in labor markets should be the VSL used to estimate the mortality benefits of adaptation policies to climate change.

Different VSLs may be justified on the grounds that the beneficiaries of adaptation policies (those whose lives are saved by the policies, for example, the elderly and persons with chronic cardiovascular and respiratory illnesses) are different from the workers (males in their prime) examined by most compensating wage studies. The nature of the risks are also different: workplace risks are voluntary, while the risks associated with thermal stresses are involuntary, which could lead to different WTP, even if the magnitude of the risk reductions were the same.

9.3.2 Other Hedonic Pricing and Consumer Behaviour Studies

An alternative revealed-preference approach for estimating the VSL is to relate the price of a product to the product's attributes, including its safety. Atkinson and Halvorsen (1990) regress the price of cars on car attributes, such as the car size, make, fuel efficiency, luxury index, and the risk of dying in an accident when driving a car of this make and model to find the implicit marginal price of risk.

Consumer behaviour studies examine the time spent by an individual in activities that increase safety, or the amount of money spent on items that reduce risk, to estimate WTP for a reduction in the risk of death. For example, Blomquist (1979) uses a probit model explaining whether a driver buckles up when driving a car to infer the VSL. Jenkins et al. (2001) examine the use of helmets when riding a bicycle. A recent meta-analysis by Blomquist (2004) concludes that the VSLs from averting behaviour studies are typically smaller than that produced by other approaches.

Both hedonic approaches and consumer behaviour studies, however, assume that individuals know perfectly both their baseline risk of dying and the risk reduction afforded by certain products or risk-reducing activities. Estimates of the VSL based on these approaches are subject to the same caveat we expressed with VSL figures derived from labor markets: without further documentation, there is no particular reason for assuming that these estimates are appropriate for valuing mortality risk reductions in other contexts.

9.3.3 Revealed-preference VSL and Climate Change

Moore (1998) reports that if temperatures in the United States increase as predicted by the IPPC, there will be significant increases in January temperatures, but summertime temperatures will experience only small changes.[7] For example, an average annual 6.7 °F increase in temperature for New York City would make its climate comparable to that of Atlanta, but summertime temperatures would not go up commensurably. He further notes that a sample of 45 metropolitan areas in the United States shows that for each increase of one degree in the average annual temperature, July's average temperature would go up by only 0.5 °F, whereas January's average temperatures would climb by 1.5 °F.

When these predicted temperature changes are combined with daily mortality figures for Washington, DC, and for 89 large counties nationwide, which are typically higher in the winter for cardiovascular and respiratory causes, as well as for complications of infectious seasonal maladies like the flu, it follows that climate change will actually result in an overall *reduction* in the annual mortality rates. Specifically, Moore estimates that there will be 37 000 to 41 000 fewer deaths for cardiovascular, respiratory, and infectious-disease causes if the average annual temperature increases by 4.5 °F.

The next step in Moore's analysis is to estimate a wage regression relating wage rates by occupation across metropolitan areas to temperature, population, population growth between 1980 and 1990, and other variables that might potentially influence the wage rate (through the amenity value of living in that area), such as racial composition of the population, crime rate, air pollution etc. The regression confirms that workers prefer warmer climates to cooler ones, and they also prefer climates with substantial seasonal changes in temperatures.

Based on the wage regression, and assuming that the effect of temperature on wage rates reflects the workers' valuation of reduced mortality, Moore predicts that if it were warmer by 4.5 °F and the current precipitation patterns stayed the same, there would be total benefits worth $ 30– $ 100 billion annually. (This pegs the VSL at $ 0.8–2.5 million, where all figures are expressed in 1994 dollars).

---

[7] A comparison of predictions across climate models places the projected average warming in the United States for the 21st century in the range of 5 to 9 °F (2.8 °F to 5 °C) (US Global Change Research Program, National Assessment Synthesis Team, 2001). This assessment also confirms the prediction that winter warming is expected to be more pronounced than summer warming.

In Maddison and Bigano (2003), the amenity effects of climate are captured into two markets: the housing market and the labour market. A pleasant climate attracts workers, raises housing prices, and drives down wage rates. The amenity effect of climate is, thus, decomposed into its effect on wages $(\partial w/\partial C)$, minus its effect on housing prices $(\partial h/\partial C)$, leading Maddison and Bigano to run regressions using data from Italy where the dependent variable is expected after tax income net of housing costs at the province level. The regressors include climate variables, geographical and year dummies, and population density.

Their regressions indicate that, absent any changes in the precipitation patterns, Italians do not attach a significant value to marginal change in January temperatures, and that they would be prepared to pay about € 325–370 per household per year to *avoid* a one-degree increase in July temperatures. Clearly, these results are very different from those for the United States obtained by Moore.

We use the results of the Maddison and Bigano analyses to estimate the benefits of adaptation policies that would have avoided heat-related deaths in Summer 2003. For the purpose of this example, we focus on excess deaths in the city of Rome, which were estimated to be equal to 1094 for the period June–August 2003, with 319 deaths occurring during the second heat-wave (July 10–30). Maddison and Bigano report that for a household in Rome the marginal value of avoiding an increase of one degree in the July average mean temperature is about € 367 per year. In July 2003, the average mean temperature was a 28.0 °C, which is 2.8 °C above normal average mean temperature for that month (25.1 °C).[8] We estimate the total number of households in Rome to be 1 038 461 (2.7 million, the population of Rome, divided by 2.6 persons per household[9]), making the value of the temperature disamenity equal to € 1 067 123 077 per year.[10] Assuming that the value of the disamenity reflects entirely the excess deaths due to the heat-wave, this implies a value per statistical death avoided of € 3 345 213.

---

[8] The average mean temperature for June 2003 was 27.4 °C, which is five degrees warmer than normal (22.4 °C). In August, the average mean temperature was 29.1 °C, which is four degrees above normal (25.0 °C)

[9] Figure from http://www.guidagenitori.it/guidagenitori/
home.jsp?openDocument=2271&parent1=186&parent2=289&do cs=289

[10] This figure is equal to the number of households in Rome multiplied by the marginal price of a one-degree increase in July temperatures, multiplied by 2.8 degrees, the increase in July temperature over the Rome average

9.4   Contingent Valuation Studies

Contingent valuation studies have the potential to overcome some of the limitations implicit in many revealed preference studies. For example, respondents can be educated about their baseline risk levels, and can be told exactly the extent of the risk reduction they are to value. In addition, a CV study can be tailored to the specific type of risk being considered and to address the issue of latency, which arises when the risk reduction takes place in the future, but decisions about it and payment for it must be made now (see, for example, Johannesson et al., 1996; Krupnick et al., 2002). Compensating wage studies, by contrast, are ill-suited to value future risks, since they generally focus on the risk of dying in a workplace accident in the next year. They also do not lend themselves to situations with lifetime risks.

Special care has to be taken, however, with risk communication, which is generally difficult, and with the fact that respondents often struggle to grasp small risks. Visual aids and graphical renditions are often used to help respondents process risks. Risk ladders, pie charts, dots, and grids of squares are commonly used risk communication devices.

In a CV survey, the risk reduction can be delivered to the respondent in one of two possible ways: through a public program, or as a private risk reduction. From the perspective of the researcher, the choice between these two alternatives depends on the purpose for which the study is conducted, as well as on practical considerations. For example, when examining the mortality benefits of policies for adaptation to climate change, it would seem straightforward to base the valuation exercise on a public programme that reduces the risk of dying. However, researchers worry that in such a context the responses to the payment questions might be motivated by altruistic considerations, which raises the issue of double-counting of benefits. Economic theory (Jones-Lee, 1991, 1992) predicts the circumstances under which double-counting occurs, showing that it depends crucially on two factors: the type of altruism (paternalistic or non-paternalistic) and the assumption made by an individual about other people's payments.

In practice, however, it is difficult to recognize exactly what type of altruism motivates a respondent, and to tell respondents what they are to assume about other people's payments (Johannesson et al., 1996). These difficulties are the reason why many recent studies have focussed on private risk reductions, asking individuals if they would purchase a product, medical intervention or safety device which brings about a specified reduction in risk at the cost of € X. This approach is likely to work well when a specific cause of death can be directly linked with a product, as in the case of skin cancer and sunscreen (Dickie and Gerking, 1996). Individuals may, however, question the effectiveness of the product or worry about side effects.

Economic theory predicts that WTP should be increasing in the baseline risk and in the size of the risk reduction. Researchers developing a CV questionnaire about mortality risk reductions must decide whether the risks therein should be the risks the respondents subjectively believes he is facing, or whether risks should be objectively stated to the respondent. In making this decision, researchers should keep in mind that it may be difficult for respondents to estimate their own subjective risks. In addition, both WTP and subjective baseline risks and/or risk reductions may be influenced by common, unobservable individual factors. This would make the risk reductions endogenous with WTP in a regression of WTP on the former, a problem that must be addressed with instrumental variable estimation techniques.

Even working with objective risks, however, has its problems, in the sense that individuals may not accept the risks stated to them. In an earlier survey eliciting WTP for a private mortality risk reduction in the United States, for example, Alberini et al. (forthcoming) find that over a quarter of the respondents did not accept the risk stated to them. When this happens, the regression equation relating WTP to the risk reduction is affected by an error-in-variable problem, which makes the estimates of the coefficient on the risk change biased, and invalidates the estimate of the VSL. In some cases, it is possible that, especially when the risk reduction is a private good, some respondents prefer their own risk reducing activities to the product described in the scenario, and end up ascribing the latter's risk reduction a relatively low value.

Economic theory suggests that WTP for a risk reduction should increase with the size of the risk reduction. Hammitt and Graham (1999) report that many CV studies fail to detect a significant relationship between WTP and the size of the risk reduction, and Corso et al. (2001) explore the possibility that such failure could be due to poor risk communication.

9.5   What Factors Influence the VSL?

9.5.1 Effect of Age

In the transportation safety and environmental policy contexts, it has been noted that deaths occur disproportionately in certain age groups. For example, the majority of the people dying in road traffic accidents are young males, whereas epidemiological evidence from the United States (Pope et al., 1995) indicates that the over 75% of the lives saved by the Clean Air Act are those of persons 65 years old and older. Likewise, heat-waves have been linked with increased premature mortality among the elderly. This has led to the question whether the VSL should be adjusted for age.

Proponents of such an adjustment argue that the VSL should be lower for older persons because they have a shorter remaining lifetime. To see how this claim compares with economic theory, consider the life cycle model, according to which an individual at age j receives expected utility V_j over the remainder of his lifetime:

$$V_j = \sum_{t=j}^{T} q_{j,t}(1+p)^{j-t} U_t(C_t) \tag{3}$$

where V_j is the present value of the utility of consumption in each period, $U_t(C_t)$, times the probability that the individual survives to that period, $q_{j,t}$, discounted to the present at the subjective rate of time preference p. T is the maximum lifetime. The specific expression of the budget constraint of the individual depends on the assumptions about opportunities for borrowing and lending. If, for example, it is assumed that the individual can borrow and lend at the riskless rate r, but never be a net borrower, and that the individual's wealth constraint is binding only at T, the VSL at age j is equal to

$$\text{VSL}_j = (1 - D_j)^{-1} \sum_{t=j+1}^{T} q_{j,t}(1+p)^{j-t} \frac{U_t(C_t)}{U_t'(C_t)} \tag{4}$$

where D_j is the probability of dying at age j.[11]

If the term $\dfrac{U_t(C_t)}{U_t'(C_t)}$ is constant with respect to age, then it can be brought outside of the summation in equation (4), implying that WTP is proportional to the discounted remaining life years. If, in addition, the discount rate is zero, then WTP for a reduction in the risk of dying is indeed strictly proportional to remaining life years.

In sum, adjusting VSL for age relies on two restrictive assumptions: (i) that the utility divided by marginal utility does not vary with age, and (ii) that the discount rate is zero. There is no particular reason to believe that these assumptions should be true in practice. For example, if the marginal utility of consumption increases with age, then it is no longer appropriate to assume that WTP is proportional to remaining life years.

In a theoretical exercise, Shepherd and Zeckhauser (1982) assume that the utility function is of the form C^β, and show that for plausible values of β WTP for a risk reduction has an inverted-U shape that peaks when the individual is in his 50s. Jones-Lee (1989) finds some empirical support for this notion, as do Johannesson et al. (1997). Both studies report that WTP for a risk reduction of a specified size is at its highest when the individual is approximately 50 years old. Willingness to pay is lower among younger

---

[11] VSL at age j is defined as the willingness to pay for a marginal change in D_j, the probability of dying at age j

and older individuals, which is consistent with the quadratic relationship predicted by Shepherd and Zeckhauser.[12]

Krupnick et al. (2002) find that WTP declines (by about 30%) only for the oldest age group in their sample of residents of Hamilton, Ontario, and report of a similar pattern for a national sample of United States residents, although in the latter case the effect is not statistically significant. These findings recently led the Office of Management and Budget to discontinue its prior practice of discounting VSL for age in policy analyses by United States government agencies (Skrzycki, 2003). In a subsequent application of the Alberini et al. survey instrument in the United Kingdom, France and Italy, no evidence was found of an association between willingness to pay and the age of the respondent (Alberini et al., 2004).

9.5.2 Effect of Health Status on the VSL

Because the risk of dying during a heat-wave is greater for persons with chronic cardiovascular and respiratory illnesses, it is important to see if WTP depends on a person's health status. In equation (4), persons with chronic illnesses would be argued to have a higher D (the probability of dying in their j-th year of age), and lower probabilities of surviving to future ages. However, it is not clear how the remaining terms in equation (4) depend on health status, implying that theory does not offer predictions about the effect of impaired health on the VSL.

Krupnick et al. (2001) and Alberini et al. (forthcoming) find that, if anything, people with chronic cardiovascular and respiratory illnesses are willing to pay slightly *more*, rather than less, to reduce their own risk of dying. This find is in sharp contrast with the practice of some government agencies (e.g. in the United States, the Food and Drug Administration) and in medical decision-making of relying on quality-adjusted life years (QALY), whereby survival with a chronic illness would be deemed as less desirable than survival in perfect health.

We are aware of only one CV study that focussed on a population that is considered at high risk during heat-waves: Johannesson et al.'s 1991 survey of hypertensive patients. Patients with high blood pressure were recruited at a clinic in Sweden. The survey questionnaire asked these persons to report their subjective baseline risk of dying from heart diseases and other complications associated with hypertension, and to estimate the risk reduction afforded by the medication they took on a regular basis. These persons were subsequently asked to report their WTP to continue taking the medication. The estimates of WTP showed good internal validity, suggest-

---

[12] It is interesting that the shape of the relationship between age and WTP was the same in the studies, despite their focus on completely different type of risks (transportation risks in the Jones-Lee study versus risks that can be reduced through a medical intervention in the Johannesson et al. study)

ing that the likely beneficiaries of the adaptive policies examined in this chapter should be willing to pay to implement these policies.

9.5.3 Latency

In contrast with revealed preference approaches, contingent valuation allows the researcher to directly elicit WTP now for a risk reduction that takes place in the future. Economic theory holds that if actuarially fair annuities are available, WTP at age a for a risk reduction occurring at age a+t is equal to:

$$\text{WTP}_{a,a+t} = \pi_{a,a+t} \cdot (1+\rho)^{-t} \cdot \text{WTP}_{a+t,a+t} \tag{5}$$

where $\text{WTP}_{a+t,a+1}$ is WTP at age a+t for a risk reduction beginning at age a+t, and $\pi_{a,a+t}$, is the probability of surviving from age a to age a+t (Cropper and Sussman, 1990). Since $\pi_{a,a+t}$ and $(1+\rho)^{-t}$ are less than one, WTP for a future risk reduction should be less than WTP for an immediate risk reduction of the same size if the person were (a+t) years old. It should be emphasized that the actual discount factor to be used in equation (5) may or may not be equal to the market interest rate, depending on the budget constraint faced by the individual and his or her access to credit.

To our knowledge, only few studies have sought to elicit WTP for latent risks. One of them (Johannesson et al., 1996) values an extension in life expectancy to be experienced when the individual reaches the age of 75. The others were conducted in Canada, the United States, the United Kingdom, France and Italy using a standardized questionnaire (Krupnick et al., 2002; Alberini et al., 2004; Alberini, 2004). The implicit discount rate is estimated to be 8% (s.e. 0.7%) in Canada, 4.5% (s.e. 0.55%) in the United States, 10% (s.e. 0.91%) in the United Kingdom, 5% (s.e. 1.07%) in France, and 6% (s.e. 0.99%) in Italy.

9.5.4 Cohort and Country Effects on VSL

Climate change is a complex phenomenon that will be experienced over a relatively long time horizon. Modelling scenarios often cover the next 30 or 50 years, or extend even beyond the 21st century. The present value of the benefits of saving lives over such a long period would be calculated as:

$$\sum_{t=1}^{T} \text{VSL}_t \cdot L_t \cdot (1+\delta)^{-1} \tag{6}$$

where δ is an appropriate discount rate and VSL is allowed to change over time. There are at least three reasons why VSL might change over time.

First, as people become wealthier with economic growth, it seems reasonable to believe that their WTP for a reduction in the risk of death may increase. Second, preferences can change over time, and the rate at which people are willing to trade off income for risk reductions may change with economic growth. [13] Third, the VSL in equation (6) may reflect the different composition by age in the population.

Finally, consider European Union-wide policies and programmes. The benefits of these programs would be experienced by different countries, raising the question whether a different VSL should be used for each country. Economic theory and empirical studies suggest that wealthier persons – and hence, wealthier countries – should hold greater WTP amounts for the same risk reduction, which would imply higher VSLs. (The income elasticity of WTP in mortality risk studies is usually of the order of 0.2–0.5.) However, when computing Europe-wide benefits, should a different VSL be used for each country, depending on the country's income, or should just one VSL value be used for all countries? Economic considerations might suggest that we allow for variation in taste and income across countries, but political considerations may dictate that only one VSL be used for all countries.

9.5.5 What VSL Figures?

In its guidelines for the estimation of the benefits of environmental policies (US EPA, 2000), the United States Environmental Protection Agency recommends using a VSL of $ 6.1 million (1999 dollars). To arrive at this figure, the Agency compiled VSL values from 26 studies, mostly compensating wage studies. The US EPA does not adjust the VSL for age, futurity of the risk, and cancer, but it does adjust it for growth in income.

The European Commission uses a baseline central VSL of € 1.4 million. It does adjust for age – the central VSL figure is multiplied by 0.70 for persons aged 70 and older – and it does apply a "cancer premium." In other words, if the cause of death is cancer, the VSL is inflated by 50% to account for the morbidity and the dread associated with this health endpoint.

The EC does not adjust the VSL to account for growth in income, but it does apply a discount factor of 4% for latent risk reductions. All of these adjustments results in a final VSL of € 1.0 million.

---

[13] Costa and Kahn (2002) estimate compensating wage studies for different years in the United States. Using Census micro-data and fatality risk figures from the Bureau of Labor Statistics for 1940, 1950, 1960, and 1980, they conclude that the quantity of safety has increased over time, and that the compensating differential has increased

Since there are no official VSL figures for Italy or the Czech Republic, we have developed and administered a contingent valuation survey instrument to samples of respondents in these two countries.[14] The survey instrument is standardized, save for the language and other minor adaptations, and is self-administered by the respondent using the computer. We describe the questionnaire and the results of the Italy survey in the next section. The reader is referred to Alberini, Scasny, Kohlova and Melichar, chapte 10 this book for a full report on the Czech Republic study.

9.6 The Value of Adaptation to Heat-Waves: A Survey of the Italian Public

9.6.1 Cardiovascular and Respiratory Mortality Risk Questionnaire

Heat-waves are typically associated with excess deaths for cardiovascular and respiratory causes. Our Thermal Stresses (TS) questionnaire, therefore, elicits WTP for reductions in the risk of dying for these causes from a sample of Italian citizens.

The risk reduction is of a private nature. Our questionnaire is self-administered by the respondent using the computer. This allows us to tailor risks and scenarios to the respondent's individual circumstances, such as his or her age, gender, and health status.

In addition, it avoids interviewer bias, and the responses to the questions are automatically entered in an Excel spreadsheet, which reduces data entry cost and time.

The questionnaire is divided into seven sections. In Section 1, after querying the respondent about gender and age, we ask the respondent if he or she has ever been diagnosed to have certain cardiovascular and respiratory illnesses (including heart disease, chronic obstructive pulmonary disease, and emphysema), and cancer. We also ask people to tell us a little about the health and longevity of other family members, to assess their current and future health, and to report a subjectively assessed life expectancy.

In Section 2, we ask questions assessing their health over the last four weeks, as well as any physical mobility limitations and psychological well-being. Our questions are adapted from the Short Form 36 (SF36) questionnaire, which is widely used in medical research to assess physical and emotional health.

---

[14] We considered conducting a compensating wage study of the Italian labour market, but earlier research by Barone and Nese (2002) using Eurostat data finds no evidence of a significant relationship between wage rates and objectively measured workplace risks

Section 3 provides a simple probability tutorial, leading to the explanation of one's chance of dying, which is expressed as X in 1000 over 10 years, and is graphically depicted using a grid of 1000 squares. White squares represent survival, while blue squares represent death. Respondents are then tested for probability comprehension.

In Section 4, we wish to acquaint respondents with the concept that it is possible to reduce one's risk of dying, and that many people do so on a routine basis. For example, we tell respondents that a pap smear can reduce the risk of dying of cervical cancer (in women) and that blood pressure medication reduces the risk of dying of a heart attack. We then introduce cardiovascular and respiratory illnesses, and allow the respondent to learn more about them by reading a glossary which is launched by double-clicking a link on the screen. The respondents are then asked questions about treatments or actions they take against cardiovascular and respiratory illnesses, and their cost.

In Section 5, we present the chance of dying for a person of the respondent's age, gender, and health status. This is shown using blue squares in the grid of 1000 squares. We highlight the chance of dying for cardiovascular and respiratory illnesses using orange squares, emphasizing that these risks increase with age.

Section 6 presents the hypothetical risk reduction scenario. The respondent is randomly assigned to one of two versions of the questionnaire. In version 1, we tell respondents that a new medical test is available that would be safe and without side effects, and would reduce the respondent's chance of dying by X in 1000 over the next 10 years, where X ranges from 1 to 22, depending on the respondent's age and gender. The risk reduction is expressed both in absolute and percentage terms. The experimental design for the baseline risk and risk reduction is displayed in Table 1.

We chose a medical test because in the initial focus groups we found that our participants were skeptical of "products" that would deliver the risk reduction and were much more comfortable with a medical test, especially if the latter was approved by the national health care system. Respondents are told that this test would have to be done and paid for every year for 10 years to be effective. The payment question is in a dichotomous choice format with one follow-up. The bid amounts are shown in Table 2. The payment mechanism is a co-pay, modelled after the routine charge for medical tests within the Italian national health care system.

Respondents assigned to version 2 of the questionnaire are simply asked to imagine that it were possible to reduce their risk by a certain amount, and are subsequently queried about their WTP using dichotomous-choice questions. Our focus groups indicated that people accept such an abstract risk reduction, and that this approach is likely to get them to focus more sharply on the size of the risk reduction, without being distracted by other details. In both versions of the questionnaire, we include several questions that attempt to find out the reasons why our subjects are, or are not, willing to pay the stated amounts.

Table 1. Baseline risks and risk reductions assigned to respondents in the survey

Males				Females			
Age	Baseline risk (all causes of death)	Baseline risk (cardio-vasc. and respira-tory)	Risk reduction (cardio-vasc. and respira-tory)	Age	Baseline risk (all causes of death)	Baseline risk (cardio-vasc. and respira-tory)	Risk reduction (cardio-vasc. and respira-tory)
30–34	12	2	1	30–34	5	2	1
35–39	15	4	2	35–39	8	2	1
40–44	23	6	3	40–44	13	4	2
45–49	37	11	5	45–49	20	5	2
50–54	62	18	3 or 6	50–54	38	7	3
55–59	105	34	5 or 8	55–59	49	13	4
60–64	177	64	5 or 10	60–64	80	25	4 or 5
65–69	297	122	5 or 12	65–69	138	54	5 or 8
70–74	478	225	12 or 22	70–74	247	118	8 or 12

Table 2. Bid amounts

Initial bid (euro per year)	If yes	If no
110	250	70
250	500	110
500	950	250
950	1200	500

Next, we move on to Section 7, which describes, and elicits WTP for, a risk reduction that would take place X years from now (where X varies with the respondent's age), when the respondent is older. To make this question meaningful to the respondents, we show them that the chance of dying for any cause and for cardiovascular causes increases as one becomes older. Baseline risks and risk reductions used in this valuation task are reported in Table 3.

In Section 8, we ask questions intended to investigate the intertemporal rate of preference of the respondent, and his or her financial risk aversion. Section 9 concludes the survey with the usual socio-demographic questions, and with debriefing questions about the respondent's interpretation of the questions.

We wish to emphasize that climate change is never mentioned to the respondent in this survey. We chose to do so for two reasons. First, we wished to keep the risk reduction a private good, because it is difficult to identify the altruistic components of WTP, and to account for them appro-

Table 3. Future risk reductions

Current age	Age when the risk reduction would start	Risk reduction for males	Risk reduction for females
30–34	50 or 70	6 if at 50, 12 if at 70	3 if at 50, 8 if at 70
35–39	50 or 70	6 if at 50, 12 if at 70	3 if at 50, 8 if at 70
40–44	60 or 75	10 if at 60, 12 if at 75	5 if at 60, 12 if at 75
45–49	60 or 75	10 if at 60, 12 if at 75	5 if at 60, 12 if at 75
50–54	65 or 75	12 if at 65, 12 if at 75	8 if at 65, 12 if at 75
55–59	75	12	12
60–64	N/A	N/A	N/A
65–69	N/A	N/A	N/A
70–74	N/A	N/A	N/A

priately to avoid double-counting. Second, linking risk changes to public policies for emissions reductions or adaptation to climate change would require that we provide information about them. In our opinion, doing so would have resulted in an excessively heavy cognitive burden, which prompted us to choose a context-free risk reduction.

9.6.2 Sampling Plan and Survey Administration

The survey was administered in five cities in Italy on 31 May–9 June 2004, resulting in 801 completed questionnaires. The original sampling plan called for 800 interviews, split into 2 groups: 600 people recruited from the general population and 200 people recruited from those suffering of chronic cardiovascular and respiratory diseases, as shown in Table 4.

Table 4. Sampling plan for the Italy survey

Region	City	Sample size from general population	Sample size for chronic group	Total sample size
Northwest	Milan	100	30	130
	Genova	60	20	80
	Total	160	50	210
Northeast	Venice	120	40	160
Central region	Rome	120	40	160
South and Islands	Bari	200	70	270
Total		600	200	800

Respondents were recruited by the CIRM Institute, a Milan-based survey firm, among the residents of five cities: Venice, Milan, Genoa, Rome and Bari. The sample is stratified by age, with an equal number of respondents in each of three broad age groups (30–44, 45–59 and 60–75). The sample is comprised of a roughly equal number of men and women.

The survey was self-administered by respondents using the computer at centralized facilities.[15] In all facilities, two interviewers were present at the time to welcome the respondents, introduce to them the survey and provide assistance if necessary.

Prior to the final survey, a pilot study of 20 respondents was conducted in Venice on 24–25 May 2004 in order to fine-tune the questionnaire and to test every aspect of the survey. (In a pilot study, a draft questionnaire similar to the one that will be used in the final survey is administered to respondents under conditions similar to those of the final survey.) A training session for interviewers was conducted prior to the final survey to inform them about the questionnaire structure and to anticipate problems that might arise in the course of the administration of the final survey.

---

[15] The survey questionnaire was programmed in Microsoft Visual Basic 6 (VB6), Service pack 4. The programme was interactive, in the sense that answers to questions about gender and health demographics were used to customize subsequent risk questions. Follow-up questions and debriefing questions also depended on the answers given by the respondents to earlier questions. Creating the program involved the development of about a hundred forms and two modules for general procedures. In general, the development of the cCASHh programme was conducted in four steps: (i) development of the visual portion of the application (screen with which the respondents interact); (ii) addition of VB6 programming language which joins visual elements and automates the programme; (iii) testing of the application to identify and remove bugs; and (iv) compilation of the application. In order to control access to data, the link with the ACCESS database where the answers to the questions are stored was carried out by means of ADO technology (*ActiveX Data Objects*). For additional flexibility, the link with ACCESS is easily modifiable from outside by editing the connection string in the *mortalit.ini* file. The versatility of the project is also evident in the *lingua.ini* file which can also be amended from the outside. The possibility of adapting the language which appears on the screen without having to re-program the application is an enormous advantage when working in an international context. Finally, from a practical point of view, as the use of a server was not possible in all the places in which the interviews were administered, the questionnaire-administering program was installed on several PCs at the selected facilities, each of which was connected to a database. The operating system for all PCs was Windows2000. Respondents were given access as *users* and were protected by a procedure which did not allow them to access *query* data in the database. The queries were devised to extract data quickly according to certain criteria

Table 5. Descriptive statistics of the respondents

Variable	Valid observa-tion	Mean	Standard Deviation	Minimum	Maximum
Abstract (dummy)	801	0.50	0.50	0	1
Male (dummy)	801	0.48	0.50	0	1
Age (years)	801	50.50	13.52	30	77
Married (dummy)	801	0.70	0.46	0	1
Separated or divorced (dummy)	801	0.07	0.25	0	1
Widow/widower (dummy)	801	0.06	0.24	0	1
Never married (dummy)	801	0.17	0.38	0	1
College degree (dummy)	801	0.11	0.45	0	1
Household size	801	2.89	1.22	1	7
Household income (euro/yr)	677	20 956	8 624	6 000	60 000
Chronic group (dummy)	801	0.25	0.43	0	1

9.6.3 Data

Descriptive statistics of our survey respondents are displayed in Table 5. As is consistent with the experimental design, half of the respondents were assigned to the abstract risk reduction, and the remainder were asked to consider a medical exam that would reduce risk.

The sample is relatively well-balanced in terms of gender, with only a slight prevalence of women. Almost 70% of the respondents are married. Divorced, separated, and widowed persons account for about 13% of the sample, and about 17% of the respondents are single. The average respondent is 50 years old. Regarding household income, 16% of the respondents failed to answer the income question. The average income among those respondents who did report income information is € 20 956 a year. Finally, 25% of the respondents belongs to the chronic oversampling group.

9.6.4 Responses to the Payment Questions

In this paper, attention is restricted to the immediate risk reduction. Our first order of business is to make sure that respondents comply with the economic paradigm by checking that the percentage of "yes" responses to the initial payment question declines with the bid amount. As shown in Table 6, this is indeed the case. The percentage of "yes" responses is about 66% at the lowest bid amount included in the study, and about 41.3% at the highest.

Table 6. Percentage "yes" responses to the initial payment question (immediate risk reduction)

Initial bid	N	% yes
110	158	65.89
250	155	52.26
500	304	42.76
950	184	41.3

Next, we consider the sequences of responses to the initial and follow-up payment questions. As is often the case in contingent valuation surveys, the most frequently observed pair of responses is "no"-"no" (NN) (40.07%), followed by "yes"-"yes" (28.71%). YN and NY combinations account for 19.75% and 11.24% of the sample, respectively.

To obtain estimates of the VSL, we begin with forming intervals around the respondent's unobserved true WTP amount using the responses to the payment question. To illustrate, suppose that an individual was offered an initial bid of € 250 for the specified risk reduction ΔR_i, which he declined to pay. He was then queried about € 110, which he accepted to pay. We interpret his NY responses to imply that his true willingness to pay for ΔR_i lies between € 110 and € 250.

Next, we divide the lower and upper bound for WTP by ΔR_i to obtain a lower and upper bound for the respondent's VSL. Assuming that VSL is a random variate with cdf $F(y;\lambda)$, where λ is the vector of parameters that indexes the distribution, the respondent's contribution to the likelihood is:

$$F(\text{VSL}_H; \lambda) - F(\text{VSL}_L; \lambda) \tag{7}$$

where VSL_L and VSL_H denote the lower and upper bound of the interval around the respondent's VSL. Finally, the log likelihood function of the sample is:

$$\sum_{i=1}^{n} \log[F(\text{VSL}_{Hi}; \lambda) - F(\text{VSL}_{Li}; \lambda)] \tag{8}$$

where the subscript i denotes the respondent ($i = 1, 2, \ldots, n$). The parameters λ are estimated by the method of maximum likelihood.

We experimented with the Weibull, lognormal, normal and exponential distribution for VSL, finding that the Weibull and the lognormal were comparable in terms of fit of the data, and that they outperformed normal and exponential. In what follows, we therefore work with the Weibull distribution. Using the Weibull distribution with shape θ and scale σ, equation (8) is simplified to:

$$\sum_{i=1}^{n} \log\left[\exp\left(-(\text{VSL}_{Li}/\sigma)^{\theta}\right) - \exp\left(-(\text{VSL}_{Hi}/\sigma)^{\theta}\right)\right] \tag{9}$$

Mean VSL is computed as $\hat{\sigma} \cdot \Gamma\left(\dfrac{1}{\hat{\theta}} + 1\right)$, where $\Gamma(\bullet)$ is the gamma function, while median VSL is equal to $\hat{\sigma} \cdot [-\ln(0.5)]^{1/\hat{\theta}}$, the hats denoting the maximum likelihood estimates.

Using with the Weibull distribution, we estimate mean VSL to be € 3 767 200 per year (standard error € 465 000), while median VSL is € 891 110 per year (s.e. € 89 800). The former figure is very similar to the one we obtained using Maddison and Bigano's results. The latter is closer to, but lower than, DG Environment's central estimate of the VSL.

9.6.5 VSL Regressions

In this section, we report the results of regressions relating VSL to experimental treatment variables and individual characteristics of the respondents. We adopt an accelerated-life model based on a Weibull hazard. In equation (9), σ is replaced by an individual-specific scale $\sigma_i = \exp(x_i\beta + z_i\gamma)$, while the shape parameter θ is common to all respondents. This is equivalent to specifying the following regression equation for the underlying log VSL*:

$$\log \mathrm{VSL}_i^* = x_i\beta + z_i\gamma + \varepsilon_i \tag{10}$$

where x denotes individual characteristics of the respondents, z denotes experimental treatment variables, β and γ are vectors of coefficients, and ε is a type I extreme value error term with scale θ.

In the specification of this paper, z is a dummy indicator (ABSTRACT) that takes on a value of 1 if the respondent was assigned to the abstract risk reduction, and 0 if the individual was asked to think of a medical test that reduces the chance of dying. The vector x includes gender, age and age squared, whether the respondent suffers from chronic cardiovascular and respiratory illnesses (CHRONIC), income divided by the number of family members (PCAPPINC), a missing income dummy (MISSINC)[16], whether the respondent is married (MARRIED), the number of dependent children of ages 18 and younger (DEPENDENTS), and a college education dummy (COLLEGE). We also include city dummies to capture possible differences in the cost of living.

---

[16] Specifically, we created a dummy, MISSINC, that takes on a value of one if the respondent did not answer the income question. If so, PCAPPINC is recoded to zero. The recoded PCAPPINC and MISSINC must be included in the regression equation together. The coefficient of MISSINC, if significant, captures any systematic differences in VSL among those respondents who do and do not report household income. The coefficient on PCAPPINC must be interpreted as the effect of income, conditional on knowing what the respondent's income is

Table 7. Interval-data accelerated life Weibull model of VSL (N = 801)

Variable	Coefficient	t Statistic
Intercept	13.2637	12.54
Abstract	0.3784	2.76
Chronic	0.2369	1.39
Age	0.0756	1.70
Age squared	−0.0012	−3.00
pcappinc (1000)	0.0242	1.19
Missinc	0.2081	0.80
Married	0.1541	0.89
Dependent children of ages 18 and younger	0.0195	0.57
College	0.1508	0.63
Male	−0.4183	−3.04
City dummies	yes	
Log likelihood	−1041.25	

The results of this run are reported in Table 7. One striking result is that the coefficient on the ABSTRACT dummy is significant at all the conventional levels. The magnitude of this coefficient implies that the VSL of individuals who were assigned the abstract risk reduction delivery mechanism are 46% higher than those of the other respondents, all else the same.

Another striking result is that the coefficient on age is positive and significant at the 10% level, while that on age squared is negative and significant at the 1% level. This would seem to suggest a quadratic relationship, but in practice the magnitude of the coefficient on the quadratic term swamps out the effect of the linear term in age. As shown in Figure 1 for a representative female resident of Rome who has the average income of the sample, is in good health, is married and has two children, the VSL is highest at age 30, and declines as she gets older. The decline in the VSL as age increases is dramatic. [17]

We did not detect a statistically significant relationship between VSL and income per family member, marital status and family composition of the sample. All else the same, however, the VSL of males is 35% less than that of a woman. Finally, our respondents with documented cardiovascular and respiratory illnesses are not willing to pay less than healthier respon-

---

[17] For good measure, we experimented with an alternative specification of the econometric model, where age and age squared are replaced with four age dummies (AGE4049, AGE5059, AGE6069 and AGE70PLUS). The coefficient estimates from this run confirm that VSL declines monotonically with age, and that in the oldest age group the VSL is only about 15–20% of the VSL of the persons in the youngest age group (30–39)

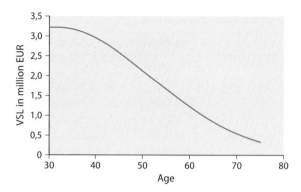

Fig. 1. Relationship between VSL and age

dents: if anything, their WTP is 23% greater than that of healthier respondents, but this effect is not statistically significant at the conventional levels. This result is against the prescriptions of the QALY approach, which would place a lower value on the lives of people with chronic illnesses, on the grounds of these individuals' lower quality of life.

9.7 Conclusions

This chapter has presented approaches for estimating the mortality benefits of adaptation to heat-waves. We have presented examples of estimates based on earlier studies, as well as the design, administration and results of an original contingent valuation survey of Italian residents. Our survey questionnaire asks people to report information about their willingness to pay for a specified reduction in the risk of dying for cardiovascular and respiratory causes. We ask people to value private risk reductions, without mentioning climate change and adaptation policies, to avoid possible double-counting of the benefits. The VSL figures elicited from this study can be used for estimating the mortality benefits of adaptation policies that save lives during sustained heat-waves.

The responses to the WTP questions in our survey are broadly consistent with the economic paradigm and suggest that people understood the commodity being valued. The mean VSL from our study is about € 3.7 million, while the mean VSL is about € 0.9 million. Our respondents' VSL depends systematically on gender and age, and is much lower among older people. We do not find evidence of a significant association between VSL and income per household member, marital status or family size. We suspect that our failure to detect such an association could be due to the likely correlation between age and income. If so, the strong effect of age on the VSL might be the combination of a true age effect and of an income effect. We will further explore this possibility in future refinements of this research.

References

Abdalla CW, Roach BA, Epp DJ (1992) Valuing Environmental Quality Changes Using Averting Expenditures: An Application to Groundwater Contamination. Land Economics 68(2):163–169

Alberini A (2004) Willingness to Pay for Risk Reductions: The Effects of Age, Health Status, and Latency. Paper presented at North Carolina State University, Durham, April 2004

Alberini A, Krupnick A (2002) Valuing the Health Effects of Pollution. In: Tietenberg T, Folmer H (eds) The International Yearbook of Environmental and Resource Economics 2002/2003. Edward Elgar, Cheltenham, United Kingdom

Alberini A, Cropper M, Krupnick A, Simon N (2002) Willingness to Pay for Future Risk: How Much does Latency Matter? Paper presented at the Second World Congress of Environmental and Resource Economists, Monterey, CA, June 2002

Alberini A, Cropper M, Krupnick A, Simon N (forthcoming) Does the Value of a Statistical Life Vary with Age and Health Status? Evidence from the U.S. and Canada. Journal of Environmental Economics and Management 48(1):769–792

Alberini A, Cropper M, Fu T-T, Krupnick A, Liu J-T, Shaw D, Harrington W (1997) Valuing Health Effects of Air Pollution in Developing Countries: The Case of Taiwan. Journal of Environmental Economics and Management 34:107–126

Alberini A, Hunt A, Markandya A (2004) Willingness to Pay to Reduce Mortality Risks: Evidence from a Three-country Contingent Valuation Study. FEEM working paper 111.04, Milan, July

Alberini A, Scasny M (2004) What is the Value of a Statistical Life in the Czech Republic? Evidence from a Contingent Valuation Survey. Draft paper, University of Maryland, College Park, October

Arabsheibani GR, Marin A (2000) Stability of the Estimates of the Compensation for Danger. Journal of Risk and Uncertainty 20(3):247–269

Atkinson SE, Halvorsen R (1990) The Valuation of Risks to Life: Evidence from the Market for Automobiles. The Review of Economics and Statistics 72(1):133–136

Barone A, Nese A (2002) Rischio sul lavoro e premio salariale in Italia. Lavoro e relazioni industriali

Blomquist G (1979) Value of Life Saving: Implications of Consumption Activity. Journal of Political Economy 96(4):675–700

Blomquist G (2004) Self-Protection and Averting Behavior, Values of Statistical Lifes, and Benefit Cost Analysis of Environmental Policy. Review of Economics of the Household 2:89–110

Carson RT, Wright J, Carson N, Alberini A, Flores N (2002) A Bibliography of Contingent Valuation Studies and Papers. La Jolla, California

Costa DL, Kahn ME (2002) Changes in the value of life, 1940–1980. NBER Working Paper 9396, Cambridge, December

Cropper M, Sussman F (1990) Valuing Future Risks to Life. Journal of Environmental Economics and Management 19:160–174

Dickie M, Gerking S (1996) Formation of Risk Beliefs, Joint Production and Willingness to Pay to Avoid Skin Cancer. Review of Economics and Statistics 78(3):451–463

Dickie M, Ulery V (forthcoming) Parental Altruism and the Value of Avoiding Acute Illness: Are Kids Worth More than Parents. Journal of Environmental Economics and Management

Gegax D, Gerking S, Schulze W (1991) Perceived Risk and the Marginal Value of Safety. The Review of Economics and Statistics 73(4):589–596

Gragg M, Kahn M (1997) New Estimates of Climate Demand: Evidence from Location Choice. Journal of Urban Economics 42:261–284

Johannesson M, Johansson P-O (1996) To Be, or Not to Be, That Is the Question: An Empirical Study of the WTP for an Increased Life Expectancy at an Advanced Age. Journal of Risk and Uncertainty 13:163–174

Johannesson M, Johansson P-O, Lofgren K-G (1996) On the Value of Changes in Life Expectancy: Blips Versus Parametric Changes. Journal of Risk and Uncertainty 15(3):221–239

Johannesson M, Johansson P-O, O'Conor RM (1996) The Value of Private Safety Versus the Value of Public Safety. Journal of Risk and Uncertainty 13:263–275

Jenkins R, Owen N, Wiggins L (2001) Valuing Reduced Risks to Children: The Case of Bicycle Safety Helmets. Contemporary Economic Policy 19(4):397–408

Jones-Lee MW, Hammerton M, Phillips PR (1985) The Value of Safety: Results of a National Sample Survey. The Economic Journal 95:49–72

Jones-Lee MW (1989) The Economics of Safety and Physical Risk. Basin Blackwell, Oxford

Jones-Lee MW (1991) Altruism and the Value of Other People's Safety. Journal of Risk and Uncertainty 4(2)

Jones-Lee MW (1992) Paternalistic Altruism and the Value of a Statistical Life. The Economic Journal 102:80–90

Krupnick A, Alberini A, Cropper M, Simon N, O'Brien B, Goeree R, Heintzelman M (2002) Age, Health and the Willingness to Pay for Mortality Risk Reductions: A Contingent Valuation Study of Ontario Residents. Journal of Risk Uncertainty 24:161–186

Lanoie P, Pedro C, Latour R (1995) The Value of a Statistical Life: A Comparison of Two Approaches. Journal of Risk and Uncertainty 10:235–257

Leigh JP (1995) Compensating Wages, Value of a Statistical Life, and Inter-Industry Differentials. Journal of Environmental Economics and Management 28(1):83–97

Loehman E, De VH (1982) Application of Stochastic Choice Modeling to Policy Analysis of Public Goods: A Case Study of Air Quality Improvements. The Review of Economics and Statistics LXIV(3):474–480

Maddison D, Bigano A (2003) The Amenity Value of the Italian Climate. Journal of Environmental Economics and Management 45:319–332

Moore TG (1998) Health and Amenity Effects of Global Warming. Economic Inquiry, 471–488

Pope CA, Thun MJ, Namboodiri MM, Dockery DD, Evans JS, Speizer FE, Health CW Jr (1995) Particulate Air Pollution as a Predictor of Mortality in a Prospective Study of U.S. Adults. American Journal of Respiratory Critical Care Medicine 151:669–674

Sandy R, Elliott RF (1996) Unions and Risk: The Impact on the Level of Compensation for Fatal Risk. Economica 63(250):291–309

Sandy R, Elliott RF, Siebert WS, Wei X (2001) Measurement Error and the Effect of Union on the Compensating Differentials for Fatal Workplace Risks. Journal of Risk and Uncertainty 23(1):33–56

Shepard DS, Zeckhauser RJ (1982) Life-Cycle Consumption and Willingness to Pay for Increased Survival. In: Jones-Lee MW (ed) The Value of Life and Safety. North Holland, Amsterdam, the Netherlands

Siebert WS, Wei X (1994) Compensating Wage Differentials for Workplace Accidents: Evidence for Union and Non-union Workers in the U.K. Journal of Risk and Uncertainty 9(1):61–76

Skrzycki C (2003) Under Fire, EPA Drops the 'Senior Death Discount'. Washington Post, (May 13, 2003)

Timmins C (2003) If You Can't Take the Heat, Get Out of the Cerrado... Recovering the Equilibrium Amenity Cost of Non-Marginal Climate Change in Brazil. Environmental Economics

US Environmental Protection Agency (1998) Regulatory Impact Analysis for Ozone and Particulate National Ambient Air Quality Standards., Washington, DC

US Environmental Protection Agency (1999) The Benefits and Costs of the Clean Air Act Amendments of 1990–2010. Report to the US Congress, Washington, DC, November

US Environmental Protection Agency (2000) Guidelines for Preparing Economic Analyses, EPA-R-00-003. Washington, DC, September

US Global Change Research Program, National Assessment Synthesis Team (2001) Climate Change Impacts on the United States. Cambridge University Press, Cambridge

Viscusi WK (1993) The Value of Risks to Life and Health. Journal of Economic Literature 31(4):1912–1946

10 The Value of a Statistical Life in the Czech Republic: Evidence from a Contingent Valuation Study*

Anna Alberini, Milan Scasny, Marketa Braun Kohlova, Jan Melichar

10.1 Introduction

The Value of a Statistical Life (VSL) is a key input when one wishes to compute the mortality benefits of environmental policies or other safety regulations that save lives.

The VSL is defined as the rate at which individuals are prepared to trade off income for risk reductions, and is the figure that must be multiplied by the expected number of lives saved by the policy to estimate its mortality benefits.

The VSL is usually estimated using one of three main approaches. The first is by observing the compensating wage differentials that workers must be paid to take riskier jobs (see Viscusi and Aldy, 2003, for a recent literature review). The second approach examines other behaviours where people weigh costs against risks (Blomquist, 2004), and the third is through contingent valuation surveys where respondents report their willingness to pay (WTP) to obtain a specified reduction in mortality risks.[1]

The European Commission has recently adopted an interim VSL of about 1 million € for policy analysis purposes, with adjustments for the age of the beneficiaries of environmental policy and for whether the cause of death is cancer (in which case a premium of 40% is applied).[2] These figures do not vary with the income of the beneficiaries. In principle, they

* We are grateful to Bettina Menne of the World Health Organization, Rome; Charles University Environment Center for financial and logistical support; Marketa Sychrovska for administrative assistance; Jan Melichar and Jan Urban for their help in administering the survey in the Czech Republic; Aline Chiabai for her role in the development of the Italian survey instrument, and Pippo Nocella for implementing the computer programme for the administration of the Italian survey

[1] See Alberini (2004) for an examination of recent contingent valuation studies used for US environmental policy purposes

[2] See http://europa.eu.int/comm/environment/enveco/others/
recommended_interim_values.pdf

could be adopted to estimate the mortality benefits of policies that promote adaptation to climate change.[3]

Without further documentation, however, there is no reason to believe that these figures, which were developed by experts on the basis of previous studies, most of which were conducted in Western Europe, reflect the preferences of people living in the countries that recently joined the European Union. Differences in incomes and in the taste for income and risk may imply different VSLs.[4]

Another problem with the DG Environment figures is that they are based on transferring the VSL from other contexts (e.g. transportation or workplace safety) and/or that they are based on individuals' WTP for all causes of death, rather than for causes specifically associated with environmental or climate exposures.

This paper reports the results of an original contingent valuation survey in the Czech Republic where residents of three cities (Prague, Brno and Ostrava) were asked to report information about their WTP for reductions in their own risk of dying for cardiovascular and respiratory causes. By focussing on these causes of death, we obtain WTP figures that are immediately applicable to the climate change and environmental policy contexts. The VSL is then obtained by dividing WTP by the risk reduction being valued.

To circumvent the problem of possible double-counting of benefits, we ask respondents to consider a private reduction in their own risk of dying. Our WTP questions adopt the dichotomous-choice format, in that respondents are asked whether or not they would pay a specified Czech Koruna (CZK) amount. To refine information about WTP, this is followed by a higher (lower) CZK amount if the respondent agreed (declined) to pay the initial amount.

---

[3] Because extreme weather events, heat-waves, and the spreading of certain vector-borne diseases are considered potential, and worrisome, consequences of climate change, these policies may include heat, storm and other weather advisories, emergency procedures, development of health care facilities, training of health professionals, preparation of evacuation plans etc.

[4] Costa and Kahn (2002) suggest that the compensating wage differentials observed in the labor market – and hence, they conclude, the VSL – has increased over time in the US, as has the quantity of workplace safety. Using Census micro-data and fatality risk figures from the Bureau of Labor Statistics for 1940, 1950, 1960, and 1980, they estimate the implied elasticity of VSL with respect to per capita GNP to be 1.5 to 1.7. A meta-analysis of compensating wage studies by Viscusi and Aldy (2003) pegs the income elasticity of VSL to be 0.5–0.6, and certainly less than one. DeBlaeij et al. (2000) conduct a meta-analysis of the WTP to reduce transport risks, finding a considerable higher income elasticity of 1.33. Liu et al. (1997) compare estimates of the VSL from compensating wage studies in Taiwan based on 1982–1986 data with predictions based on VSL-income relationship from developing countries

Our experimental design allows us to answer four questions that are important for climate change and environmental policy. First, we survey individuals of ages 30–75, so that we can answer the question whether the VSL varies with the age of the beneficiary of the policy – an issue previously explored by Shepherd and Zeckhauser (1982), Jones-Lee (1976), Johannesson et al. (1997) and Alberini et al. (2004a) – and with their health status.

This question is particularly important in the climate change context, because the mortality effects of heat-waves and thermal stresses fall disproportionately on the elderly and persons with chronic conditions.

Second, we vary the size of the risk reduction to the respondent to see if WTP increases with the size of the risk reduction, as prescribed by the economic theory, and if so, by how much. Third, we ask people to value both an immediate and a future risk reduction, as is consistent with the type of exposure changes implied by many environmental and climate change adaptation policies. Fourth, we experiment with two alternative mechanisms for delivering the risk reduction: a completely abstract risk reduction, and one that would be delivered by a new medical test.

Using the survey data, we estimate mean VSL to be 40.16 million CZK (€ 1.27 million at the current exchange rate, € 2.86 million at the Purchasing Power Parity [PPP]), while median VSL is 18.52 million CZK (€ 0.58 million at the current exchange rate, € 1.32 million at the PPP). These estimates increase by roughly 5% to 15% if we increase household income to levels that are closer to the population's household incomes of the three cities than the incomes in our sample. The VSL is lower for older people, but not for individuals with cardiovascular or respiratory illnesses.

The remainder of this paper is organized as follows. Section 10.2 defines the VSL, and presents advantage and disadvantages of using contingent valuation. Section 10.3 describes the structure of the questionnaire. Section 10.4 presents the experimental treatments used in our study, and Section 10.5 the sampling plan and the administration of the survey. Section 10.6 describes the data, and section 10.7 the VSL estimates. In section 10.8, we discuss the results of our VSL regressions, while in Section 10.9 the conclusions are given.

10.2 Use of Contingent Valuation with Mortality Risks

Contingent valuation is a method of estimating the value that a person places on a good. The approach asks people to directly report their willingness to pay (WTP) to obtain a specified good, rather than inferring them from observed behaviours in regular marketplaces.

Contingent valuation can be used to value a reduction in a person's chance of dying, and the VSL implied by this person's WTP. We begin by defining the VSL, which is the rate at which a person is prepared to trade off income for risk reduction. Formally, assume that the individual's utility is $U(y)$, where y is aggregate consumption, and that he has a probability p of dying at the end of this period. Assuming that the state-dependent utility of consumption when the individual is dead is zero, expected utility is equal to

$$E(U) = (1 - p)U(y) \tag{1}$$

The VSL is defined as the rate at which an individual is prepared to trade off income for risk changes, while keeping expected utility the same:

$$\text{VSL} = \frac{dy}{dp} = \frac{1}{1 - p} \frac{U(y)}{U'(y)} \tag{2}$$

Although the VSL is a derivative, in a contingent valuation survey individuals are asked to report their WTP for a specified finite risk chance Δp, and the VSL is approximated as WTP/Δp.

If the researcher's goal is to estimate the VSL, contingent valuation has several advantages over revealed-preference approaches. For example, it is flexible, in that it can be adapted to any desired context, and it is not limited to workplace risks. Moreover, respondents can be informed about their baseline risk and are told what risk reduction they are to value, so that the researcher does not have to rely on the (untested) assumption that perceived risks are identical to objective risks, as is usually the case in compensating wage and consumer behaviour studies. Finally, by assigning the risk changes to be valued exogenously to the respondent, the researcher avoids, by construction, problems like endogeneity of risk and wages, and poorly observed measures of risk.

One potential disadvantage of contingent valuation is that it imposes a heavy cognitive burden on the respondents (see Carson, 2000). Visual aids are usually deployed to communicate risks to the respondents (Corso et al., 2001), and respondents are sometimes given practice questions about probabilities.

The most widely used approach to eliciting information about the respondent's WTP is the so-called dichotomous-choice format. A dichotomous choice payment question asks the respondent if he would pay $X to obtain the good. There are only two possible responses to a dichotomous choice payment question: "yes" and "no." The dollar amount $X is varied across respondents, and is usually termed the bid value.

When dichotomous choice questions are used, the researcher does not observe WTP directly: at best, he can infer that the respondent's WTP amount was greater than the bid value (if the respondent is in favour of the programme) or less than the bid amount (if the respondent votes against the plan), and form broad intervals around the respondent's WTP amount. To estimate the usual welfare statistics, it is necessary to fit binary data models (Cameron and James, 1987).

10.3 **Structure of the Questionnaire**

Heat-waves are typically associated with excess deaths for cardiovascular and respiratory causes. For this reason, the questions at the heart of our questionnaire elicit WTP for reductions in the risk of dying from these causes. Our questionnaire is self-administered by the respondents using the computer, and was developed by translating into the Czech a questionnaire that was previously administered to a sample of Italian residents (see Alberini et al., 2004b).

The questionnaire is divided into seven sections. In section 1, we query the respondent about gender and age, and if he or she has ever been diagnosed to have certain cardiovascular and respiratory illnesses (including heart disease, chronic obstructive pulmonary disease and emphysema), and cancer. We also ask people to tell us a little about the health and longevity of other family members, to assess their current and future health, and to report a subjectively assessed life expectancy.

In section 2, we ask questions about the respondent's health over the last four weeks, physical mobility limitations (if any) and psychological well-being. Our questions are adapted from the Short Form 36 (SF36) questionnaire, which is widely used in medical research to assess physical and emotional health. Section 3 provides a simple probability tutorial, leading to the explanation of one's chance of dying, which is expressed as X in 1000 over 10 years, and is graphically depicted using a grid of 1000 squares, a commonly used risk communication device. White squares represent survival, while blue squares represent death. Respondents are then tested for probability comprehension.

In section 4, we acquaint respondents with the concept that it is possible to reduce one's risk of dying, and that many people do so on a routine basis. For example, we tell respondents that a pap smear can reduce the risk of dying of cervical cancer (in women) and that blood pressure medication reduces the risk of dying of a heart attack. We then introduce cardiovascular and respiratory illnesses, and allow the respondent to learn more about them by reading a glossary which is launched by double-clicking a link on the screen. The respondents are then asked questions about treatments or actions they take against cardiovascular and respiratory illnesses, and their cost.

In section 5, we present the chance of dying for a person of the respondent's age, gender, and health status. This is shown using blue squares in the grid of 1000 squares. We highlight the chance of dying for cardiovascular and respiratory illnesses using orange squares, emphasizing that these risks increase with age. For example, cardiovascular and respiratory causes account for a small fraction of the total risk of dying among persons of age up to 40 years, but for 50% among 70-year-olds. These mortality risks were taken from the official population statistics of the Czech Republic.

Section 6 presents the hypothetical risk reduction scenario. People are asked to value a reduction of X in 1000 in their chance of dying for cardiovascular and respiratory 9 causes over the next 10 years, where X ranges

Table 1. Baseline risks and risk reductions assigned to respondents in the survey (immediate risk reduction)

Age	Male				Female			
	FILL1	**FILL2**	**FILL3**	**FILL4**	**FILL1**	**FILL2**	**FILL3**	**FILL4**
30–34	17	2	1	50%	8	2	1	50%
35–39	30	4	2	50%	13	2	1	50%
40–44	51	8	4	50%	22	4	2	50%
45–49	81	16	5	31%	35	6	2	33%
50–54	126	30	3	10%	55	9	3	32%
55–59	189	59	6	10%	86	13	4	30%
60–64	274	94	5	5%	141	35	4	11%
65–69	387	160	8	5%	240	72	7	10%
70–74	539	274	12	4%	397	155	7	5%
75–79	714	487	12	2%	611	342	7	2%

from 1 to 12, depending on the respondent's age and gender. The experimental design for the baseline risk and risk reduction is displayed in Table 1, where the baseline risk of dying for all causes is denoted as FILL, the baseline risk of dying for cardiovascular and respiratory causes if denoted as FILL2, and the risk reduction in the latter is denoted as FILL3.

During the focus groups we conducted as part of the initial questionnaire development work in Italy, we noticed that people of ages 30 to 40 would regard their risk reductions – which were usually of about 1 in 1000 – as negligible, although these risk reductions in many cases account for half of the baseline risk. Accordingly, we decided that the questionnaire should explicitly remind respondents of the percentage risk reductions implied by the absolute risk reductions. The percentage risk reduction is denoted as FILL4 in Table 1.

Respondents are told that to obtain the risk reduction described in the questionnaire, they would have to pay a given amount of money every year for 10 years. The payment question is in a dichotomous choice format with one follow-up.

In section 7 we describe and elicit WTP for a risk reduction that would take place X years from now (where X varies with the respondent's age), when the respondent is older. To make this question meaningful to the respondents, we show them that the chance of dying for any cause and for cardiovascular causes increases as one gets older.

Baseline risks and risk reductions used in this valuation task are reported in Table 2, where they are denoted as FILL5 and FILL7, respectively.

In section 8, we ask questions intended to investigate the intertemporal rate of preference of the respondent, and his or her financial risk aversion. Section 9 concludes the survey with the usual socio-demographic ques-

Table 2. Baseline risks and risk reductions assigned to respondents in the survey (future risk reduction)

Age	FILL5	FILL7	
	Future age	Male	Female
30–34	50 and 70	3	3
		12	7
35–39	50 and 70	3	3
		12	7
40–44	60 and 75	5	4
		12	7
45–49	60 and 75	5	4
		12	7
50–54	65 and 75	8	7
		12	7
55–59	75	12	7

tions, and with debriefing questions about the respondent's interpretation of the questions.

We wish to emphasize that climate change is never mentioned to the respondent is this survey. We chose to do so for two reasons. First, we wished to keep the risk reduction a private good, because it is difficult to identify the altruistic components of WTP, and to account for them appropriately to avoid double-counting. Second, linking risk changes to public policies for emissions reductions or adaptation to climate change would require that we provide information about them. In our opinion, doing so would have resulted in an excessively heavy cognitive burden, which prompted us to choose a context-free risk reduction.

10.4 Experimental Treatments

Our study involves a total of four experimental treatments. The first is that our respondents are randomly assigned to one of two mechanisms for delivering the risk reduction: a completely abstract risk reduction, and a medical test. In the abstract risk reduction version of the questionnaire, respondents are simply told to suppose that it were possible to reduce their risks. Our focus groups indicated that people accept such an abstract risk reduction, and that this approach is likely to get them to focus more sharply on the size of the risk reduction, without being distracted by other details. The purpose of the experimental treatment is, therefore, to see if people truly focus on probabilities better in this fashion.

Table 3. Bid amounts

BID 1	BID 2 if BID 1 = yes	BID 3 if BID 1 = no
1 500 Kč	3 500 Kč	950 Kč
3 500 Kč	7 000 Kč	1 500 Kč
7 000 Kč	13 000 Kč	3 500 Kč
13 000 Kč	17 000 Kč	7 000 Kč
In EURO using the exchange rate (31.5 CZK/€)		
48 €	111 €	30 €
111 €	222 €	48 €
222 €	413 €	111 €
413 €	540 €	222 €
In EURO by PPP (14.0 CZK/€)		
107 €	250 €	68 €
250 €	500 €	107 €
500 €	929 €	250 €
929 €	1 214 €	500 €

Table 4. Annual payment (for ten years) offered as an alternative to an immediate payment

Immediate payoff	Annual payment	Implicit discount rate
140 000 CZK	16 000 CZK (1 150 € by PPP)	2.5%
(4 444 € by exchange rate	21 000 CZK (1 500 € by PPP)	8.1%
10 000 € by PPP)	23 000 CZK (1 650 € by PPP)	10.2%

Regarding the medical test, we had previously found that people were comfortable with this vehicle for the risk reduction, as long as the medical test was approved by the national health care system. In this case, the payment mechanism is a co-pay, modeled after the routine charge for medical tests within the Czech national health care system. Respondents are told that the medical test would have to be done, and paid for, every year for ten years, for it to be effective. They are also told that the test would be safe and have no side effects.

The second experimental design is that respondents are randomly assigned to one of four possible bid sets, which we report in Table 3.

Our third experimental treatment is that respondents younger than age 55 are randomly assigned to one of two possible ages at which the future risk reduction is supposed to start, and the risk reduction itself takes one of two possible sizes. Respondents aged 56–59 are assigned only on future age when the risk reduction begins and one risk reduction.

Our fourth experimental treatment refers to section 8 of the questionnaire, and varies the annual payment to be paid annually to the respondent for ten years as an alternative to an immediate payment of 140 000 CZK. The annual payments are reported in Table 4, along with the implied discount rate. All of these amounts were obtained as the 13 purchase power parity equivalents of the corresponding euro figures used in the Italian survey.

10.5 Sampling Plan and Administration of the Questionnaire

The survey was self-administered by the respondents using the computer in three cities of the Czech Republic – Prague, Brno and Ostrava.[5] These were selected on the grounds of population and income, and to ensure variability in the population risk factors.

For example, Prague, the capital of the Czech Republic and its largest city, has relatively high age- and gender-specific mortality rates, and both cardiovascular and respiratory mortality rates are relatively high. By contrast, Brno has relatively low overall mortality rates, low mortality rates for respiratory causes, but somewhat high mortality rates for cardiovascular causes. Finally, Ostrava, an industrial city in the Northeast of the Czech Republic, has high overall and respiratory mortality rates, but low mortality rates for cardiovascular causes. These characteristics are summarized in Table 5.

We began with translating the questionnaire developed by Alberini et al. (2004 b) for Italy into the Czech, and by testing it in three focus groups in Prague and Karlovy Vary, a city in western Bohemia. Evidence from the focus group was used to revise slightly the questionnaire to improve clarity, where needed.[6,7]

---

[5] Charles University Environment Center (CUEC) was commissioned to translate the questionnaire, pretest, and implement the survey in the Czech Republic. The Czech Republic team is comprised of Milan Scasny (team leader), Marketa Braun Kohlova, Jan Melichar and Jan Urban, plus three researchers involved in the recruitment and administration of the survey. The project started in March 2004 and was completed by September 2004

[6] The focus groups were attended by 5–7 participants. We used paper handouts in this phase of the research

[7] For example, in section 4, people found our question about checkups and medical tests to reduce one's chance of dying somewhat confusing. Clarity and comprehension was greatly improved when we added a response category that allowed people to tell us that they did undertake such actions, and that they did so because their doctor had specifically prescribed them. Likewise, we felt it was necessary to add a question to ascertain whether respondents were familiar with certain preventive actions (e.g. aspirin therapy) to reduce the chance of dying for cardiovascular causes. All questionnaire amendments were discussed mainly by email and telephone, and during three work meetings held in Brussels on 13 March 2004, in Washington on 24–25 May 2004, and in Budapest on 26–27 June 2004

Table 5. Characteristics of the three cities

Prague (capital; central Bohemia)	– Capital city
	– Has relatively high mortality overall, high both for respiratory and cardiovascular causes
	– There is a wide variation in income, and the wealthiest people in the CR live in Prague
	– Lowest unemployment rate
Brno (second largest city in the CR; Moravia; Southeast CR)	– Former capital of Moravia province; economy based mostly on agriculture
	– Low mortality overall, low respiratory causes, relatively high cardiovascular cause in the region
	– Relatively high unemployment rate in the region
Ostrava (Northeast CR)	– Industrial city; the income of the region is among the lowest among the regions of the CR. However, the city does not have low income
	– Highest overall mortality rates, high respiratory deaths, low cardiovascular mortality rates
	– High unemployment rate (16% in the region, one of the highest in the CR)

The self-administered computer questionnaire was conceived as an internet-based application and developed using PHP language. It was then installed on three servers,[8] to be accessed by individual computers at the three facilities in the cities where the survey was conducted. Each of the individual computers had high-speed internet access.

The computer questionnaire was pre-tested in Prague, on 4 August 2004. The respondents were recruited by the professional survey company, Factum Invenio (www.factum.cz), following the sampling quotas specified for the final sample.[9] The pretest included 55 respondents and was conducted at the same location as for the final survey (the university computer lab in downtown Prague). The room included 17 computers, all of them dedicated to the pre-test.

The sampling plan for the final survey called for quota sampling of the 30–45, 46–60, and 61–75 age groups in roughly equal proportions. We also wished to have an equal number of men and women, and a sample that

---

[8] The URLs of these servers were http://www.czp.cuni.cz/mortalita (the official web site of CUEC belonging to Charles University server), http://vydra.ff.cuni.cz/jd/mortalita (the server of the computer center of Faculty of the Arts, Charles University in Prague), and http://mortalita.zabukem.cz (a private server)

[9] As explained below, there were three equally represented age groups, an even number of men and women, and that variation in income was ensured by sampling by quotas on education

Table 6. Administration of the survey

City	Address	Days	Date
∎ **Prague** N = 351	Charles University in Prague, Faculty of Arts and Philosophy, Celetna 20, Prague 1	8 days	30 August – 5 Sept plus 9 Sept 2004 10 a.m. – 8 p.m.
∎ **Brno** N = 296	Masarykova Univerzita in Brno, Fakulty of Mediciny, Komenského aq. 2, Brno	9 days	31 August – 8 Sept 10 a.m. – 7:15 p.m.
∎ **Ostrava** N = 307	Ostravská univerzita, Centre of information technologies, Bráfova 5, Ostrava	12 days	30 August – 10 Sept 10 a.m. –7:15 p.m.

was similar to the population in terms of educational attainment and financial circumstances of the household.

The data were collected in Prague, Brno and Ostrava at facilities with 10 personal computers within University buildings. Table 6 summarizes locations, sample sizes and dates when the data collection took place.

Throughout the administration of the final version of the questionnaire, we made sure that assistance was available for those respondents who needed help. This proved necessary only in a few cases. Most people, even those who were not familiar with computers, quickly learned how to use the mouse and how to type in open-ended answers when needed.

Our survey respondents were recruited by Centrum Výzkumu Veřejné Mínky (CVVM) Public Opinion for Research Centre, a survey firm within the Sociological Institute, Academy of Sciences of the Czech Republic (http://www.cvvm.cz). They were offered a token of 100 CZK (3.2 €) in Prague, and 80 CZK at the other locations, for participating in the study. The payment was made upon completion of the questionnaire. Our final sample is comprised of 954 observations.

10.6 Data

10.6.1 Individual Characteristics of the Respondents

In this section, we present descriptive statistics for the purpose of examining whether the final sample is consistent with the sampling plan. Our final sample size was 954, with 36.79% of our respondents in Prague, 31.03% in Brno and 32.18% in Ostrava.

Table 7 reports descriptive statistics for one experimental treatment variable, the abstract risk reduction scenario v. medical test, and for individual characteristics of the 17 respondent. As shown in Table 7, half of the respondents were given the abstract risk reduction scenario. The sample is

Table 7. Descriptive statistics of the sample

Variable label	N	Mean	Stand. Dev.	Minimum	Maximum
▌Abstract risk reduction	954	0.50	0.50	0	1
▌Male	954	0.47	0.50	0	1
▌Age	954	50.75	12.84	30	75
▌Cardio	954	0.20	0.40	0	1
▌Pressure	954	0.18	0.38	0	1
▌Cholesterol	954	0.31	0.46	0	1
▌Diabetes	954	0.33	0.47	0	1
▌Lungs	954	0.13	0.33	0	1
▌Cancer	954	0.04	0.19	0	1
▌Married	954	0.54	0.50	0	1
▌Household size	953	2.74	1.51	1	25
▌Dependents	953	0.49	0.90	0	9
▌Household income in 1000 CZK	876	247.51	129.22	108	660
▌Works full-time	954	0.51	0.50	0	1

Table 8. Marital status of the respondents

Category	Frequency	Percent of the sample
▌Married	511	53.56
▌Divorced or separated	238	24.96
▌Widow/widower	82	8.60
▌Never married	121	12.68
▌Missing	2	0.21

balanced in terms of gender, with only a slight prevalence of women. The average age is almost 51 years, as required by the sampling plan. [10]

Regarding the health status of the respondents, 20% of the sample suffers from heart disease or has had a heart attack, 18% has high blood pressure, 31% has high cholesterol, and 13% suffers from a chronic respiratory illness, like asthma, COPD, or emphysema. About one-third of the sample reports being a diabetic. Only about 4% of the sample has or has had cancer.

About 54% of the respondents are married, almost a quarter of the sample is divorced or separated, 9% is comprised of widows or widowers, and 13% never married (Table 8). Returning to Table 7, the average household size is 2.74 people, while annual household income is on average 247510

---

[10] The older age group is relatively less represented in Prague and Ostrava, where it accounts for about 24% of the sample

CZK. This is slightly less than the average household income in the population (which as of 2003 is 290 000 CZK in Prague, 264 000 CZK in Brno and 278 000 CZK in Ostrava), and is confirmed by comparing the cumulative distribution of household income in the sample and population, as we do in Figure 1.[11]

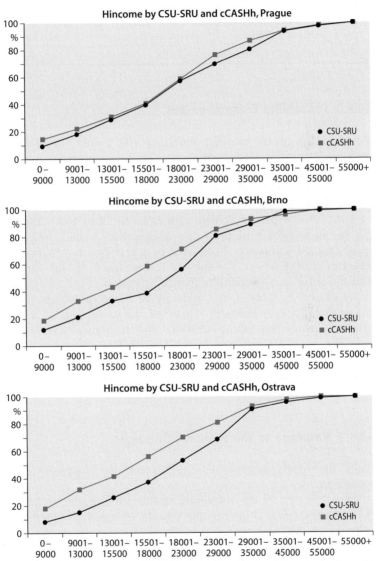

Fig. 1. Comparison of the household income (HINCOME) for three cities under examination with CSU-CRU, Household Budget Survey by Czech Statistic Office, for the year 2003

[11] Czech Statistical Office, 2003. Median household income ranges from 252 000 CZK (in Prague) to 267 000 CZK (in Ostrava)

Table 9. Comprehension of risks. Absolute and relative frequencies of the respondents who ...

	Person A	Person B	Indifferent between A and B	Missing
▌In the probability quiz, choose ...	157 [16.46%]	794 [83.23%]	N/A	3 [0.31%]
▌In the probability choice question, choose ...	681 [71.38%]	136 [14.26%]	135 [14.15%]	2 [0.21%]

10.6.2 Probability Comprehension

When asking people to value mortality risk reductions, it is important to make sure that the concept of chance has been communicated effectively to them. We included two questions meant to test for probability comprehension in section 3 of the questionnaire. First we ask respondents to indicate which of two people has the higher chance of dying – person A, with risk of 5 in 1000, or person B, with risk equal to 10 in 1000. Then we ask people to tell us which of these two people they would prefer to be. These questions are similar to those used in a previous mortality risk survey by Alberini et al. (2004a). We dub the former question our "probability quiz" and the latter our "probability choice" question.

As shown in Table 9, the majority of the respondents (83%) answered the probability quiz correctly. However, the percentage of respondents who failed this question (about 17%) is slightly higher than we have seen in previous studies conducted in the United States and Canada (Alberini et al., 2004a), and in the United Kingdom, France and Italy (Alberini et al., 2004c). Likewise, the majority of the respondents chose the person with the lower risk of dying.

10.6.3 Response to the Payment Questions

Table 10 reports descriptive statistics for the responses to the payment questions.

About 56.19% of the responses to the initial payment questions were a "yes." Similar proportions of the sample reported a sequence of two "yes" (32%) and two "no" responses (29%), respectively, to the payment questions. The relative frequency of "yes"-"no" responses is also relatively high (24%), while "no"-"yes" pairs account for 15% of the sample.

We report the percentage of respondents that answered "yes" to the payment questions within the subsample that received a given bid set in Table 11. The table shows clearly that the percentage of "yes" responses declines with the initial bid, which means that responses are consistent with the

Table 10. WTP response patterns

Pattern	Percent of the sample
Yes (yes to the initial payment question)	
YY	31.87
YN	24.32
NY	14.99
NN	28.72

Table 11. Number of respondents assigned to initial bid amounts and percentage of yes responses

	Bid amount (CZK)			
	1500	3500	7000	13 000
Number of respondents	235	245	241	233
Percentage of yes	72.34	67.35	47.72	36.91

Table 12. Percentage of yes responses by FILL3 level

Yes	FILL3 (FILL3)									
	1	2	3	4	5	6	7	8	12	Total
Yes	47.51	53.01	58.77	61.54	52.63	50.00	61.04	75.00	70.45	
Total	181	166	114	182	114	48	77	28	44	954

economic paradigm. It also shows that the bids were selected judiciously, in the sense that they cover a broad range of the distribution of WTP. Finally, Table 11 suggests that the median WTP lies between 3500 and 7000 CZK.

In Table 12, we check for patterns in the percentage of "yes" responses by FILL3.

We were expecting this percentage to increase with FILL3 – the larger the risk reduction, the more one should be willing to pay for it – but the relationship is not monotonic. In retrospect, we suspect that there could be two reasons for this result. First, WTP may be less than proportional to the risk reduction. Second, people may be influenced by the relative risk reduction, and not just by the absolute risk reduction. In future research, we will explore these effects.

10.7 VSL Estimates

To obtain estimates of the willingness to pay for the risk reductions, we combine the responses to the initial and follow-up payment question, and form intervals around the respondent's unobserved WTP amount. To further elaborate on our approach, suppose that a respondent answered "yes" to the initial bid amount of 7000 CZK and was subsequently asked whether he would pay 13000, which he declined to do. We infer that the respondent's WTP falls in the interval between 7000 and 13000 CZK. This respondent's implied VSL, therefore, falls between 7000/FILL3 and 13000/FILL3 CZK.[12]

Next, we assume that VSL follows a certain distribution in the population, and specify an interval-data model of the responses, which is estimated by the method of maximum likelihood. Formally, assuming that VSL is a random variate with cdf $F(y,\lambda)$, where λ is the vector of parameters that index the distribution, the log likelihood function of the sample is:

$$\sum_{i=1}^{n} \log[F(\text{VSL}_{Hi}; \lambda) - F(\text{VSL}_{Li}; \lambda)] \tag{3}$$

where VSL_{Li} and VSL_{Hi} are the lower and upper bound, respectively, of the interval around the respondent's unobserved VSL amount.

As shown in Table 13, we tried different two-parameter distributions for VSL, finding that the lognormal and the Weibull fit the data best. The Weibull model pegs mean VSL at 40.16 million CZK (s.e. 2.73 million CZK). This is equivalent to € 1.27 million when the exchange rate is used, and € 2.86 million at PPP.[13] Median VSL is 18.52 million CZK (s.e. 1.14 million

Table 13. Goodness of fit for various distributions, double-bounded interval-data model of the responses to the dichotomous-choice payment question and follow-up. N=953 valid observations

Distribution assumed for VSL	Log likelihood	Parameter 1	Parameter 2
Normal	−1506.78	Mean 27.16 [s.e. 1.57]	Scale 42.98 [s.e. 1.68]
Lognormal	−1325.36	μ 2.81 [s.e. 0.06]	σ 1.63 [s.e. 0.06]
Weibull	−1329.81	Scale 31.42 [s.e. 1.80]	Shape 0.69 [s.e. 0.03]
Exponential	−1382.09	Scale 32.60 [s.e. 1.30]	Shape restricted to 1

---

[12] This presumes that WTP is proportional to the size of the risk reduction, as economic theory predicts to be the case for small risk changes (Hammitt and Graham, 1999)

[13] The current exchange rate is 31.5 CZK to the euro, and the PPP is 14.03 CZK to the euro

CZK), which corresponds to € 0.58 million at the current exchange rate (€ 1.32 million based on a PPP conversion).[14]

10.8 Interval Validity of the Responses

To test the internal validity of the responses, we run regressions relating VSL to experimental treatment variables and individual characteristics of the respondents. We use an accelerated life model with a Weibull baseline hazard. Formally, this implies that

$$\log \mathrm{VSL}_i^* = \mathbf{x}_i \boldsymbol{\beta} + \varepsilon_i \tag{4}$$

where VSL* is the unobserved VSL, and $\mathbf{x}$ is a vector of covariates. The error term is a type I extreme value variate with scale θ. This means that VSL* is a Weibull with scale $\sigma_i = \exp(\mathbf{x}_i \boldsymbol{\beta})$ and shape parameter θ, and that the log likelihood function is:

$$\sum_{i=1}^{n} \log[\exp(-(\mathrm{VSL}_{Li}/\sigma_i)^{\theta}) - \exp(-(\mathrm{VSL}_{Hi}/\sigma_i)^{\theta})] \tag{5}$$

Our vector of covariates includes a dummy for the abstract experimental treatment (ABSTRACT), age group dummies, a gender dummy (MALE), an education dummy (COLLEGE), annual household income expressed in thousand CZK (INC), a dummy equal to one if the respondent is married (MARRIED), and continuous variables for household size (HHSIZE) and number of children under the age of 18 (DEPENDENTS). To capture health status, we use a number of dummies denoting whether the respondent has heart disease (CARDIO), high blood pressure (PRESSURE), high cholesterol (CHOLESTEROL), is a diabetic (DIABETES), has a chronic respiratory condition (LUNGS), and has or has had cancer (CANC).

The results for this specification of the Weibull accelerated life model are shown in Table 14.[15] The table shows that several covariates are significant predictors of the respondent's VSL. All else the same, WTP – and hence the VSL – is about 67% higher among those subjects who were assigned to the abstract risk reduction, but 34% lower for men.

---

[14] The lognormal model produces a substantially larger estimate of mean VSL, 60.29 million CZK (s.e. 7.25 million CZK) (€ 1.90 million, or € 4.29 million with a PPP conversion). Median VSL is 16.60 million CZK (s.e. € 0.99 million) (€ 0.53 million at the regular exchange rate, 1.18 with PPP conversion)

[15] City dummies included in the model to account for differences in the cost of living in different cities do not have any further explanatory power for WTP, and are thus excluded from the specification reported in this paper

Table 14. Weibull interval-data model of VSL

Number of observations	875
Noncensored values	0
Right Censored values	278
Left censored values	255
Interval censored values	342
Missing values	79
Name of distribution	Weibull
Log likelihood	−1139.264505

Parameter	DF	Estimate	Standard error	95% Confidence limits		Chi-square	Pr > ChiSq
Intercept	1	3.1747	0.1825	2.8170	3.5324	302.61	<0.0001
Abstract	1	0.5166	0.0994	0.3217	0.7114	26.98	<0.0001
Male	1	−0.4194	0.1001	−0.6157	−0.2231	17.54	<0.0001
Age 40–49	1	−0.6605	0.1406	−0.9360	−0.3849	22.07	<0.0001
Age 50–59	1	−0.7447	0.1540	−1.0464	−0.4429	23.39	<0.0001
Age 60–69	1	−0.8577	0.1717	−1.1942	−0.5211	24.95	<0.0001
Age 70 plus	1	−1.4688	0.2175	−1.8952	−1.0425	45.59	<0.0001
College	1	0.0122	0.1166	−0.2164	0.2407	0.01	0.9169
Inc	1	0.0022	0.0005	0.0012	0.0031	20.24	<0.0001
HHSIZE	1	0.0079	0.0488	−0.0877	0.1035	0.03	0.8714
Dependents	1	0.0470	0.0764	−0.1027	0.1967	0.38	0.5386
Cardio	1	0.1222	0.1330	−0.1385	0.3829	0.84	0.3583
Lungs	1	0.0656	0.1518	−0.2320	0.3632	0.19	0.6656
Pressure	1	−0.1868	0.1396	−0.4604	0.0867	1.79	0.1807
Cholesterol	1	0.2288	0.1214	−0.0092	0.4668	3.55	0.0596
Diabetes	1	0.0968	0.1158	−0.1301	0.3238	0.70	0.4030
Cancer	1	0.3451	0.3248	−0.2914	0.9816	1.13	0.2879
Weibull shape	1	0.8572	0.0373	0.7871	0.9335		

The most striking result, however, is that the VSL declines monotonically with age. The youngest respondents in the sample have the highest VSL values, but by the time one is 70, the VSL is only about 23% of the VSL of respondents in the youngest age group. This finding is comparable to evidence from the Italian study based on the same questionnaire, and is in sharp contrast with Johannesson et al. (1997) and Alberini et al. (2004 a).

We do not find any evidence of a statistically significant association between the VSL and the educational attainment of the respondent, but household income is, as predicted by economic theory, positively and significantly associated with a person's WTP, and hence his or her VSL. At the average level of household income in the sample (about 247 500 CZK), the median VSL is 18.30 million CZK (€ 0.58 at the current exchange rate).

If we raise income to bring it to levels closer to the average household income of Brno (264 000 CZK), the median VSL is 19.32 million CZK (€ 0.61 million), a 5.6% increase, and if we raise it to the average household income of Prague (290 000 CZK), the median VSL is 21 million CZK (€ 0.66 million), a 15% increase. If we then further raise household income from 290 000 CZK to 365 000 CZK (a 25% increase) raises the VSL from € 0.66 million to € 88 million (a 33% increase).

Controls for household size and number of dependent children, however, do not have any additional explanatory power. We find little evidence of an association between the health status dummies and the VSL.[16] The coefficients on most of them, however, are positive, dispelling the notion that people in poor health would be willing to pay less to increase their chance of survival.

10.9 Conclusions

We have conducted a contingent valuation survey in three cities of the Czech Republic to elicit people's WTP for reductions in their risk of dying for cardiovascular and respiratory illnesses. These causes of death are typically associated with heat-waves and with environmental exposures (e.g. air pollution).

Our questionnaire, which is self-administered by the respondents using the computer, is virtually identical – except for the language and other minor amendments required for adaptation to the Czech context – to a questionnaire previously developed and administered in Italy (Alberini et al., 2004 b). Our study incorporated several experimental treatments in this study.

In this paper, attention is restricted to the immediate risk reduction. We compute the VSL as WTP divided by the size of the risk reduction given to the respondent. We obtain WTP responses that imply a mean VSL of 40.16 million CZK (€ 1.27 million at the current exchange rate, € 2.86 million at the PPP), while median VSL is 18.52 million CZK (€ 0.58 million at the current exchange rate, € 1.32 million at the PPP).

VSL is higher among wealthier people, and we calculate the income elasticity of VSL to be about one. A striking result is that the VSL declines

---

[16] The likelihood ratio test of the null that the coefficients on the health status dummies are jointly equal to zero is 9.34 (p value = 0.16)

dramatically with the age of the respondent. For example, all else the same, a person in his 70s has a VSL that is about only one-quarter than that of a 30-year old. Finally, people who suffer from chronic illness or conditions that could lead to heart diseases are willing to pay no less than the respondents without these ailments.

To our knowledge, this is the first effort to obtain VSL values that could be used directly for valuing the mortality benefits of environmental and climate change policies in an eastern European country that has recently joined the European Union. The approach used in this paper involves a broader population than that studied in Johannesson et al. (1991) and can be compared directly with the results of the companion study in Italy (Alberini et al., 2004 b).

References

Alberini A (2004) Willingness to Pay for Mortality Risk Reductions: A Reexamination of the Literature. Final report to the US Environmental Protection Agency under Cooperative Agreement 015-29528, College Park, July

Alberini A, Chiabai A, Nocella G (2005) Valuing the Mortality Effects of Heat Waves. In: Menne B, Ebi K (eds) Climate Change and Adaptation Strategies for Human Health. Steinkopff, Darmstadt, pp 345–371

Alberini A, Cropper M, Krupnick A, Simon N (2004a) Does the Value of a Statistical Life Vary with Age and Health Status? Evidence from the United States and Canada. Journal of Environmental Economics and Management 48(1):769–792

Alberini A, Hunt A, Markandya A (2004b) Willingness to Pay to Reduce Mortality Risks: Evidence from a Three-country Contingent Valuation Study. FEEM Working paper 111.04, Milan, July

Blomquist GC (2004) Self-protection and Averting Behavior, Values of Statistical Lives, and Benefit Cost Analysis of Environmental Policy. Review of Economics of the Household 2:69–110

Cameron TA, James MD (1987) Efficient Estimation Methods for 'Closed-Ended' Contingent Valuation Surveys. Review of Economics and Statistics 69(2):269–276

Carson RT (2000) Contingent Valuation: A User's Guide. Environmental Science Technology 34:1413–1418

Corso P, Hammitt JK, Graham JD (2001) Valuing Mortality Risk Reductions: Using Visual Aids to Improve the Validity of Contingent Valuation. Journal of Risk and Uncertainty 23(2):165–184

Costa DL, Kahn ME (2002) Changes in the Value of Life, 1940–1980. NBER Working Paper 9396, Cambridge, December

deBlaeij A, Florax RJG, Rietvield P (2000) The Value of Statistical Life in Road Safety: A Meta-analysis. Working paper TI 2000-089/3, Tienbergen Institute, Amsterdam

Hammitt JK, Graham JD (1999) Willingness to Pay for Health Protection: Inadequate Sensitivity to Probability? Journal of Risk and Uncertainty 8:33–62

Johannesson M, Johansson P-O, Löfgren K-G (1997) On the Value of Changes in Life Expectancy: Blips Versus Parametric Changes. Journal of Risk and Uncertainty 15:221–239

Johannesson M, Jonsson B, Borquist L (1991) Willingness to Pay for Antihypertensive Therapy – Results for a Swedish Pilot Study. Journal of Health Economics 10:461–474

Jones-Lee MW (1976) The Value of Life: An Economic Analysis. Martin Robertson, London

Liu JT, Hammitt JK, Liu J-L (1997) Estimated Hedonic Wage Function and Value of Life in a Developing Country. Economic Letters 57:353–358

Shepherd DS, Zeckhauser RJ (1982) Life-Cycle Consumption and Willingness to Pay for Increased Survival. In: Jones-Lee MW et al (eds) The Value of Life and Safety. North-Holland, Amsterdam, The Netherlands

Viscusi WK, Aldy JE (2003) The Value of a Statistical Life: A Critical Review of Market Estimates throughout the World. Journal of Risk and Uncertainty 27(1):5–76

11 Implications of the SRES Scenarios for Human Health in Europe

Pim Martens, Irene Lorenzoni, Bettina Menne

11.1 Introduction

Good health is a fundamental resource for social and economic development. Higher levels of human development mean that people live longer and enjoy more healthy years of life. While the health of the 879 million people in the WHO European Region has in general improved over time, inequalities between the 52 Member States in the Region[1] and between groups within countries have widened. In addition to the east–west gap in health, differences in health between socioeconomic groups have increased in many countries.

The most important causes of the burden of disease in the Region are non-communicable diseases (NCDs – 77% of the total), external causes of injury and poisoning (14%) and communicable diseases (9%). In 2002, NCDs caused 86% of the 9.6 million deaths and 77% of the 150.3 million DALYs in the Region. They originate from complex interactions of genetics, behaviour and the environment and, thus, require long-term planning and treatment. In addition, injuries are a particular problem for young people. Further, poverty and under funded services create a double burden of non-communicable and communicable diseases for some countries in the eastern half of the Region. This double burden is partly responsible for the health gaps between and within countries. Just seven leading risk factors – tobacco, alcohol, high blood pressure, high cholesterol, overweight, low fruit and vegetable intake and physical inactivity – are mainly responsible for the differences between countries in the burden of disease due to seven leading conditions ischaemic heart disease, unipolar depressive disorders, cerebrovascular disease, alcohol use disorders, chronic pulmonary disease, lung cancer and road-traffic injury.

In the European Region, roughly two distinct epidemiological patterns of communicable disease can be observed: the western European countries, with steady decline or stabilization of communicable disease morbidity and mortality over the last decades; and the countries in the eastern part of the Region, where during the 1990s, episodes characterized by outbreaks of communicable disease were observed, although some progress has been

---

[1] The WHO European Region is composed by European Union Countries, Central and Eastern European Countries (CCEE) and Newly Independent States (NIS)

achieved in the last four years. A further distinction can be made between many CCEE and the NIS: the former group of countries show moderate incidence rates, compared to high and increasing morbidity and mortality rates of selected communicable diseases in the NIS. These rates are associated with socioeconomic factors, the state of public health spending and public health infrastructure. Factors contributing to increases in communicable diseases include population displacements, environmental change and political unrest.

Income poverty has spread in NIS: previously affecting only a small part of their population (3.3% in 1987–1988), it currently takes its toll on about a half. In those NIS countries for which data are available, 46% of the population (168 million people) now live below a poverty line of US$ 4 a day. In EU countries, about 37 million people, i.e. about 10% of the total population, live under the income poverty line of less than 50% of the median income. In general, large and increasing numbers of people in European societies today are at risk of experiencing poverty sometime in their lives. Throughout the 1990s, income inequality increased in all western European countries for which such data are available, except for Denmark, i.e. in Austria, Belgium, Finland, Germany, Ireland, Italy, the Netherlands, Norway, Sweden, Switzerland, and the United Kingdom.

The European Region has been subject to health care reforms and re-organizations, in particular after 1990. It is difficult to match the demand for better health with a stable or declining expenditure on health by the public sector. A significant trend emerging in the last few years is the reversal in the fortunes of market-oriented policies for health system reform, including solutions such as privatization, corporatization, reliance on user charges or fee-for-service compensation schemes, provider or purchaser competition and advanced forms of decentralization. The countries which pioneered these solutions in the early 1990s already experienced their limitations, and have often smoothened their policies, or discontinued them completely: examples of this trend are the dismantling of the "Internal Market" and the shift in emphasis from General Practitioners fund holding to Primary health care groups after 1997.

This introduction reflects some health trends in Europe during the last decade. The central question regarding the future developments of global health is: will the current and anticipated socioeconomic trends and global environmental changes be consistent with a sustainable and healthy life for humankind? And more specifically, what will future prospects for Europe be? In this chapter we attempt to describe possible European futures, based on published scenario work, and attempt to estimate what European health could look like in the 21st century.

11.2 Scenarios

Scenarios are descriptions of possible futures that reflect different perspectives on past, present and future developments with a view to anticipating the future [1]. Scenarios do not provide predictions of future outcomes; rather, they outline (quantitatively, qualitatively or both) various possible futures in order to provoke reflection and debate about events and circumstances that may need to be faced. Even contemplating different possible future outcomes may induce people to reconsider and modify their present activities in view of a different, preferred outcome. For scenarios to be credible, they should be transparent and internally consistent views of the world [1].

Scenario analysis has evolved significantly over the past decades. In their early days, scenarios were used primarily as planning and forecasting tools, displaying a rather mechanistic and deterministic worldview. Currently scenario analysis supports a more open form of exploration [2].

Scenarios as aids to decision-making can be useful tools to:
- *articulate our key considerations and assumptions*: scenarios can help to imagine a range of possible futures if we follow a key set of assumptions and considerations
- *blend quantitative and qualitative knowledge*: scenarios are powerful frameworks for using both data and model-produced output in combination with qualitative knowledge elements
- *identify constraints and dilemmas*: exploring the future often yields indications for constraints in future developments and dilemmas for strategic choices to be made
- *expand our thinking beyond the conventional paradigm*: exploring future possibilities that go beyond our conventional thinking may result in surprising and innovative insights.

Depending on how scenarios are produced and developed, they can have several limitations:
- *lack of diversity*: scenarios are often developed from a narrow, disciplinary based perspective, resulting in a limited set of standard economic, technological and, to a lesser extent, environmental assumptions
- *extrapolations of current trends*: many scenarios do have a 'business-as-usual' character, assuming that current conditions will continue for decades, excluding surprises
- *inconsistent*: sets of assumptions made for different sectors, regions, or issues are often not made consistent with each other
- *not transparent*: key assumptions and underlying implicit judgements and preferences are often not made explicit. Also, not which factors are exogenous and which ones are endogenous, and to what extent societal processes are autonomous or influenced by concrete policies.

Nowadays, scenarios as powerful exploratory tools can be used to paint pictures of possible future outcomes based on changing assumptions. In a sense, they can be used to promote considerations on capacity building for adaptation and mitigation in the face of climate change, by broadening and deepening the mindset of stakeholders involved in a process of exploring possible futures. To the best of our knowledge, scenarios which relate human health to global and environmental changes are scarce. Some examples do exist, however, of national attempts at envisioning changes to the health system using scenario-based approaches (see [3]). And some global scenario studies exist that describe changes in human health in a globalizing world [4, 5]. In the sections below, we outline scenarios that have been developed for the climate change community and attempt to link them to the health of the European population.

It has been argued that, when assessing the impacts of climate change on health, wider circumstances and ranges of uncertainty (see [6]) should be considered, including:

▌ future emissions of greenhouse gases
▌ assumptions underlying global climate models that are used to generate scenarios
▌ other non-climate factors that may affect health in the future.

11.2.1 The IPCC SRES Scenarios

The Intergovernmental Panel on Climate Change (IPCC) [7] has published scenarios produced via a broad consultative process for estimating emissions of greenhouse gases over the coming century. This set of scenarios (known as the IPCC SRES scenarios) focuses on changes in economic, technological, demographic trends and energy use as major drivers for global climate change. Specifically, the scenarios explore the global and regional dynamics that may result from changes at a political, economic, demographic, technological and social level. The scenarios do not include additional climate initiatives, although their storylines do account for other 'non-climate' change policies on greenhouse gas emissions. The distinction between classes of scenario was broadly structured by defining them *ex ante* along two dimensions. The first dimension relates to the extent both of economic convergence and of social and cultural interactions across regions; the second ranges from the tendency towards satisfying economic objectives to the prevalence of environmental and equity objectives. This process therefore led to the creation of four scenario 'families' or 'clusters', each containing a number of specific scenarios.

The first 'family' of scenarios (A1 and B1) emphasizes successful economic convergence and social and cultural interaction, while the second (A2 and B2) focuses on diverse regional developments. And whereas the 'A' storylines (A1 and A2) emphasize economic development and leave only a subsidiary role for environmental and social concerns, the 'B' storylines (B1 and B2) reverse these priorities.

The first group of scenarios [A1] is characterized by fast economic growth, low population growth and the accelerated introduction of new, cleaner and more effective technologies. Under this scenario, social concerns and the quality of the environment are subsidiary to the principal objective: the development of economic prosperity. Underlying themes combine economic and cultural convergence, and the development of economic capacity with a reduction in the difference between rich and poor, whereby regional differences in per capita income decrease in relative (but not necessarily absolute) terms.

The second group of scenarios [A2] also envisages a future in which economic prosperity is the principal goal, but this prosperity is then expressed in a more heterogeneous world. Underlying themes include the reinforcement of regional identity with an emphasis on family values and local traditions, and strong population growth. Technological changes take place more slowly and in a more fragmented fashion than in the other scenarios. This is a world with greater diversity and more differences across regions.

In the third group [B1], striving for economic prosperity is subordinate to the search for solutions to environmental and social problems (including problems of inequity). While the pursuit of global solutions results in a world characterized by increased globalization and fast-changing economic structures, this is accompanied by the rapid introduction of clean technology and a shift away from materialism. There is a clear transformation towards a more service and information-based economy.

The fourth group [B2] sketches a world that advances local and regional solutions to social, economic and ecological problems. This is a heterogeneous world in which technological development is slower and more varied, and in which considerable emphasis is placed on initiatives and innovation from local communities. Due to higher than average levels of education and a considerable degree of organization within communities, the pressure on natural systems is greatly reduced.

Invariably temperature changes by the end of the 21st century, as projected by the IPCC SRES, will be the most pronounced under an A2 scenario, smallest under the B1 scenario and more intermediate under the A1 and B2 scenarios.

11.2.2 Socioeconomic Scenarios for Europe

The IPCC SRES scenarios are mainly concerned with sectoral and societal choices affecting greenhouse gas emissions. Other scenarios in existence include aspects of environmental change which extend well beyond climate change (such as, for instance, the UK Technology Foresight scenarios – see [8]).

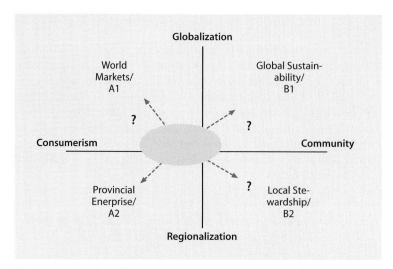

Fig. 1. The preliminary IPCC SRES storylines (A1 to B2) compared to the Foresight Scenario scenarios [9]

Work carried out before the IPCC SRES scenarios were officially published found a correspondence between the IPCC SRES scenarios and the UK Technology Foresight Scenarios storylines (see [9, 10]). Despite not being exactly the same, these storylines are comparable because they are based on a similar logic: both types of scenarios represent future changes framed by two orthogonal axes which 'frame' four future views of the world (see Fig. 1). In addition, both types of scenarios have a common origin (see [11]).

Based on the above scenarios correspondence and the logic of the four quadrants, Jordan et al. [12] (as part of the ACACIA project) elaborated a series of future socioeconomic scenarios for Western Europe[2] in the 21st century. These were intended to provide the basis for reflection on how sectors and institutions may cope with impacts of climate change. Analysts claim that if scenarios are sufficiently malleable and structured, these can inform decisions in the present by informing and encouraging reflection. The scope is to show how indeed the future is not a given, but is indeed shaped and moulded by choices and actions carried out in the present (see Box 1).

---

[2] In Jordan et al.'s study [7], Western Europe was intended to include the 15 Member States of the European Union (EU), five countries with imminent accession to the EU (Hungary, Poland, Czech Republic, Slovenia, Estonia), Cyprus, Bulgaria, Lithuania, Slovakia, Romania. Note that this definition of Western Europe is geographically more restrictive than the 'Europe' considered by the WHO

▌ Box 1: Possible futures for Western Europe (2020–2080) (adapted from [12])

▌ **A1/World Markets (WM)**: It is envisaged that enlargement of the EU beyond its current 15 member states (MS) will be instrumental to maximize economic growth by opening up new markets and access to resources in the former Eastern Block. Trade among MS increases (favoured by the Euro), at the expense of environmental quality. Paradoxically, although accelerated trade favours economic interdependence among states, social cohesion remains weak, as emphasis is placed on individual consumerist gains rather than societal welfare. Voluntary agreements and market-based instruments are preferred to regulation. The burden of urban environmental degradation is left to bear upon the poorer sectors of society – the richer wealthier citizens find refuge in the countryside increasing the demand for housing in 'healthier' areas. The liberalization of the energy markets favour generation of supply to meet demand. Profligate use of energy relies on fossil fuels as the backbone of energy provision; car and air transport continue to be the preferred means of individual mobility.

▌ **A2/Provincial Enterprise (PE)**: Under this scenario, worries about the costs of enlargement and potential federalization of the EU slow down these processes. Power and political disputes among Member States erupt; this hampers integration of European policies. Societal values become more focused on the locality, although still oriented towards consumerism and the market. Thus, states favour retaining sovereignty rather than transferring it to the EU – potential new member states are not assisted in gaining the standards adopted in existing EU countries, with consequent degradation of the natural environment. The single European market disintegrates as nations gradually erect barriers to safeguard their economies, slowing down the volume of international trade. These protectionist outlooks dampen the original strive of the EU towards social and political integration. International environmental problems are disregarded in favour of what is considered to be important nationally/regionally. Governments provide little or no assistance to degraded urban environments, which become home to the poorer sections of society, while the wealthier take refuge elsewhere. Regional/national energy resources (mainly fossil fuels) are heavily relied upon in this scenario – consumer demand is still high despite the international disarray in provision.

▌ **B1/Global Sustainability (GS)**: In this world, the EU widens its geographical borders to include new entrants providing they meet social and environmental standards. Thus, enlargement proceeds at a slower pace than under WM. At the same time, nations delegate sovereignty to the EU which becomes more federalistic, aiming to guarantee a more democratic, multilevel governance in which nations delegate sovereignty. Important functions remain in the hands of public bodies, ever more accountable to society and the public as they work within their remit of promoting sustainability: the welfare of society becomes more important than private profit. The energy and transport sectors remain subject to state intervention to ensure sustainability targets. Higher environmental quality is promoted at national and international levels. The private sector strives to meet demands for 'green consumerism' by 'cleaning up its act'. Strong social and environmental standards are imposed: cities grow as centres of culture. Long-haul tourist destinations are replaced by recreation closer to home.

Train and public transport replace the need for energy-intensive and individually comfortable modes of transport. The economy undergoes a radical change to favour local and regional activities that bring social and environmental well-being.

▌ **B2/Local Stewardship (LS):** Devolution of power to national and sub-national levels is favoured as a means to effectively address daily affairs; local authorities start exerting significant influence in the management of regional issues as they are seen to be closer to citizens and therefore more accountable. The EU is left to perform subsidiary functions that cannot be managed wholly on a national/regional scale, such as dealing with global environmental problems. Enlargement of the EU is invariably delayed. Discord and animosity between member states fray international trade within and outside the EU. The notion of economic self-sufficiency becomes increasingly important as the public demands, on subnational levels, high environmental quality linked to communitarian lifestyles in smaller towns. People dedicate some of their time voluntarily to improve their lifestyles and environments. Local production and leisure activities are preferred. 'Green' forms of energy are favoured. Disadvantaged areas are assisted for improvement. The Common Agricultural Policy is dissolved, as it cannot survive in a multi-layered political system where local prerogatives prevail. Invariably this leads to varying levels of environmental policy.

11.3 Future Health Scenarios for the European Union

Let us now consider what prospects the population of Europe may expect in the future, based on the socioeconomic and climate scenarios outlined above. These do not, by any means, span the range of all the almost infinite combinations of future possibilities. Rather, they fall within the wider framework of the ACACIA project and the IPCC; in other words, they are mainly focused on changes that may occur in Western Europe throughout the 21st century.

11.3.1 A1/World Markets (WM) Scenario Group

Under an A1/World Markets scenario, where population will continue to decline due to low fertility rates but where economic considerations and private profit prevail, it is foreseeable that variations in individuals' financial assets may increase health divides in European Member States. Thus, individuals who are part of organizations and institutions which may afford to pay medical cover for their employees, or who have invested in a private medical insurance, will be able to afford the best health care available at increasing prices. Rapid technological development and improvement will favour growth of the biotechnology and pharmaceuticals sectors in parallel with increasing reliance of private health care. At the same time, poorer fractions of society, living in more degraded inner city areas and not being able to afford treatment due to lack of state health support, may be subject to a wider variety of illnesses and diseases, in particular with regard to chronic long lasting diseases. Some of these may be do emerging strains of infectious diseases and lack of surveillance systems. Furthermore, according the 'storyline' of this scenario, only limited action to adapt to global environmental changes (such as climate) will take place with support from private funds. It is conceivable that these will be allotted to strategic areas of concern to particular vested interests (e.g. urban areas, strategic ports and commercial facilities).

11.3.2 A2/Provincial Enterprise Scenario Group

Health in the developed countries will to a large extent be left to individual choice and less of an issue for public policy. Western Europe's developed regions will increasingly invest in education and better welfare. Globally, however, the gains in health brought about by increasing economic growth and technology may be partly offset by erosion of social and environmental capital, the global division of labour, the exacerbation of the rich-poor gap between and within countries, and the accelerating spread of consumerism.

It is possible that Europe will not fare much better than other countries worldwide in terms of threats to health. Under this type of scenario, severe environmental degradation coupled with resource depletion and erosion of community support will undermine any resources devoted to disease prevention and control. The emergence of 'new' diseases for which it may not be possible to produce cures or vaccines relying exclusively on regional/ state expertise, knowledge and resources may put a great proportion of the population at risk. Certainly the state of the populations' health may be increasingly endangered by the risk of even more pronounced climate change than in a WM situation. Single parts of Europe will be affected in different ways. Even when the effects of climate change may prove beneficial for the majority of the population, other parts will be more negatively affected, necessitating urgent adaptive measures. Whereas an increased risk of food poisoning throughout warmer months of the year as a consequence of climate change could affect the population at large, an increase in winter temperatures may reduce the incidence of winter mortality of people affected by circulatory diseases and respiratory conditions [13].

11.3.3 B1/Global Sustainability Scenario Group

A central element in this scenario is a high level of environmental and social consciousness, combined with a global approach to sustainable development. Accordingly, in the developed world, life expectancy could be increased by an improved social structure (and the concomitant benefits to individual lifestyles), and by investments focused on decreasing pressure on ecological systems via the sustained management of resources. Extensive welfare nets could prevent poverty-based social exclusion.

In Western Europe, healthcare systems would be properly adjusted to an ever ageing and older population. It is conceivable that under these conditions, health for citizens in Europe shall mirror those in other developed countries worldwide, thanks to a combination of improved environmental quality, improved services for all members of its population and modified health care to cater for the needs of an increasingly ageing population. Furthermore, the 'limited' climate change projected for this scenario (due to decreased reliance on fossil fuels and increasing use of renewable resources operating in modified societal and economic systems) would not pose high threats of significant risk of 'alien' diseases affecting the population.

11.3.4 B2/Local Stewardship Scenario Group

Under this scenario, it is possible that increasing concern for environmental and social sustainability will spur local governments to enact policy-based solutions to environmental and health problems. Environmentally aware citizens will exercise a growing influence on national and local policy. There will be shift to local decision-making, with a high priority being given to human welfare, equality and environmental protection. Education and welfare programs will be widely pursued, thus, reducing mortality, and, to a lesser extent, fertility. However, the rate of implementation will vary across regions and countries. Increased expenditure on 'health' and 'environment' will be implemented first in richer countries and it will take time for poorer countries to follow.

For Western Europe as a whole, a B2 Local Stewardship world may mean improved health (in terms of provision and conditions) for part of its population, depending on when they are resident. In fact, if Europe develops into a diversity of national/regional entities relying on economic self-sufficiency and the voluntary contributions of their residents to improve their conditions, health standards and services may in part reflect the idiosyncrasy and self-determination of the situation. Undoubtedly the concern for community welfare and amelioration of the environment will provide more robust conditions than under an A2 scenario. Changes to climatic conditions will also have to be dealt with under this scenario: temperature and rainfall changes will be similar to those projected for an A1/WM situation. However, the different characteristics of this 'world' based on more integrated regional services, 'care in the community' patterns and limited travel may provide some of the desired backup to deal with changing circumstances.

11.4 **Conclusions**

The processes of globalization in today's world – which span socioeconomic change, demographic change and global environmental change – oblige us to broaden our conception of the determinants of population health. In recent decades, health conditions have improved and life expectancy has increased in almost all countries. Within many populations or communities, however, the prospects for health are being adversely affected by the diminution of social capital – i.e. the widening gaps between rich and poor and the weakening of public health systems. Furthermore, the biosphere's capacity to sustain healthy human life is beginning to be impaired by the loss of natural capital, e.g. changes in the climate system [14].

We must, therefore, be increasingly alert to the impact on health of larger-scale socioeconomic processes and systemic environmental disturbances. Our vulnerability to disease could easily be increased by changes in environmental conditions or a breakdown in public health services. As the European populations age, the burden imposed by chronic disease will inevitably escalate. Increasingly, the factors that affect health are transcending national borders and, as they do so, the health needs of diverse countries are beginning to converge. At the same time, global processes are increasingly influencing national health systems. In a world where nations and economies are increasingly interdependent, ill health in any population affects all people, rich and poor alike.

This chapter has sought to provide an example of how storyline scenarios available for Europe can be used to imagine and project possible changes to population health and health provision in the region. This exercise was intended to provide an initial indication of how similar scenarios could be developed. In order to lay the foundations of more complete and complex projections, and provide a more in-depth analysis, it would be opportune to combine qualitative (descriptive) elements with more quantitative ones in a scenario approach. To this end, it would be worthwhile widening the scope of the assessment beyond the boundaries of Western Europe, to include CCEE and NIS countries. We would recommend that this approach follows two phases. First, collection and analysis of data relating to basic social and economic indicators (e.g. GDP, population growth, state of the environment, growing and declining sectors, morbidity and mortality) of these countries. This phase would serve to understand the current health situation in the countries considered. Second, these baseline data could be used in conjunction with the logic of scenarios (either the accepted SRES or others, or new ones to reflect the reality and major issues of the area under consideration) to envision various possibilities of what the future may hold in terms of health provision and access in such a vast geographical area, composed of many national, political and cultural realities. For this effort to be credible and meaningful, it should result from a transparent process involving scientists, health practitioners, and stakeholders, using the data available combined with imagination and common sense. On a global scale, progress has already been made (see e.g. [15]).

Beware, however. Scenarios can provide the aid to decision-making, not the solution. As such they should not be entirely relied upon, but combined with other approaches. Managing health effectively will require a microapproach, taking into account the social, cultural and behavioural determinants of health (beliefs and practices that account not just for illness and poor health, but also for good health). But this microapproach will only ensure sustained better health in combination with a macroapproach. A fall in mortality rates will be brought about not only by rapid macroeconomic development, but also by policies designed to satisfy the basic needs of the majority of the population. Rapid progress through the health transition urgently requires substantial investments in education and a restructuring of health systems so that they provide better access to poor people. Simultaneously, international action needs to be undertaken to ensure that impacts of global environmental changes on health will be minimal (e.g. reductions of the emission of greenhouse gases to reduce the impacts on health caused by the anticipated climate change the coming decades).

However, the major determinants of ill health are beyond the direct control of health services [16]. Therefore, new integrated programmes are needed that combine protection of the global environment, and economic and social policies more effectively.

References

1. Parry M, Carter T (1998) Climate impact and adaptation assessment. Earthscan Publications Ltd, London, UK
2. Berkhout F, Hertin J, Jordan A (2002) Socio-economic futures in climate change impact assessment: using scenarios as 'learning machines'. Global Environmental Change 12(2):83–95
3. Murray G (2003) Alternative future paths for Ontario hospitals. In: Dopson S, Mark AL (eds) Leading health care organisations. Palgrave MacMillan, London pp 196–213
4. Martens P (2002) Health transitions in a globalising world: towards more disease or sustained health? Futures 37(7):635–648
5. Martens P, Huynen MMTE (2003) A future without health? The health dimension in global scenario studies. Bulletin of the World Health Organization 81(12):896–901
6. Kovats RS, Martens P (2000) Health. In: Parry ML (ed) Assessment of the potential effects of climate change in Europe: the Europe ACACIA project. Jackson Environment Institute, University of East Anglia, Norwich, pp 227–242
7. IPCC (2000) Emissions scenarios. Cambridge University Press, Cambridge
8. Berkhout F, Eames M, Skea J (1999) Environmental futures. UK Office of Science and Technology/Foresight Programme
9. Lorenzoni I et al (2000) A co-evolutionary approach to climate change impact assessment: part I. Integrating socio-economic and climate change scenarios. Global Environmental Change 10(1):57–68

10. Lorenzoni I et al (2000) A co-evolutionary approach to climate change impact assessment: part II. A scenario-based case study in East Anglia (UK). Global Environmental Change 10(1):145–155
11. Schwartz P (1992) Art of the long view: scenario planning – protecting your company against an uncertain future. Century business, London
12. Jordan A et al (2000) Europe in the new millennium. In: Parry ML (ed) Assessment of the potential effects of climate change in Europe: the Europe ACACIA project. Jackson Environment Institute, University of East Anglia, Norwich, pp 35–45
13. Langford IH, Bentham G (1995) The potential effects of climate change on winter mortality in England and Wales. International Journal of Biometeorology 38:141–147
14. McMichael AJ, Beaglehole R (2000) The changing global context of public health. Lancet 356:495–499
15. Huynen MMTE, Martens P, Hilderink HBM (2005) The health impacts of globalisation: a conceptual framework. Globalization and Health 1(14):1–12
16. Woodward A et al (2000) Protecting human health in a changing world: the role of social and economic development. Bulletin of the World Health Organization 78(9):1148–1155

12 Conclusions

Bettina Menne, Kristie L. Ebi

Europe's climate is changing, with increases in average temperature and changes in the hydrologic cycle experienced in recent decades. There is consensus among climate scientists that human activities are at least partly responsible for the climate change that occurred during the twentieth century (Houghton et al., 1996). The inherent inertia in the climate system means that climate will continue to change for many decades after stabilization of greenhouse gas emissions. That is, climate change will continue even as effective, worldwide mitigation of greenhouse gas emissions is achieved. With only limited measures implemented to curb greenhouse gas emissions, atmospheric concentrations continue to rapidly rise, thus, committing the Earth to more, and more intense, climate variability and change over the coming decades. Therefore, adaptation is needed to reduce population vulnerability to climate-sensitive diseases even as national and international programmes implement policies and measures to achieve mitigation goals.

Climate variability and change will affect all European nations, from increasing extreme weather events such as floods and droughts, to providing weather conditions suitable for the spread or retreat of a variety of climate-sensitive diseases. Public health can design and implement policies and measures to reduce projected impacts. The approaches taken will include modifying existing programmes to take increasing climate variability and change into account; implementing disease-specific policies and measures from other countries and regions when a disease changes its geographic range; re-implementing programmes that have been neglected; and developing new policies when new threats arise (Ebi et al., 2005).

Although these approaches hold promise for the reduction of current and future burdens of climate-sensitive diseases, relatively little has been done in public health to incorporate projections of climate variability and change into policy planning. Barriers include a limited sense that climate change matters in public health, few exposure-response relationships for climate-sensitive diseases that can be used to measure current and project future impacts, and the political will to tackle the problem (adapted from Last, 1998).

As detailed throughout this volume, the cCASHh project has improved our understanding of the current effects of weather and climate variability on population health in Europe, projected future health impacts, identified populations who are particularly vulnerable to current and future climate, reviewed current adaptation approaches and identified barriers to improving adaptive capacity, and estimated the benefits of specific adaptation options. This chapter summarizes the results by answering the following questions:

- Which methods can be used to assess the health impacts of and adaptation to climate change?
- What are the climatic risks Europe might be facing?
- What can be learned from observed health impacts and vulnerabilities?
- What strategies, policies and measures are currently available to reduce impacts?
- What are the costs?
- What are the projected future health impacts?
- Which policy responses need to be strengthened or developed?
- What factors influence adaptive capacity?
- What additional research is needed?

12.1 Which Methods Can Be Used to Assess the Health Impacts of and Adaptation to Climate Change?

As discussed in Chapter 3, no single method is appropriate to assess all climate-related health risks and to identify effective adaptation policies and measures. Approaches to climate impact assessments are often characterized as top-down or bottom-up (Carter et al., 1994; Feenstra et al., 1998). Top-down approaches begin with an understanding of system behaviour, and then project how the system could change under different scenarios of climate change in order to estimate impacts. Bottom-up approaches focus on quantifying current risks to a particular system, which are then aggregated to estimate future impacts under climate change scenarios. The relative contribution of top-down and bottom-up approaches in vulnerability assessments depends on the purpose and scope of the study and the availability of relevant data and other resources. The cCASHh project primarily used bottom up quantitative and qualitative assessment techniques, based on the assessment guidelines from Kovats et al. (2003). The methods used are theoretically transferable to other regions, but would need to be adapted to the specific questions raised and the social, political, economic, and other context of the region or country conducting the assessment. Limited information was available through the cCASHh project on adaptation options.

12.2 What Are the Climatic Risks Europe Might Be Facing?

Clear trends in European climate that have emerged over the last 50 years include increases in minimum and maximum temperatures, changes in precipitation characteristics in certain locations, and increases in the number of extreme events such as high temperatures, heavy precipitation and persistent dryness. These changes have coincided with the increase in global mean temperature. Although there is still debate concerning attribution of these trends to anthropogenic climate change, there is mounting evidence that Europe's climate may be responding to increases in the intensity of the global greenhouse effect.

Mean climatic changes that can be expected in a warmer Europe include a possible expansion of Mediterranean and maritime temperate climates with a commensurate reduction in continental temperate climates. Drying is likely over southern Europe, while northern Europe is likely to become milder and wetter. In addition, there will be a shift in the climatology of extreme events, although there is high uncertainty about how much climate variability may increase over the next decades. Changes in extreme events may be experienced as changes in the frequency of events and/or as changes in their intensity or magnitude. Return intervals for heat-waves, periods of prolonged dryness and drought, heavy precipitation, storminess, and storm surges are projected to decrease. This increasing climate variability will challenge public health systems and increase the burden of climate-sensitive diseases unless policy-makers take a proactive, anticipatory approach to designing policies and measures to reduce current and future burdens of climate-sensitive diseases (McGregor, Chapter 2 in this book; McGregor, 2005).

12.3 What Can Be Learned from Observed Health Impacts and Vulnerabilities?

The cCASHh project assessed the health impacts of heat-waves and floods, as well as the distribution and occurrence of vector-, rodent- and food-borne diseases.

12.3.1 Heat-waves

There is a direct relationship between mortality and temperature that differs by climatic zone and geographic area. High ambient temperature is associated with mortality from heat stroke, cardiovascular, renal and respiratory diseases, metabolic disorders etc. Few studies have examined the impacts of heat-waves on morbidity; there is some evidence for increased emergency admissions for renal disease, and for increased respiratory disease in the elderly and children. Although everyone is at risk, the most vulnerable are those over the age of 65 years, particularly if they have certain chronic diseases, take certain medications and are not physically fit. Analyses in Budapest, Hungary, confirmed that the timing of heat-waves is important, with events in early summer associated with higher mortality than late season heat-waves. The potential toll of heat-waves was dramatically demonstrated in the 2003 heat-wave that caused 27000 to 40000 excess deaths.

12.3.2 **Flooding**

Floods are the most common natural disaster causing loss of life and economic damage in Europe. The main areas of Europe prone to frequent flooding are the Mediterranean coast, the dyked areas of the Netherlands, the Shannon callows in central Ireland, the north German coastal plains, the Rhine, Seine and Loire valleys, some coastal areas of Portugal, the Alpine valleys, the Po valley in Italy, and the Danube and Tisza valleys in Hungary (Estrela et al., 2001). The frequency of great floods increased during the twentieth century, underscoring the need for measures to prevent their negative health impacts.

Adverse health impacts arise from a combination of some or all of the following factors: characteristics of the flood event itself (depth, velocity, duration, timing etc.); amount and type of property damage and loss; whether flood warnings were received and acted upon; the victims' previous flood experience and awareness of risk; whether or not flood victims need to relocate to temporary housing; the clean-up and recovery process, and associated household disruption; degree of difficulty in dealing with builders, insurance companies etc.; pre-existing health conditions and susceptibility to the physical and mental health consequences of a flooding event; degree of concern over a flood recurrence; degree of financial concern; degree of loss of security in the home; and degree of disruption of community life (Ebi et al., Chapter 5 in this book). Adverse health impacts associated with flooding include direct physical effects, such as drowning and injuries, and indirect effects caused by other systems damaged by the flood (such as water- and vector-borne diseases, acute or chronic effects of exposure to chemical pollutants released into floodwaters, food shortages etc.). The risk of communicable disease outbreaks following flooding is small in industrialized countries due to effective water treatment and sewage pumping, safe drinking water, and public health infrastructure. However, national ministries and governments may need to take additional action on a case-by-case basis.

The physical effects largely manifest themselves within weeks or months following flooding, and are largely related to the shock of the flood, damp or dusty living conditions, and the recovery process.

Mental health effects directly attributable to the experience of being flooded or indirectly during the recovery process tend to be much longer lasting and can be worse than the physical effects.

In general, those most at risk include the disadvantaged, children, and older adults. However, there is a lack of detailed data on those affected by flooding in Europe, making conclusions uncertain.

12.3.3 Vector- and Rodent-borne Disease

The associations of weather and climate were assessed with the following diseases: leishmaniasis, Lyme borreliosis, tick-borne encephalitis (TBE), malaria, Hantavirus infection, and West Nile virus infections. Although there is considerable evidence that climate is an important determinant of the geographic range of vectors, data are insufficient to conclude that recent climate change has resulted in increased disease.

It is likely that climate change has already led to changes in *I. ricinus* populations in Europe. In general, existing data are not reliable enough to allow comparisons over time and space of changes in tick prevalence and disease incidence on a pan-European level. However, studies from specific areas based on particularly reliable long-term data sets have shown that ticks have spread into higher latitudes (observed in Sweden) and altitudes (observed in the Czech Republic) in recent decades and have become more abundant in many places (Tälleklint and Jaenson, 1998; Daniel et al., 2003; Lindgren et al., Chapter 6.2 in this book).

Also over the past few decades, the incidence of Lyme borreliosis and other tick-borne diseases have increased in some European countries. Although this may be the result of improved reporting and other factors, studies from regions with reliable long-term surveillance data show that changes in transmission intensity and seasonality have occurred that may be due to the climatic factors that help determine tick abundance (Lindgren, 1998; Lindgren and Gustafson, 2001; Daniel et al., 2004).

There is no convincing evidence that recent changes in climate have affected the re-emergence of malaria in the WHO European region, but studies suggest a possible link between temperature increase and increases in the geographic range and intensity of outbreaks of malaria (Kuhn, Chapter 6.4 in this book).

While the evidence is not compelling that sandfly and visceral leishmaniasis distributions in Europe have altered in response to recent climate change, this hypothesis cannot be discounted (Lindgren et al., Chapter 6.1 in this book).

12.3.4 **Foodborne Disease**

Results of analyses conducted for the cCASHh project show that, in general, cases of salmonella increase by 5–10% for each degree increase in weekly temperature, for ambient temperatures above about 5 °C. The effect of temperature is most apparent in the week before illness, indicating that inappropriate food preparation and storage around the time of consumption is most important. Rates of salmonellosis are declining in most countries in Europe, suggesting that improvement of current measures will be an effective adaptation to controlling salmonella under warmer climate conditions. There was no evidence of a strong role of temperature in the transmission of campylobacteriosis. Some notable outbreaks of waterborne disease (such as cryptosporidiosis) have been associated with heavy rainfall events. Waterborne diseases and other foodborne diseases were not examined because of a lack of data (Kovats et al., in Chapter 7).

12.4 **What Strategies, Policies and Measures Are Currently Available to Reduce Impacts?**

Although the 2003 heat-wave and the 2002 flood do not prove that the world is getting hotter or the weather more extreme, the impacts of these events highlighted some of the shortcomings in the existing public health responses, such as the lack of mechanisms to predict the events, to rapidly detect the health impacts, and the lack of knowledge of effective prevention measures (WHO and EEA, 2004).

12.4.1 **Heat-Waves**

Prevention of heat-waves includes early warning systems, rapid health system preparedness and response, urban planning, and housing improvements. A survey of meteorological agencies on heat warnings found that, with the exception of Lisbon and Rome, no warning systems were in place before 2003. A survey following the 2003 heat-wave found that cities in Italy, Spain, France, the United Kingdom, Germany, and Hungary have started or developed warning systems. A survey of Ministries of Health found that few have put heat-wave prevention and response plans into place that include strategies for identifying vulnerable subgroups, health monitoring, population advice, and financial incentives to encourage vulnerability reduction. An important gap is the lack of formal collaborative agreements with the national weather services (Kosatsky, in Chapter 8.1). Further actions and research on this topic have been endorsed in the European Environment and Health action plan and are partially addressed in integrated research projects (Ebi et al., in Chapter 8).

Results from the cCASHh project provided information on appropriate public health measures during a heat-wave; however, the effectiveness of measures needs to be evaluated. In 2004, France, Italy, Portugal and the United Kingdom began to include heat-related mortality in computerized surveillance systems.

12.4.2 Flooding

A survey to Ministries of Health found that although all of the respondent countries had emergency intervention plans, none had strategies to prevent long-lasting health impacts of flooding or financial incentives for citizens to increase their ability to resist the effects of floods. Items commonly included in the flood impact prevention plans are assessment of environmental (flood plains) and social (ill or disadvantaged populations) vulnerabilities, communications strategies, and zoning regulations. European flood warning is available through several systems; however, the public health community is rarely informed in advance. Increased cooperation across countries through international river commissions should facilitate improved warnings of flooding risks, including implementation of the UNECE and EC guidance on flood prevention. But if the projected increase in the rate and intensity of flooding events occurs, some regions will need to assess the siting of housing and infrastructure, and strengthen insurance provision and public education.

12.4.3 Vector- and Rodent-borne Disease

The measures currently available to control vector- and rodent-borne diseases are disease specific; however, they can be classified into the following: diagnosis and treatment, vaccination, vector control, reservoir host control, information and health education, and disease surveillance and monitoring. The level of implementation of these measures depends on the severity of disease, the number of people affected, the level of standardization of diagnostic procedures, the effectiveness of control measures, technological advances, and health system awareness. At the EU level, currently the following climate-sensitive diseases are monitored by the recently established networks, with different degrees of standardization and completeness: campylobacteriosis, cryptosporidiosis, leptospirosis, salmonellosis, meningococcal disease, pneumococcal infections, cholera, malaria and viral hemorrhagic fevers. Leishmaniasis and Hantavirus are not monitored. In some instances, the coordination and collaboration between vector control activities, including veterinary surveillance, and the health sector could be strengthened.

12.4.4 **Foodborne Disease**

The most important mechanisms to prevent food- and waterborne diseases are surveillance and monitoring, microbiological risk assessment, risk management and risk communication. The effectiveness of these programmes varies across countries, providing opportunities for decreasing current burdens of foodborne diseases. Including projected climate change into surveillance and monitoring should increase future adaptive capacity if increasing temperatures increase the rate of diseases such as salmonella.

12.5 **What Are the Costs?**

The contingent valuation surveys (Chapters 9 and 10) on reducing the risk of dying in a heat-wave, estimated the value of a statistical life (VSL) for Italy and the Czech Republic. The mean VSL was about € 3.7 million for Italy and about € 0.9 million in the Czech Republic. The VSL depended on gender, age and health status, and was much lower among older adults but higher in ill adults.

12.6 **What Are the Projected Future Health Impacts?**

12.6.1 **Heat-Waves**

Climatologists now consider it "very likely" that human influence on the climate system at least doubled the risk of a heat-wave such as that experienced in 2003 (Stott et al., 2004). This could mean that the 2003 heat-wave-associated mortality could have been an early signal of the health impacts of climate change. Both short- and long-term adaptation policies and measures are needed to reduce vulnerability to heat-waves.

12.6.2 **Floods**

Although the extent to which climate change will affect the frequency and intensity of extreme weather events is still a subject of debate in the scientific community, there is no question that there are many anthropogenic influences that affect the impacts of floods. Changes in land use and hydrology create multiplying effects when the natural or "ecological" protection has disappeared. Examples are reduced wetland buffering areas, straightening of rivers, forestry fragmentation and logging (Kirch, Menne et al., 2005). Many locations are in flood prone areas or along the coasts, which guarantees that flooding will continue to cause adverse physical and mental health impacts.

12.6.3 Vector- and Rodent-borne Disease

The distribution of *Ixodes ricinus*, tick-borne encephalitis, and Lyme borreliosis are likely to continue to change in response to changes in temperature and precipitation. Latitudinal and altitudinal shifts of *Ixodes ricinus* ticks have been observed. To date, attempts to predict *I. ricinus* or Lyme borreliosis distributions have largely been limited to statistical "pattern-matching" models, with different results. Only one attempt to simulate the impact of climate change on the distribution of TBE in Europe has been reported. The simulations suggested TBE would invade progressively higher altitudes and latitudes in Europe during the 21st century, but that TBE would be eliminated from more southern currently endemic countries (such as Hungary, Switzerland, Croatia, Slovenia) by the 2020s (Randolph, 2001).

Based on the results of the extended reviews in this book it seems most probable that climate change in Europe will:

- facilitate the spread of Lyme borreliosis and TBE into higher latitudes and higher altitudes
- contribute to an extended and more intense Lyme borreliosis and TBE transmission season in some areas and
- diminish the risk of Lyme borreliosis, at least temporarily, in locations with repeated droughts or severe floods.

There is considerable potential for climate-driven changes in Leishmaniasis distribution. Sandfly vectors already have a wider range than that of *L. infantum*, and imported dogs infected with *L. infantum* are common in central and northern Europe (Gothe et al., 1997). Once conditions make transmission suitable in northern latitudes, these imported cases could act as a source of infections to permit the development of new endemic foci. Although only a small percentage of human *L. infantum* infections in Europe currently cause visceral Leishmaniasis, the large population with sub-clinical infections, as well as HIV, remain a concern as they are at risk from immunosuppression. An additional public health threat is the ability of several of the European sandfly species to be potent vectors of viruses. Climate-induced changes in sandfly abundance may thus increase the risk for the emergence of new diseases in the region.

The WHO (1999) pointed out that "the increasing number of cases of imported malaria has raised the question of the risk of the re-introduction of malaria into some areas of Europe". Given the remarkably small number of secondary cases in contrast to the high number of imported cases in most of Europe (see Chapter 6), it appears that under current conditions this risk is extremely low. Probably the greatest risk is in those eastern European countries, where per capita health expenditure is relatively low (e.g. US$ 64/year in Moldova in 2000), so that health services are less efficient at detecting and treating imported malaria cases.

Although several models, by focusing on the biological determinants of the potential limits of malaria transmission, predicted a potential increase of malaria for Europe (Martens et al., 1999; Lindsay and Thomas, 2001; van Lieshuit, 2004), Rogers and Randolph (2000) and Kuhn (2003) projected that the risk of malaria for Europe is very low under current socioeconomic conditions. However, none of the models address possible changes in the environment, changes in socioeconomic levels, and changes in the efficiency of national health services across Europe to rapidly detect and treat malaria cases. Given the climatic associations with malaria transmission described above, the question remains whether climate change could lead to a substantial risk of malaria re-emergence in Europe.

Further research is needed to understand the potential role of climate change on other vector-borne diseases, including Hantavirus and West Nile virus.

12.6.4 Food- and Waterborne Disease

It was estimated that temperature influences the transmission of infection in about 35% of cases of salmonellosis in England and Wales, Poland, the Netherlands, the Czech Republic, Switzerland and Spain (Kovats et al., 2004b). Diseases associated with water are varied and cover multiple environmental pathways. At present, there is insufficient evidence to estimate the role of rainfall extremes in disease outbreaks in Europe, and it is not possible to assess whether climate change would have an overall impact on the burden of waterborne disease in Europe.

12.7 Which Policy Responses Need to Be Strengthened or Developed?

The present state of public health reflects the success or otherwise of the policies and measures designed to improve lifestyle, education, the environment, health care services and systems, and the economy. Evidence over the past few years clearly demonstrates that additional policies and measures are needed at the pan European, national and local levels to reduce the burden of climate-sensitive diseases. The design and implementation of incremental adaptation can draw on the considerable experience in public health with responding to both gradual long-term changes (such as disease rates changing over decades) and stimuli with severe consequences over short time scales (such as disease outbreaks or extreme events) (Ebi et al., 2005).

Because climate change will continue for the foreseeable future and because adaptation to these changes will be an ongoing process, active management of the risks and benefits of climate change needs to be incorporated into the design, implementation and evaluation of disease control strategies and policies across the institutions and agencies responsible for maintaining and improving population health (Ebi et al., 2005). In addition, understanding the possible impacts of climate change in other sectors could help decision-makers identify situations where impacts in another sector, such as water or agriculture, could adversely affect population health.

Incremental changes in current measures may be needed if changes in climatic and socioeconomic variables create favourable conditions for increasing the range or intensity of climate-sensitive diseases, or if thresholds are crossed. Thresholds could be crossed either because a disease was close to its boundary conditions or because there was a sudden and/or large change in prevailing weather conditions. If a disease emerges in a country outside its normal range, then measures implemented in countries where the disease is endemic may be modified to reflect the local context. In some cases, new diseases or conditions will emerge that will require the development of new policies and measures.

Responses at the Pan European Level

Examples of responses at the pan European level include recommendations on extreme weather events in the 4th Ministerial Conference, recognition of the potential impacts of climate change in the Convention on the Protection and Use of Transboundary Watercourses and International Lakes, and development of surveillance systems.

Recommendations on extreme weather events were elaborated by the cCASHh team and discussed at a preparatory meeting to the 4th Ministerial Conference for Environment and Health, in Bratislava, February 2004, with representatives of Ministries of Health and the Environment, international organizations, and the European Commission. The recommendations were further adopted at the fourth Ministerial Conference for Environment and Health (Menne, 2005). In summary:

> it was recognized that there is a need to improve the understanding of the regional and national burden of disease due to weather and climate extremes, to identify effective and efficient interventions, such as early warning systems, surveillance mechanisms and crisis management mechanisms, enhance effective and timely coordination and collaboration among public health authorities, meteorological services and agencies (national and international), emergency response agencies and civil society as well as improving public awareness of extreme weather events, including actions that can be taken at individual, local, national and international levels to reduce impacts.

Within the Convention on the Protection and Use of Transboundary Watercourses and International Lakes of the Economic Commission for Europe, there is a recommendation to implement evidence-based measures to limit the impacts of flooding on human health, the environment, and ecosystems. Currently decision-making before, during and after floods is fragmented throughout several bodies in some of the European countries. This differentiation of power in decision-making may make it difficult to for decisions to be reached rapidly, particularly in an emergency situation.

Because current prevention measures may not be sufficient to cope with potential changes in the range or intensity of climate-sensitive diseases, the cCASHh project identified one pan European response is the improvement of active disease surveillance systems, including the capacity of rapid detection of new diseases and effective communication of risks and responses. Since 1999, the European Commission has managed a Communicable Disease Network. However, there is a need for a substantial reinforcement of this system if the EU is to be in a position to control communicable diseases effectively and to take into account early signals of changes and threats, including how climate change could affect disease rates.

Responses at National Levels

National, comprehensive flood and heat-wave mitigation plans can save lives. In the case of floods, risk managers should: develop maps of potential flood risks, including the location of chemical and nuclear plants and other sources of potentially hazardous materials; analyze vulnerability of communities potentially affected by a flood; develop an inventory of resources that could be mobilized during and following a flood; and establish a regional or national coordination mechanism that includes the health sector to deal with floods.

In general, the less familiar a population is with a particular health risk, the less effective the risk is controlled. The 2003 heat-wave and the increasing rates of Lyme borreliosis and salmonella raise the question of how familiar communities and populations are with the risks. Therefore, there is a need to better communicate preventive actions that can be taken by members of the population to reduce their risk of climate-sensitive diseases.

12.8 What Factors Influence Adaptive Capacity?

Good health is essential to sustainable development. While the health of the 879 million people in the WHO European Region has, in general, improved over time, inequalities between the 52 Member States in the Region and between groups within countries have increased. In addition to the east–west gap in health, differences in health have increased between socio-economic groups within many countries.

The survey of experts concerning the determinants of individual countries' adaptive capacity found that income, equality, type of health care system, and quick access to information were the most important factors thought to influence adaptive capacity. With the information gathered, an index of adaptive capacity was constructed for the countries in the WHO European Region. The results demonstrated that there is a wide range of adaptive capacity, from the lowest in Albania, to the highest in Luxembourg. On average, the Russian Federation scores poorly, the ten newest members of the European Union are in the mid-range, and most of the 15 original members have the highest adaptive capacity scores, owing to their high incomes, universal health care coverage, and high access to information.

12.9 **What Additional Research is Needed?**

The cCASHh project increased our understanding of the potential impacts of climate change on health in Europe. More, however, needs to be understood in order to more effectively protect population health. Some of the knowledge gaps are summarized in Table 1. In general, greater understanding is needed of the current relationships between weather/climate and disease outcomes, how other drivers of those outcomes interact with weather, how the range and intensity of climate-sensitive diseases might change with a changing climate, and what response options can be implemented to reduce the current and projected burdens of disease (including identification of where and when response options should be implemented).

For example, many knowledge gaps exist of the health impacts of heat-waves, including characterization of the relationship between heat exposure and a range of health outcomes, understanding interactions between harmful air pollutants and hot weather, analysis of the morbidity associated with heat-waves, how to effectively communicate the necessary information to at-risk groups, and how to motivate appropriate behavioural changes during heat-waves. In order to reduce the health impacts of future heat-waves, fundamental questions need to be addressed, such as can a heat-wave be predicted, can it be detected, can it be prevented, and what can be done. Criteria need to be developed for how to identify regions that are particularly vulnerable.

Understanding the impacts of and responses to floods is limited by the lack of a comprehensive European database on flood impacts. Mapping of flood risks would be a useful first step. In addition, there is no common understanding on what best measures are needed. Many health care and public institutions and industries are located in flood prone areas. The health sector needs to be involved into activities targeted to develop early

Table 1. Research gaps in understanding projected health impacts of and adaptations to climate variability and change

Subject	Gaps	Audience
Vulnerability and adaptation assessments	▓ Better understanding of the associations between weather/climate and health outcomes ▓ Development of indicators for detecting early effects of climate change on human health ▓ Health scenarios ▓ Improved downscaled climate scenarios, including more accurate projections of extreme events ▓ Monitoring and evaluation of implemented adaptation options to determine their effectiveness ▓ Appropriate incorporation of health impacts into integrated assessment models ▓ Quantification of the ancillary benefits of climate policies, including stabilization of greenhouse gases, on human health, particularly the benefits of reducing outdoor air pollution ▓ Integration of health into global monitoring mechanisms for the early detection of changes ▓ Effective communication of climate-related health risks to individuals and communities ▓ Identification of additional adaptation options to reduce current and future vulnerability to climate-sensitive diseases	Health sector; climate change community; insurance sector; international law; remote sensing communities
Extreme weather events	▓ Better understanding of the health impacts of extreme events ▓ Identification of thresholds when population vulnerability will significantly increase ▓ Estimation of the costs of adverse health outcomes associated with extreme weather events ▓ Estimation of the costs and benefits of reducing the health impacts of extreme weather events	Health sector, climate change community; insurance sector; international law; economics; international disaster community; civil protection agencies

Table 1 (continued)

Subject	Gaps	Audience
Heat-waves	█ Better understanding of the impacts on morbidity, particularly in children and the elderly █ Better understanding of the socio-economic and clinical risk factors for heat-related morbidity and mortality █ Understanding the synergies between heat-waves and air pollution, and the most vulnerable groups █ Effectiveness of public health interventions for reducing heat-related morbidity and mortality, including the effectiveness of heat event warning systems █ Identification of lessons learned from community intervention programmes █ Development of more effective risk communication █ Surveillance programs for early detection of population health impacts █ Estimation of the risks and benefits of air conditioning █ Better understanding the impacts on occupational health	Health sector; weather services; insurance sector; energy sector; economics; international disaster community; civil protection agencies
Floods	█ Analysis of flood morbidity and mortality using routine data sources or pre-existing cohorts █ Better understanding of the impacts of floods on European health care systems █ Estimation of the effectiveness of flood early warning systems █ Estimations of the costs and benefits of preventing morbidity and mortality from floods	Health sector; river basis authorities; insurance sector; international environmental law; international disaster community; civil protection agencies
Water- and foodborne diseases	█ Better understanding of the impacts of climate change on domestic water supply in Europe █ Better understanding of how weather and climate could affect the transmission of pathogens █ Better understanding of the associations between weather (particularly extreme events), water quality, and measurable health outcomes █ Incorporation of projected climate variability and change into food safety regulations and standards	Health sector; regulatory agencies; international law

Table 1 (continued)

Subject	Gaps	Audience
Vector-borne diseases	▪ Development of quantitative models of vector-borne disease transmission ▪ Better understanding of the factors that determine the geographic distribution of vectors (including rodents), and how distributions may change with climate change ▪ Better understanding of the role of climate in vector-host relationships, particularly in endangered ecosystems ▪ Creation of additional international integrated surveillance networks, including for potentially emerging or re-emerging pathogens ▪ Standardization of methods for vector control activities, including monitoring and evaluation of results ▪ Development of scenarios that incorporate the drivers of vector-borne diseases	Health sector; climate change community; international health regulation
Other infectious diseases	▪ Better understanding of the potential risk of introduction of new or re-emerging diseases	Health sector; security; international health regulation

warning systems. As with heat-waves, there are questions about how to best communicate the risks and stimulate responsible behaviours. Cost-benefit analyses would further help decision-makers allocate resources.

The effects of climate change on vector-, rodent-, food- and waterborne diseases are connected to other societal and developmental factors, such as migration, population densities, land-use, and ecological changes as well as trade of food and goods. An understanding of these complex mechanisms is needed.

12.10 Conclusions

The cCASHh project has provided valuable information to EU member states regarding the potential impacts of climate variability and change on human health, but many uncertainties remain. For example, how significant is the expected increase in a particular health risk due to global climate change, compared with current risk levels. This question is in particular important with regard to Article 2 of the UN Framework Convention on Climate Change, which concerns what constitutes dangerous climate change.

There are many uncertainties around climate projections and future risk projections. The scale of these projections is currently much broader than the scale at which adaptation measures could occur. As long as projections of regional increases in health risks are less certain, no regret adaptation strategies are needed, by for example focusing on improved monitoring, early warning, surveillance, early disease detection and research. We do not know how large the risks (in terms of additional costs) will be of taking early actions compared with the risks (in terms of additional disease burden) of acting later. However, this does not preclude from taking actions in anticipation of the health impacts of climate change. Numerous research questions have been identified and must be prioritized to answer the most important gaps. Research on future adaptation options might need to address costs, benefits, effectiveness and barriers of implementation. The lower the expected reduction of health impacts through adaptation now and the higher the health risks in the near future, the more important it is to act sooner rather than later.

References

Carter TR, Parry ML et al (1994) IPCC Technical Guidelines for Assessing Climate Change Impacts and Adaptations. Part of the IPCC Special Report to the First Session of the Conference of the Parties to the UN Framework Convention on Climate Change. Dept of Geography, University College London

Daniel M, Danielova V et al (2003) Shift of the tick Ixodes ricinus and tick-borne encephalitis to higher altitudes in central Europe. Eur J Clin Microbiol Infect Dis 22(5):327–328

Daniel M, Danielova V et al (2004) An attempt to elucidate the increased incidence of tick-borne encephalitis and its spread to higher altitudes in the Czech Republic. Int J Med Microbiol 293(Suppl 37):55–62

Ebi K et al (2005) Integration of public health with adaptation to climate change. Lessons learned and new directions, Taylor and Francis, 295 pages

Estrela M, Dimas, Marcuelle, Rees, Cole, Weber, Grath, Leonard, Ovesen, Feher (2001) Sustainable water use in Europa. Environmental Issue Report. EEA, Copenhagen, European Environment Agency, p 21

Feenstra J, Burton I et al (eds) (1998) Handbook on Methods for Climate Change Impact Assessment and Adaptation Strategies. UNEP, Nairobi, Kenya

Gothe R, Nolte I et al (1997) Leishmanioses of dogs in Germany: epidemiological case analysis and alternative to conventional causal therapy. Kleintierpraxis 25(1):68–73

Houghton J, Ding Y et al (eds) (2001) Climate Change 2001: The Scientific Basis. Contribution of Working Group I to the Third Assessment Report of the Intergovernmental Panel on Climate Change. Cambridge University Press, Cambridge, UK

Kirch W, Menne B et al (2005) Extreme weather events and public health responses. Springer, Berlin Heidelberg New York, p 360

Kovats RS, Haines A et al (1999) Climate change and human health in Europe. Bmj 318(7199):1682–1685

Kuhn K, Campbell-Lendrum D et al (2003) Malaria in Britain: past, present, and future. Proc Natl Acid Sci USA 100(17):9997–10001

Kuhn K, Campbell-Lendrum D et al (2002) A continental risk map for malaria mosquito (Diptera: Culicidae) vectors in Europe. Journal of Medical Entomology 39(4):621–630

Lindgren E (1998) Climate and tick-borne encephalitis in Sweden. Conservation Ecology 2(1):5–7

Lindgren E, Gustafson R (2001) Tick-borne encephalitis in Sweden and climate change. Lancet 358(9275):16–18

Lindsay SW, Thomas CJ (2001) Global warming and risk of vivax malaria in Great Britain. Global Change and Human Health 2(1):80–84

Martens P, Kovats RS et al (1999) Climate change and future populations at risk from malaria. Global Environmental Change 9:89–107

Randolph S (2002) Predicting the risk of tick-borne diseases. International Journal of Medical Microbiology 291:6–10

Rogers DJ, Randolph SE (2000) The global spread of malaria in a future, warmer world. Science 289:1763–1765

Talleklint L, Jaenson TG (1998) Increasing geographical distribution and density of Ixodes ricinus (Acari: Ixodidae) in central and northern Sweden. J Med Entomol 35:521–526

WHO (2001) Floods: Climate Change and Adaptation Strategies for Human Health. WHO, Copenhagen, EUR/01/503 6813

WHO/EEA (2004) Report of the meeting "Extreme weather and climate events and public health responses", Bratislava, Slovakia, 9–10 Feb 2004. World Health Organization, Copenhagen

Annexes

Annex 1:
cCASHh Articles, Contributions and Reports

Ahern MJ, Kovats RS et al (2005) Global health impacts of floods: epidemiological evidence. Epidemiological Reviews 27:36

Campbell-Lendrum DH, Kovats RS et al (2004) Summing up the health impacts of climate change. Climatic Change

Daniel M, Danielova V et al (2003) Shift of the tick Ixodes ricinus and tick-borne encephalitis to higher altitudes in central Europe. Eur J Clin Microbiol Infect Dis 22(5):327–328

Daniel M, Danielova V et al (2004) An attempt to elucidate the increased incidence of tick-borne encephalitis and its spread to higher altitudes in the Czech Republic. Int J Med Microbiol 293(Suppl 37):55–62

Danielova V (2002) Natural foci of tick-borne encephalitis and prerequisites for their existence. Int J Med Microbiol 291(Suppl 33):183–186

Danielova V, Daniel M et al (2004) Prevalence of Borrelia burgdorferi sensu lato genospecies in host–seeking Ixodes ricinus ticks in selected South Bohemian locations (Czech Republic). Cent Eur J Public Health 12(3):151–156

Danielova V, Holubova J et al (2002) Tick-borne encephalitis virus prevalence in Ixodes ricinus ticks collected in high risk habitats of the south-Bohemian region of the Czech Republic. Exp Appl Acarol 26(1/2):145–151

Danielova V, Holubova J et al (2002) Potential significance of transovarial transmission in the circulation of tick-borne encephalitis virus. Folia Parasitol (Praha) 49(4): 323–325

Few R, Ahern MJ et al (2004) Floods, health and climate change: a strategic review. Working Paper, Tyndall Centre for Climate Research 63

Hajat S, Ebi KL et al (2003) The human health consequences of flooding in Europe and the implications for public health: a review of the evidence. Applied Environmental Science and Public Health 1(1):13–21

Hajat S, Kovats RS et al (2002) Impact of hot temperatures on death in London: a time series approach. Journal of Epidemiology and Community Health 56:367–372

Johnson H, Kovats RS et al (2005) The impact of the 2003 heatwave on mortality and hospital admissions in England. Health Statistics Quarterly 25:6–12

Kirch W, Menne B et al: Extreme weather events and public health responses. Springer, Berlin Heidelberg New York, 360 pages

Kovats RS (2003) Climate change, temperature and foodborne disease. Eurosurveillance Weekly 7(49)

Kovats RS (2004) Will climate change really affect our health? Results from a European assessment. Journal of the British Menopause Society 139–144

Kovats RS, Campbell-Lendrum D et al (2001) Early effects of climate change: do they include changes in vector borne diseases? Philosophical Transactions of the Royal Society B 356:1057–1068

Kovats RS, Ebi KL et al (2003) Methods of assessing human health vulnerability and public health adaptation to climate change. WHO/WMO/Health Canada, Copenhagen

Kovats RS, Edwards SJ et al (2005) Climate variability and campylobacter infection: an international study. International Journal of Biometeorology 49(4):207–214

Kovats RS, Edwards SJ et al (2004) The effect of temperature on food poisoning: a time-series analysis of salmonellosis in ten European countries. Epidemiol Infect 132(3):443–453

Kovats RS, Koppe C (2004) Heatwaves: past and future impacts on health. In: Ebi KL, Smith J, Burton I (eds) Integration of Public Health with Adaptation to Climate Change: Lessons learned and New Directions. Lisse, Taylor & Francis Group

Kovats RS, Koppe C (2005) Heatwaves past and future impacts on health. In: Ebi KL, Smith J, Burton I (eds) Integration of Public Health with Adaptation to Climate Change: Lesons learned and New Directions. Taylor & Francis Group, Lisse, The Netherlands, pp 136–160

Kovats RS, Menne B et al (2003) National assessments of health impacts of climate change: a review. In: McMichael AJ, Campbell-Lendrum D, Ebi KL et al (eds) Climate Change and Human Health: Risks and Responses. WHO, Geneva, pp 181–200

Kovats RS, Wolf T et al (2004) Heatwave of August 2003 in Europe: provisional estimates of the impact on mortality. Eurosurveillance Weekly 8(11)

Kovats RS, Edwards S et al (2003) Environmental temperatures and food-borne disease: time-series analysis in 8 European countries. Epidemiology 14(5):S15–S16

Kriz B, Benes C et al (2004) Socio-economic conditions and other anthropogenic factors influencing tick-borne encephalitis incidence in the Czech Republic. Int J Med Microbiol 293(Suppl 37):63–68

Materna J, Daniel M et al (2005) Altitudinal distribution limit of the tick Ixodes ricinus shifted considerably towards higher altitudes in Central Europe: results of the three years monitoring in the Krkonose Mountains (Czech Republic). Cent Eur J Public Health

Materna J, Danielova V et al (2003) Results of monitoring of Ixodes ricinus tick distribution on the territory of Krkonose National Park. Journal Krkonose 1(19)

McMichael AJ, DH C-L et al (eds) (2003) Climate Change and Human Healh, Risk and Responses. World Health Organization, Geneva

Menne B (2000) Floods and public health consequences, prevention and control measures. UN 2000 (MP.WAT/SEM.2/1999/22)

van Pelt W, de Wit MA et al (2003) Laboratory surveillance of bacterial gastroenteric pathogens in The Netherlands, 1991–2001. Epidemiology and Infection 130(3):431–441

van Pelt W, Mevius D et al (2004) A large increase of Salmonella infections in 2003 in the Netherlands: hot summer or side effect of the avian influenza outbreak? Euro-Surveillance Monthly 9(7/8):3–4

WHO (2001) Floods: Climate Change and Adaptation Strategies for Human Health. WHO, Copenhagen, EUR/01/503 6813

WHO (2003) Climate Change and Human Health -Risk and Responses (Summary). Geneva

WHO (2004) Report of WHO meeting on Climate and Foodborne Disease. Rome, Italy, 27–28 February 2003. In preparation. WHO, Rome

WHO (2004) The Vector-borne human infections of Europe. Their distribution and burden of public health. WHO Regional Office for Europe, Copenhagen

WHO/EEA (2004) Report of the meeting "Extreme weather and climate events and public health responses", Bratislava, Slovakia, 9–10 February 2004. World Health Organization, Copenhagen

van Lieshout M et al (2003) Global change and health scenarios for Europe: An overview and assessment of existing scenario studies. ICIS/Maastricht

Internal Reports

Penning-Rowsell EC, Wilson T (2003) The emergency planning and health impacts in Europe: an exploratory overview. Report for cCASHh project. Flood Hazard Research Centre, Enfield

Valter J, Kott I, Kriz B, Daniel M, Hostynek J: Fluctuation of the mesoclimate in the tick-populated highlands of the Klatovy Region (Southwest Bohemia)

Annex 2:
Institutions that Participated in the Research

The WHO Regional Office for Europe, European Centre for Environment and Health, Rome, Italy

The European Centre for Environment and Health (ECEH) was established as a part of WHO Regional Office for Europe, with its 52 Member States, after the first WHO Ministerial Conference on Environment and Health (1989). It comprises two divisions, one in Rome and the other in Bonn. ECEH Rome focuses on developing evidence-based strategies and tools to protect health from the harmful effects of environmental hazards. ECEH Rome actively participates in the WHO/Europe environment and health process, which is marked by a series of ministerial conferences. Activities focused on the most recent, the *Fourth Ministerial Conference on Environment and Health*, which was held in Budapest in 2004 on the theme "The Future for our Children". At the Conference, European health and environment ministers committed themselves to ensuring healthy environments for children. The Programme Global Change and Health was established in 1999, and coordinates a Europe-wide interagency network that
▮ assesses and monitors the health impact of global environmental changes
▮ evaluates policy options and advocates with Member States on preventive measures
▮ carries out capacity-building activities.

London School of Hygiene & Tropical Medicine, London, United Kingdom

The London School of Hygiene & Tropical Medicine is Britain's national school of public health and a leading postgraduate institution in Europe for public health and tropical medicine. Part of the University of London, the London School is an internationally recognized centre of excellence in public health, international health and tropical medicine with a remarkable depth and breadth of expertise. It is one of the highest-rated research institutions in the United Kingdom. The Centre on Global Change and Health is a cross-departmental initiative that brings together staff and students from a wide range of disciplines to contribute to the School's rapidly growing body of research on globalization, environmental change and health.

Stockholm University, Stockholm, Sweden

Stockholm University is a centre for higher education and research, organized into four faculties: natural sciences, humanities, social sciences and law. Its 37 000 students and 6000 employees make Stockholm University one of Sweden's largest educational establishments as well as one of the largest employers in the Stockholm area. Stockholm College began in 1878 with a series of public lectures on the natural sciences. In 1904, the college became an official degree-granting institution and in the following two decades both the Faculties of Law and Humanities were established. In 1960 Stockholm College became a state university and four years later the Faculty of Social Sciences was added.

National Institute of Public Health, Prague, Czech Republic

The National Institute of Public Health (NIPH) is a health care establishment for basic preventive disciplines – hygiene, epidemiology, microbiology and occupational medicine. Its main tasks are health promotion and protection, disease prevention and follow-up of environmental impact on the health status of the population. The main activities of NIPH comprise science and research, reference and methodological advice, providing expert opinions on the health safety of various products (e. g. cosmetics, food supplements, items of daily use etc.), systematic monitoring of the environmental impact on population health in the Czech Republic, preparation of legislation in the field of health protection, including harmonization of Czech legislation with the norms of the European Union. In the field of health promotion and disease prevention, NIPH concentrates on the most important health problems – epidemiological surveillance of severe infections (AIDS, hepatitis, newly emerging and re-emerging infections) and promotion of a healthy lifestyle (prevention of cardiovascular diseases and tumours, healthy nutrition, drug abuse prevention). The Institute plays an active role in pre- and postgraduate training of physicians and other health care workers and provides consultations to professionals working in the field.

Deutscher Wetterdienst, Freiburg, Germany

Germany's National Meteorological Service, the Deutscher Wetterdienst (DWD), is responsible for meeting meteorological requirements arising from all areas of the economy and society in Germany. Its area of responsibility is defined by the statutory tasks of providing information and performing research as laid down in the Law on the Deutscher Wetterdienst. The Department for Human Biometeorology has received increasing attention in the past few years. Pollen flight, hypersensitivity to changes in the weather, UV radiation and also the bioclimate and air quality are just some of the aspects of human biometeorology. The DWD undertakes research and development, produces forecasts and expertise and provides advice in the field of meteorological environmental conditions for human beings.

The Potsdam Institute for Climate Impact Research, Potsdam, Germany

PIK was founded in 1992 and now has a staff of about 140 people. The historic buildings of the Institute as well as the high-performance computer are located on Potsdam's Telegrafenberg campus. At PIK, researchers in the natural and social sciences work together to study global change and its impacts on ecological, economic and social systems. They examine the Earth system's capacity for withstanding human interventions and devise strategies for a sustainable development of humankind and nature. PIK research projects are interdisciplinary and undertaken by scientists from the following five departments: Integrated Systems Analysis, Climate System, Natural Systems, Social Systems and Data & Computation.

Fondazione Eni Enrico Mattei (FEEM), Milan, Italy

The Fondazione Eni Enrico Mattei (FEEM) is a non-profit, non-partisan research institution established to carry out research in the field of sustainable development.

Recognised by the President of the Italian Republic in July 1989, it has since become a leading international research centre. One of its principal aims is to promote interaction between academic, industrial and public policy spheres in order to comprehensively address concerns about economic development and environmental degradation.

The Fondazione's activities are guided by four fundamental criteria: to analyse relevant and innovative research areas; to focus on "real" world issues; to integrate multidisciplinary approaches; to create and foster international research networks.

International Centre for Integrative Studies, Maastricht, The Netherlands

ICIS addresses the increasing need for integrated analyses of environments (such as cities and regions) and complex issues (e.g. sustainable development, human health, tourism and water). Such integrative studies involve analysis of the causes, effects and the mutual interlinkages between economic, environmental, institutional and sociocultural processes associated with a specific environment or complex issue. These interdisciplinary analyses complemented with participatory processes involving stakeholders usually form the basis for the development of visions and long-term strategies. ICIS can be characterized as an institute for Integrated Assessment. The mission of ICIS is, thus, to support consciousness-raising processes with regard to integrative thinking and acting in policy circles, the business community and society at large. ICIS is thus both research-oriented and client-oriented.

Annex 3:
European Region Survey on National Preparedness and Response to Extreme Weather Events

> **Objective:** To describe the role of European Health Ministries with respect to extreme weather events (floods, heat waves, cold spells and windstorms) in terms of their mandates, programs, capacities and limitations.

The WHO European Region has experienced an unprecedented rate of warming in the recent past. During the period 1976–1999, the mean daily maximum temperature in most areas during the summer months has increased by more than 0.3 °C per decade.

Increasing variability in the European climate has led many experts to predict that extreme weather events including floods, heat-waves, cold spells, and windstorms will become both more frequent and more severe.

In the summer of 2003, heat-waves struck large areas of Western Europe and caused many unanticipated deaths in several countries. According to provisional data provided by national authorities, there were more than 14000 excess deaths in France alone.

Flooding is the most common natural disaster in the European Region. From January to December 2002, the Region suffered about 15 major floods that killed about 250 people and affected as many as 1 million. An international disaster database recorded 238 floods in the Region between 1975 and 2001. During the last decade, floods killed 1940 people and made 417000 homeless.

In recent years, preventing and responding to the effects on human health of extreme weather has become a public health action priority. The WHO Regional Office for Europe is working to place climate change higher on the global public health agenda. As a first step, there is a need to document the policy frameworks, programmes and lead agencies in relation to current public heath responses to extreme weather. The current survey, funded through the European Commission grant No. EVK2-CT-2000-00070, will allow WHO to develop national and regional profiles of extreme weather prevention activities and policies. It will, therefore, identify areas where national authorities are active and areas where more efforts are needed. WHO and its partners will be able to use these profiles to advocate for increased resources, and to share examples of best practice. In addition, WHO will be in a better position to support Ministries of Health by providing technical assistance (e.g. capacity building, methods, analyses) that better respond to local needs for the implementation and support of programmes to prevent the adverse effects on health of extreme weather.

Thank you for taking the time to complete this questionnaire.

Country name:

Respondent 1:
(Name, Title and Section)

E-Mail address:

Postal address:

Telephone number:

Facsimile number:

Date:

Please provide contact information for others who also helped respond to this questionnaire

Respondent 2:
(Name, Title and Section)

E-Mail address:

Postal address:

Telephone number:

Facsimile number:

Date:

Respondent 3:
(Name, Title and Section)

E-Mail address:

Postal address:

Telephone number:

Facsimile number:

Date:

If we have questions or require clarification, may we contact you?
☐ No
☐ Yes

How do you prefer we contact you?
☐ E-Mail
☐ Telephone
☐ Fax
☐ Post

Please return this questionnaire to:
Dr. Tom Kosatsky
Technical Officer

Global Change and Health European Centre for Environmental Health
Via Francesco Crispi 10
I-00187 Rome, Italy

Tel.: +39 064877526
Fax: +39 064877599
E-Mail: tko@who.it

---

1. Please name the section of (or the person in) your Ministry that is responsible for:
a. Protecting human health from environmental threats

_____

_____

b. Disaster preparedness

_____

_____

c. Mobilization and response during emergencies

_____

_____

(Please provide us with an organizational chart of your Ministry which shows where the sections or persons described in a, b and c are located)

---

2. Is there specific legislation that gives your Ministry authority to act during emergencies?   ☐ **No**   ☐ **Yes**

_____

_____

(If yes: Please provide us with a copy of, or reference to that legislation [name of law, section, year])

3. Does the emergency measures legislation named in 2 *specifically* mention:

  a. Earthquakes?   ☐ No   ☐ Yes

  b. Chemical spills?   ☐ No   ☐ Yes

  c. Floods?   ☐ No   ☐ Yes

  d. Cold spells?   ☐ No   ☐ Yes

  e. Heat-waves?   ☐ No   ☐ Yes

  f. Windstorms?   ☐ No   ☐ Yes

---

4. Does your Ministry have a plan (whether or not you have specific legislation) for mobilization and response during emergencies?   ☐ No   ☐ Yes

If yes:

  a. Does the plan have a specific section dealing with earthquakes?   ☐ No   ☐ Yes

  b. Does the plan have a specific section dealing with chemical spills?   ☐ No   ☐ Yes

  c. Does the plan have a specific section dealing with floods?   ☐ No   ☐ Yes

  d. Does the plan have a specific section dealing with cold spells?   ☐ No   ☐ Yes

  e. Does the plan have a specific section dealing with heat-waves?   ☐ No   ☐ Yes

  f. Does the plan have a specific section dealing with windstorms?   ☐ No   ☐ Yes

  g. In what year was the mobilization and response plan last revised? _____

---

5. Is there a co-operative agreement between your Ministry and the national weather service?   ☐ No   ☐ Yes

If yes:

  a. Does it mandate (oblige) the routine transmission of meteorological information?   ☐ No   ☐ Yes

  b. Does it mandate (oblige) the transmission of forecasts in the event of extreme weather?   ☐ No   ☐ Yes

  c. Does it mandate responsibilities for decision-making (example, declaring a weather emergency) in the event of extreme weather?   ☐ No   ☐ Yes

  d. Does it mandate responsibilities for public communication in the event of extreme weather?   ☐ No   ☐ Yes

---

6. Is there a co-operative agreement between your Ministry and the National civil protection agency (ies) that describes the roles of each with respect to natural disasters/extreme weather events?   ☐ No   ☐ Yes

---

7. Is there a co-operative agreement between your Ministry and the National armed forces that describes the roles of each with respect to natural disasters/extreme weather events?   ☐ No   ☐ Yes

8. Is there a programme to monitor population health during natural disasters?

☐ No   ☐ Yes

If yes:

a. Does it include counts of recent deaths, such as through reports from funeral homes?

☐ No   ☐ Yes

b. Does it include doctor reports?   ☐ No   ☐ Yes

c. Does it include surveillance of hospital admissions and/or emergency room consultations?   ☐ No   ☐ Yes

d. Does it include use of emergency services such as ambulance calls?

☐ No   ☐ Yes

e. Does the surveillance programme include heat-waves?   ☐ No   ☐ Yes

f. Does the surveillance programme include cold spells?   ☐ No   ☐ Yes

g. Does the surveillance programme include floods?   ☐ No   ☐ Yes

h. Does the surveillance programme include windstorms?   ☐ No   ☐ Yes

---

9. In your country, is there authority for disaster relief at the regional, state or provincial (sub-national) level?   ☐ No   ☐ Yes

---

10. In your country, do city (municipal) governments have authority for disaster relief?   ☐ No   ☐ Yes

---

11. Has the media in your country demonstrated interest in extreme weather (heat-waves, cold spells, floods, windstorms, etc.) as a health problem?

a. Much interest   ☐

b. Moderate interest   ☐

c. Little interest   ☐

d. No interest   ☐

---

12. In what year was the last major flood in your country? _____

a. What were the impacts on health (deaths, injuries, impact on health services)?

---

13. In what year was the last major windstorm in your country? _____

a. What were the impacts on health (deaths, injuries, impact on health services)?

14. In what year was the last major heat-wave in your country? _____

    a. What were its impacts on health (deaths, injuries, impact on health services)?

    _____

15. In what year was the last major cold spell in your country? _____

    a. What were its impacts on health (deaths, injuries, impact on health services)?

    _____

16. Does your country have a national plan to *prevent* the impacts of floods on health? ☐ No ☐ Yes

If yes:

    a. Does it involve an assessment of which zones are most susceptible to flooding? ☐ No ☐ Yes

    b. Does it involve an assessment of which people are most susceptible to have health problems if flooding occurs? ☐ No ☐ Yes

    c. Does it include publicity campaigns to educate people in flood preparedness? ☐ No ☐ Yes

    d. Does it include zoning regulations to prohibit building in areas prone to flooding? ☐ No ☐ Yes

    e. Does it involve financial incentives to have people make their properties resistant to floods? ☐ No ☐ Yes

(If your country has a flood prevention plan, please send us a copy.)

17. Does your country have a national plan to *prevent* the impacts of heat-waves on health? ☐ No ☐ Yes

If yes:

    a. Does it involve an assessment of which sectors of your country are most susceptible to heat-waves? ☐ No ☐ Yes

    b. Does it involve an assessment of which people are most susceptible to have health problems in the event of heat-waves? ☐ No ☐ Yes

    c. Does it include publicity campaigns to educate people in preparing themselves for heat? ☐ No ☐ Yes

    d. Does it include a heat watch-warning plan with specific advice for health protection based on the weather forecast? ☐ No ☐ Yes

    e. Does it include advice to doctors and pharmacists on preventing and recognizing health problems related to heat? ☐ No ☐ Yes

    f. Does it involve financial incentives to help people make their homes cooler? ☐ No ☐ Yes

(If your country has a heat-wave prevention plan, please send us a copy.)

18. Does your country have a national plan to *prevent* the impacts on health of cold spells?   ☐ No   ☐ Yes
(If your country has a cold spell prevention plan, please send us a copy.)

19. In what ways could your Ministry play a larger role in the prevention of adverse health impacts from flooding? _____

_____

_____

_____

_____

20. In what ways could your Ministry play a larger role in the prevention of adverse health impacts from heat-waves? _____

_____

_____

_____

_____

21. Please add any other information, which will help us to describe the prevention of and response to extreme weather events in your country. _____

_____

_____

The European Centre for Environment and Health (WHO) is considering the development of a monograph describing why and how various European Cities and Ministries have developed programmes to limit the damage to health caused by heat-waves, cold spells, flood, windstorms and other extreme weather events. Would you be interested in contributing a 4–5 page text (based on a standardized outline) to this monograph? If there is general interest, we plan to pursue this project during the course of 2005.

We (I) would not be interested ☐
We (I) would be interested ☐
If Interested, your name and contact information please:

_____

_____

Thank you for describing your Ministry's extreme weather prevention and response activities.

Annex 4:
The Health Effects of Global Climate Change:
What is Your Opinion?

Today's date:

This survey is part of a large research effort about climate change conducted by the World Health Organization and universities and research institutions in several European countries.

This questionnaire includes questions about your opinions on adaptive capacity to some of the adverse health effects of global climate change. We will focus on the effects on human health of floods and landslides, heatwaves, and vector-borne diseases.

Please keep in mind that there are no right or wrong answers to the questions in this questionnaire: We are simply interested in your professional opinions and ideas. All the information you provide is **confidential** and will be published only in summary statistical form. You will not be identified in any way.

Your participation in this survey is very important, so please help us by filling out the questionnaire now to the best of your ability.

When you are finished with the questionnaire, **please return it to Bettina Menne** at this venue.

If you cannot locate Bettina Menne, **please mail** the completed questionnaire to Aline Chiabai, FEEM, Castello 5252, 30122 Venezia, Italy. You may also **fax** the completed questionnaire to Aline Chiabai at +39 041 271-1461. If you have any **questions**, please contact Aline Chiabai by phone at +39 041 271-1467 or by e-mail at aline.chiabai@feem.it.

Thank you for participating in this survey!

Part A: Global climate change

In this section, we provide a brief summary of global climate change and on some of its possible effects on human health.

Global climate change

- Many research scientists and climatology experts predict that the average temperature on the earth will increase by 1.4 to 5.8 °C over the next 100 years.
- This is due to the release into the atmosphere of certain gases (the so-called greenhouse gases), such as by-products from burning fossil fuels and emissions from manufacturing plants and automobiles. Deforestation also contributes to the increase in greenhouse gases, since trees absorb carbon dioxide, an important greenhouse gas.

What are the possible consequences of global climate change?

- Coastal erosion and flooding due to the rising of the sea level.
- Saltwater intrusion into groundwater, which may reduce drinking water supplies in certain areas
- Damage to freshwater fisheries.
- Increased and more intense precipitation in some areas, which could cause flooding, while other areas may experience droughts and increased desertification.
- In some areas, agricultural production may increase as a result of the higher temperatures. In other areas, agricultural production may suffer as a result of the more frequent rains or severe droughts.
- Displacement of some animal and plant species in certain areas; certain animal and plant species might spread to new areas.
- More frequent heat-waves and other effects.

Many of these effects have possible consequences on human health. In this survey, we would like you to focus on the following effects on human health:

- Deaths and injuries associated with floods, mudslides and landslides due to the change in precipitation patterns.
- Deaths and hospital admissions for cardiovascular and respiratory problems due to heat-waves.
- Cases of infectious diseases that would arise because certain insects and/or rodents spread to new areas, or because the pathogens they carry may become more viable. These diseases include Lyme disease, leishmaniasis, malaria and tick-borne encephalitis.

Please continue to Part B.

Part B: What is Adaptive Capacity?

In this section, we would like you to think about three main categories of health effects associated with global climate change, namely:

- deaths and injuries due to floods, mudslides and landslides;
- deaths and hospital admissions for cardiovascular and respiratory problems during heat-waves, and
- increased cases of infectious diseases spread by insects and/or rodents.

▌**B1.** How concerned is **your organization** about the following health effects of climate change? (Please circle one number for each category of health effects.)

	Very concerned	Somewhat concerned	Not concerned at all	No position/ outside the organization's mission
1. Deaths and injuries due to floods, landslides and mudslides	1	2	3	4
2. Cardiovascular and respiratory illnesses due to heat-waves	1	2	3	4
3. Increased cases of vector-borne diseases	1	2	3	4

▌**B2.** How concerned are **you professionally** about the following health effects of climate change? (Please circle one number for each category of health effects.)

	Very concerned	Somewhat concerned	Not concerned at all	No position/ outside of professional duties
1. Deaths and injuries due to floods, landslides and mudslides	1	2	3	4
2. Cardiovascular and respiratory illnesses due to heat-waves	1	2	3	4
3. Increased cases of vector-borne diseases	1	2	3	4

Now, we would like you to think about the adaptive capacity of countries and communities to these adverse health effects. By adaptive capacity, we mean the potential ability of countries, communities and individuals to adjust to climate change (including climate variability and extremes) to moderate its adverse effects, take advantage of opportunities, or cope with the consequences.

Here, the adverse effects being considered are the number of deaths, injuries and cases of illness associated with floods and landslides, heat-waves, and vector-borne infectious diseases. Adaptive capacity influences the vulnerability of communities and regions to climate change effects and hazards.

To reduce these adverse health effects, governments may implement adaptation measures, such as public health training programmes, surveillance and emergency response systems, sustainable prevention and control programmes, land use planning, and new infrastructure.

In the next section of the questionnaire, we will ask you questions about your opinions on the factors that affect a country's adaptive capacity with respect to the three categories of health effects of climate change examined in this questionnaire.

Part C: Which Country has Greater Adaptive Capacity?

In this section of the questionnaire, we will ask you to consider two hypothetical countries, A and B. These two countries could be located anywhere in the world.

Each of these two hypothetical countries is described by a set of characteristics, such as its per capita income, population density, age structure of the population, a measure of the health status of the population (life expectancy at birth), two measures of its health care system (number of physicians per 100 000, health care system coverage), and access to information.

The two countries differ from each other in the levels of two or more of these characteristics. For example, one country may have higher per capita income, but a lower life expectancy at birth than the other.

Based on the description of the two countries, you will be asked to tell us which, A or B, you believe to have the higher adaptive capacity with respect to the health effects of floods and landslides, heat-waves, and vector-borne diseases. Please consider all of these three effects when you think of the adaptive capacity of each country.

There will be a total of four questions asking you which is the country with the higher adaptive capacity out of a pair of countries.

▌ **C1.** Let us begin with two hypothetical countries, A and B. Both countries
▌ have a relatively high population density (400 people per square km),
▌ have experienced a significant amount of deforestation in the past,
▌ have significant amounts of coastline and mountains,
▌ are moderately susceptible to floods and landslides, and
▌ have a mild, Mediterranean-type climate.

In addition, they have the following characteristics:

Characteristic	Country A	Country B
Income:		
Per capita income (in US dollars)	20 000	27 000
Inequality in the distribution of income	Low	Low
Population:		
Percentage of population older than 65 years	12%	12%
Health:		
Life expectancy at birth	79 years	79 years
Physicians per 100 000	300	300
Health care system coverage	Based on private health insurance	Universal coverage
Technology and infrastructure:		
Access to information via newspaper, television, radio, internet	High	Low

In your opinion, which country has higher adaptive capacity?
1. A ☐   2. B ☐

C2. Now we would like you to consider a different pair of countries, C and D. As before, these two countries
▮ have a relatively high population density (400 people per square km),
▮ have experienced a significant amount of deforestation in the past,
▮ have significant amounts of coastline and mountains,
▮ are moderately susceptible to floods and landslides, and
▮ have a mild, Mediterranean-type climate.

In addition, they have the following characteristics:

Characteristic	Country C	Country D
Income:		
Per capita income (in US dollars)	20 000	13 000
Inequality in the distribution of income	High	High
Population:		
Percentage of population older than 65 years	12%	18%
Health:		
Life expectancy at birth	70 years	79 years
Physicians per 100 000	300	300
Health care system coverage	Universal coverage	Based on private health insurance
Technology and infrastructure:		
Access to information via newspaper, television, radio, internet	Low	Low

In your opinion, which country has higher adaptive capacity?
1. C ☐   2. D ☐

▌ **C3.** Now we would like you to consider a different pair of countries, E and F. These two countries
▌ have a relatively low population density (200 people per square km),
▌ have experienced little deforestation in the past,
▌ have significant amounts of coastline and mountains,
▌ have rarely experienced floods and landslides, and
▌ have a relatively cold, Northern European-type climate.

In addition, they have the following characteristics:

Characteristic	Country E	Country F
Income:		
Per capita income (in US dollars)	20 000	27 000
Inequality in the distribution of income	Low	High
Population:		
Percentage of population older than 65 years	18%	12%
Health:		
Life expectancy at birth	79 years	79 years
Physicians per 100 000	400	300
Health care system coverage	Universal coverage	Based on private health insurance
Technology:		
Access to information via newspaper, television, radio, internet	High	Low

In your opinion, which country has higher adaptive capacity?
1. E ☐   2. F ☐

▌ **C4.** Now we would like you to consider another, different pair of countries, G and H.
    As before, these two countries
▌ have a relatively low population density (200 people per square km),
▌ have experienced little deforestation in the past,
▌ have significant amounts of coastline and mountains,
▌ have rarely experienced floods and landslides, and
▌ have a relatively cold, Northern European-type climate.

In addition, they have the following characteristics:

Characteristic	Country G	Country H
Income:		
Per capita income (in US dollars)	20 000	27 000
Inequality in the distribution of income	Low	High
Population:		
Percentage of population older than 65 years	12%	18%
Health:		
Life expectancy at birth	70 years	70 years
Physicians per 100 000	250	400
Health care system coverage	Universal coverage	Based on private health insurance
Technology:		
Access to information via newspaper, television, radio, internet	High	High

In your opinion, which country has higher adaptive capacity?
1. G ☐   2. H ☐

Part D: A few questions about yourself

D1. What is your gender?
　　1. Male ☐　　　　　2. Female ☐

D2. What year were you born? _____

D3. Where do you live?
　　City:_____ Country: _____

D4. What is your professional title/position? _____

D5. What is the type of your highest degree or training?
　　1. Medical　　　　　　　　　　　　　　　☐
　　2. Public health or epidemiology　　　　　☐
　　3. Engineering　　　　　　　　　　　　　☐
　　4. Economics or business administration　☐
　　5. Other　　　　　　　　　　　　　　　　☐
Please explain _____

D6. What type of organization do you work for? (Please circle one.)
　　1. A public health organization　　　　　　　☐
　　2. A private or public health care organization　☐
　　3. An emergency response agency　　　　　　　☐
　　4. A university or research institution　　　　　☐
　　5. Another government organization　　　　　　☐　→ Please
　　6. A nongovernment, non-profit organization　　☐　→ Please
　　7. A private company　　　　　　　　　　　　☐　→ Please explain____
　　8. Another type of organization　　　　　　　☐　→ Please

You are almost finished with this questionnaire. Please continue to Part E.

Part E: Just a few more questions about this questionnaire

In this section of the questionnaire, we would like to ask you some questions to help us understand how clear the material presented in this questionnaire and the questions were.

E1. Did you find the description of the consequences of climate change on page 2 adequate? (Please tick one.)
   1. Yes ☐   2. No ☐   3. Don't know ☐

E2. Was any of the information on the consequences of climate change on page 2 new to you? (Please tick one.)
   1. Yes ☐   Which? _____
   2. No ☐

E3. Was the concept of adaptive capacity explained clearly? (Please tick one.)
   1. Yes ☐   2. No ☐

E4. Did you think that the text and the tables about the countries of questions C1–C4 were clear?
   1. Yes ☐   2. No ☐

E5. In answering questions C1–C4, which did you pay most attention to?
   (Please check one.)
   1. Adaptive capacity with respect to the effects of extreme events, thermal stresses, and vector-borne diseases.   ☐
   2. Adaptive capacity with respect to the effects of extreme events only.   ☐
   3. Adaptive capacity with respect to the effects of thermal stresses only.   ☐
   4. Adaptive capacity with respect to the effects of vector-borne diseases only. ☐
   5. Adaptive capacity with respect to other effects of climate change.   ☐

   → Please explain _____

Thank you for all your help! If you have any comments or suggestions that could help us improve this survey questionnaire, please write them below.

Would you like to receive a summary report on this research project? If so, please write your name and address below.

Please **return** your completed questionnaire to Bettina Menne of the World Health Organization.

If you cannot locate Ms. Menne, please **mail** your completed questionnaire to Aline Chiabai, FEEM, Castello 5252, 30122 Venezia, Italy, or **fax** it to Aline Chiabai at +39 041 271-1461.